Welding Handbook

Seventh Edition, Volume 3

Resistance and Solid-State Welding and Other Joining Processes

The Five Volumes of the Welding Handbook, Seventh Edition

1 Fundamentals of Welding

2 Welding Processes—

   Arc and Gas Welding and Cutting, Brazing, and Soldering

3 Welding Processes—

   Resistance and Solid-State Welding and Other Joining Processes

4 Metals and Their Weldability[a]

5 Engineering, Quality Assurance, and Safe Practices[b]

a. Scheduled to replace Sec. 4, 6th Ed., in 1982.
b. Scheduled to replace Sec. 5, 6th Ed., in 1984.

Welding Handbook

Seventh Edition, Volume 3

Resistance and Solid-State Welding and Other Joining Processes

W. H. Kearns, Editor

AMERICAN WELDING SOCIETY
2501 Northwest 7th Street
Miami, Florida 33125

European Edition Distributed by
THE MACMILLAN PRESS LTD
London and Basingstoke

Published in the U.K. and distributed in Europe 1980 by
THE MACMILLAN PRESS LTD.
London and Basingstoke

ISBN 0 333 27395 8

American Welding Society, 2501 N.W. 7th Street, Miami, FL 33125

Note: By publication of this handbook, the American Welding Society does not insure anyone utilizing this handbook against liability arising from the use of such handbook. A publication of a handbook by the American Welding Society does not carry with it the right to make, use, or sell patented items. Each prospective user should make an independent investigation.

Printed in the United States of America

Contents

Welding Handbook Committee

W. L. Wilcox, Chairman   Scott Paper Company
J. R. Condra, Vice-Chairman   E. I. du Pont de Nemours and Company
W. H. Kearns, Secretary   American Welding Society
I. G. Betz   Department of the Army
R. L. Frohlich   Westinghouse Electric Corporation
J. R. Hannahs   Bowser-Morner Testing Laboratories
A. F. Manz   Union Carbide Corporation
A. W. Pense*   Lehigh University
L. J. Privoznik   Westinghouse Electric Corporation
D. D. Rager   Reynolds Metals Company
R. E. Somers   Welding Consultant

*Term expired May 31, 1979

Preface

This is the third of five volumes planned for the Seventh Edition of the *Welding Handbook*. Volume 1, *Fundamentals of Welding,* was published in 1976. It contains much of the basic information on welding, including the physics of welding, heat flow, welding metallurgy, testing, residual stresses, and distortion. Volume 2, *Welding Processes—Arc and Gas Welding and Cutting, Brazing, and Soldering,* was published in 1978. In addition to the subjects named in the title, that Volume covers electroslag and electrogas welding, stud welding, surfacing, and arc welding power sources.

Volume 3 covers the newer welding processes as well as resistance welding, thermal spraying, and adhesive bonding. Certain other processes that have rather limited applications are also included. Volumes 2 and 3 of the Seventh Edition include all of the processes originally discussed in Sections 2, 3A, and 3B of the Sixth Edition.

Some of the material in Section 1 of the Sixth Edition has not yet been covered in the Seventh Edition. Consequently, this Section should be retained, as well as Sections 4 and 5 of that Edition. Safe practices for welding, cutting, brazing, and soldering is one very important subject discussed in Section 1, Sixth Edition. That subject is handled differently in this Edition. Safety considerations for each process are included in the respective chapter. The material on safety in the chapters is not all-inclusive. The reader should determine for himself the safe practices that are adequate and appropriate for each of the processes.

An index of major subjects precedes the index for this volume. It enables a reader to quickly locate the current information on major subjects in both the Sixth and the Seventh Editions.

Metric equivalents in this volume are handled in a manner different from that used in Volumes 1 and 2. Appropriate conversion factors have been placed at the end of each chapter in this volume instead of the arbitrary conversion of US Standard units to SI units and the dual dimensioning used previously. The Welding Handbook Committee considered that information in the Handbook would be presented better if it were given in units of the particular system in which the measurements themselves were taken or reported. The information may be converted to units in the system other than the one shown by using the appropriate conversion factors.

This volume was a voluntary effort by the Welding Handbook Committee and the Chapter Committees. The Chapter Committee Members and the Handbook Committee Member responsible for a chapter are recognized on the title page of that chapter. Other individuals also contributed in a variety of ways, particularly in chapter reviews. All participants contributed generously of their time and talent, and the *American Welding Society* expresses herewith its appreciation to them and to their employers for supporting the work.

The Welding Handbook Committee expresses its appreciation to Richard French, Patricia Hoel, and Hallock Campbell for their assistance in the production of this volume.

The Welding Handbook Committee welcomes your comments on the *Handbook*. Please address them to the Editor, Welding Handbook, American Welding Society, 2501 N.W. 7th Street, Miami, FL 33125.

W. L. Wilcox, *Chairman*
Welding Handbook Committee

W. H. Kearns, *Editor*
Welding Handbook

1

Spot, Seam, and Projection Welding

Chapter Committee

W. G. EMAUS, *Chairman*
Anderson Automation, Inc.

R. A. GRENDA
Weltronic Company

P. G. HARRIS
Centerline (Windsor) Ltd.

A. E. LOHRBER
Swift-Ohio Corporation

R. MATTESON
Taylor-Winfield Corporation

F. PARKER
Newcor, Inc.

D. ROTH
Kirkhof Transformer Division,
FLX Corporation

Welding Handbook Committee Members

A. F. MANZ
Union Carbide Corporation

D. D. RAGER
Reynolds Metals Company

1

Spot, Seam, and Projection Welding

FUNDAMENTALS OF THE PROCESSES

DEFINITION AND GENERAL DESCRIPTION

Spot, seam, and projection welding are three resistance welding processes in which coalescence of metals is produced at the faying surfaces by the heat generated at the joint by the resistance of the work to the flow of electric current. Force is always applied before, during, and after the application of current to prevent arcing at the faying surfaces and, in some applications, to forge the weld metal during postheating. Generally, melting occurs at the faying surface during weld time. Figure 1.1 illustrates the three processes.

In spot welding, a nugget[1] of weld metal is produced at the electrode site, but two or more nuggets may be made simultaneously using multiple sets of electrodes. Projection welding is similar except that nugget location is determined by a projection or embossment on one faying surface, or by the intersection of parts in the case of wires or rods (cross-wire welding). Two or more projection welds can be made simultaneously with one set of electrodes.

Seam welding is a variation of spot welding in which a series of overlapping nuggets is produced to obtain a continuous gastight seam. One or both electrodes are generally wheels that rotate as the work passes between them. A seam weld can be produced with spot welding equipment but the operation will be much slower.

---

1. A nugget is the weld metal joining the parts in spot, seam, and projection welds.

A series of separate spot welds may be made with a seam welding machine and wheel electrodes by suitably adjusting the travel speed and the time between welds. Movement of the work may or may not be stopped during the spot weld cycle. This procedure is known as roll spot welding.

PRINCIPLES OF OPERATION

Spot, seam, and projection welding operations involve a coordinated application of electric current and mechanical pressure of the proper magnitudes and durations. The welding current must pass from the electrodes through the work. Its continuity is assured by forces applied to the electrodes or projections which are shaped to provide the necessary current density and pressure. The sequence of operation must first develop sufficient heat to raise a confined volume of metal to the molten state. This metal then cools while under pressure until it has adequate strength to hold the parts together. The current density and pressure must be such that a nugget is formed but not so high that molten metal is expelled from the weld zone. The time of current flow (weld time) must be sufficiently short to prevent excessive heating of the electrode faces. Such heating can bond the electrodes to the work and greatly reduce their life.

The heat required for these resistance welding processes is produced by the resistance to the passage of electric current through the workpieces. Because of the short electric current path in the work and limited weld time, high

2

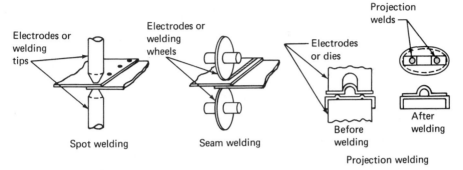

Fig. 1.1—Spot, seam, and projection welding processes

welding currents are required to develop the necessary welding heat.

Heat Generation

In an electrical conductor, the amount of heat generated depends upon three factors: (1) the amperage, (2) the resistance of the conductor, and (3) the time of current flow. These three factors affect the heat generated as expressed in the formula

$$Q = I^2Rt$$

where

Q = heat generated, joules

I = current, amperes

R = resistance of the work, ohms

t = time of current flow, seconds

The heat generated is proportional to the square of the welding current and directly proportional to the resistance and the time. Part of the heat generated is used to make the weld and part is lost to the surrounding metal.

The welding current required to produce a given weld is approximately inversely proportional to the square root of the time. Thus, if the time is extremely short, the current required will be very high. A combination of high current and short time may produce an undesirable distribution of heat in the weld zone, resulting in severe surface melting and rapid electrode deterioration.

The secondary circuit of a resistance welding machine and the work being welded are a series of resistances, and the arithmetical total of the resistance path affects the flow of current. The current flow will be the same in all parts of

the circuit regardless of the instantaneous resistance at any location in the circuit, but the heat generated at any location will be directly proportional to the resistance at that point.

An important characteristic of resistance welding is the rapidity with which welding heat can be produced. The temperature distribution in the work and electrode tips in the case of spot, seam, and projection welding is illustrated in Fig. 1.2. There are, in effect, at least seven resistances connected in series in a weld that account for the temperature distribution. For a two-thickness joint, these are:

(1) *1* and *7,* the electrical resistance of the electrode material.

(2) *2* and *6,* the contact resistance between the electrode and the base metal. The magnitude of this resistance depends upon the surface condition of the base metal and the electrode, the size and contour of the electrode face, and the electrode force. (Resistance is roughly inversely proportional to the contacting force.) This is a point of high heat generation, but the surface of the base metal does not reach the fusion temperature during the current passage due to the high thermal conductivity of the electrodes (*1* and *7*) and the fact that they are usually water-cooled.

(3) *3* and *5,* the total resistance of the base metal itself, which is directly proportional to its resistivity and thickness, and inversely proportional to the cross-sectional area of the current path.

(4) *4,* the base metal interface is the location of weld formation. This is the point of

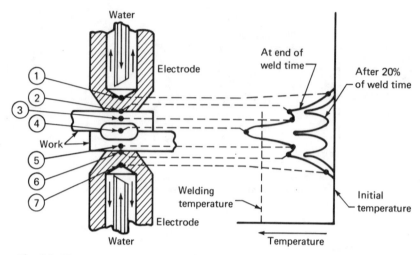

Fig. 1.2—Temperature distribution at various locations during spot, seam, and projection welding

highest resistance and, therefore, the point of greatest heat generation. Also, since heat is generated at points 2 and 6, the heat generated at this interface is not readily lost to the electrodes.

Heat is generated at all of these locations, not at the base metal interface alone. The flow of heat to or from the base metal interface is governed by the temperature gradient established by the resistance heating of the various components in the circuit. This in turn assists or retards the creation of the proper localized welding heat.

Heat will be generated in each of the seven locations in Fig. 1.2 in proportion to the resistance of each. Welding heat, however, is required only at the base metal interface, and the heat generated at all other locations should be minimized. Since the greatest resistance is located at 4, heat is most rapidly developed at that location. Points of next lower resistance are 2 and 6. The temperature rises rapidly at these points also, but not as fast as at 4. After about 20 percent of the weld time, the heat gradient may conform to the profile shown in Fig. 1.2. Heat generated at 2 and 6 is rapidly dissipated into the adjacent water-cooled electrodes 1 and 7. The heat at 4 is dissipated much more slowly into the base metal. Therefore, as the welding current continues to flow, the rate of temperature rise at plane 4 will be much more rapid than at 2 and 6. The weld-

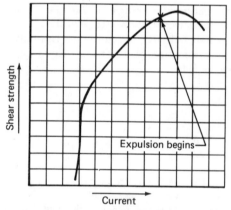

Fig. 1.3—Effect of welding current on spot weld shear strength

ing temperature is indicated by the vertical dotted line. In a well-controlled weld, the welding temperature will first be reached at numerous point contacts at the interface that melt and quickly grow into a nugget with time.

Factors that affect the amount of heat generated in the weld by a given current for a unit of weld time are: (1) the electrical resistances of the metal being welded and the electrodes, (2) the contact resistances between the workpieces and between the electrodes and the workpieces, and (3) the heat loss to the workpieces and the electrodes.

Effect of Welding Current. In the formula, $Q = I^2Rt$, current has a greater effect on the generation of heat than either resistance or time. Therefore, it is an important variable to be controlled. Two factors that cause variation in welding current are (1) fluctuations in power line voltage and (2) variations in the impedance of the secondary circuit with ac machines. Impedance variations are caused by changes in circuit geometry or by the introduction of varying masses of magnetic metals into the secondary loop of the machine. Direct current machines are not significantly affected by magnetic metals in the secondary loop and are little affected by circuit geometry. In addition to variations in welding current magnitude, current density may vary at the weld interface. This can result from shunting of current through preceding welds and contact points other than at the weld. Increase in electrode face area or projection size in the case of projection welding will decrease current density and welding heat. This may cause a significant decrease in weld strength.

A minimum current density flowing for a finite time is required to produce fusion at the interface. Sufficient heat must be generated to overcome the losses to the adjacent base metal and the electrodes. Weld nugget size and strength increase rapidly with increasing current density. Excessive current density will cause weld metal expulsion, cavitation, weld cracking, and lower mechanical strength properties. Typical variations in shear strength of spot welds as a function of current magnitude are shown in Fig. 1.3. In the case of spot and seam welding, excessive current will (1) overheat the base metal and result in deep indentations in the parts and (2) cause overheating and rapid deterioration of the electrodes.

Effect of Weld Time. The rate of heat generation must be such that welds with adequate strength will be produced without excessive electrode heating and rapid deterioration. The total heat developed is proportional to weld time. Essentially, heat is lost by conduction into the surrounding base metal and the electrodes; a very small amount is lost by radiation. These losses increase with increases in weld time and temperature of the metal, but they are essentially uncontrollable.

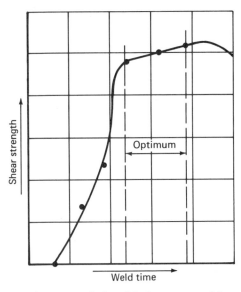

Fig. 1.4 — Relationship between weld time and spot weld shear strength

The temperature distribution curves in Fig. 1.2 show that the temperatures at the seven locations are increasing with time. Some minimum time is required to reach melting temperature at some suitable current density. If current flow is continued, the temperature at plane *4* will far exceed the melting temperature, and the internal pressure may expel weld metal from the joint. Generated gases or metal vapor may cause the expulsion of minute metal particles, called "spitting." If the work surfaces are scaly or pitted, the same thing may happen at *2* and *6*. Excessively long weld time will have the same effect of excessive amperage on the base metal and electrodes. Furthermore, the weld heat-affected zone will extend farther into the base metal. In most cases, the heat losses at some point during an extended welding interval will equal the heat input; temperatures will stabilize. An example of the relationship between weld time and spot weld shear strength is shown in Fig. 1.4, all other conditions remaining constant.

To a certain extent, weld time and amperage may be complementary. The total heat may be changed by adjusting either the amperage or the weld time. Heat transfer is a function of time, and the development of the proper nugget size requires a minimum length of time, regardless of amperage.

When spot welding heavy plates, welding current is commonly applied in several relatively short impulses without removal of electrode force. The purpose of pulsing the current is to gradually build up the heat at the interface between the workpieces. The amperage needed to accomplish welding can rapidly melt the metal if the heat time is too long, resulting in expulsion (spitting). Welding temperature can best be achieved in thick plates of moderate thermal conductivity such as steels by a series of heat pulses.

Effect of Welding Pressure. The resistance R in the heat equation is influenced by welding pressure through its effect on contact resistance at the interface between the workpieces. Welding pressure is produced by the force exerted on the weld by the electrodes. Electrode force is considered to be the net dynamic force of the electrodes upon the work, and it is the resultant pressure produced by this force that affects the contact resistance.

Pieces to be spot, seam, or projection welded must be clamped tightly together at the weld location to enable the passage of the current. Electrode force has a significant effect on the total resistance between the electrodes and, therefore, the amperage flowing through the weld. Everything else being equal, as the electrode force or welding pressure is increased, the amperage will also increase up to some limiting value. The effect on the total heat generated, however, may be the reverse. As the pressure is increased, the contact resistance and the heat generated at the interface will decrease. To increase the heat to the previous level, amperage or weld time must be increased to compensate for the reduced resistance.

The surfaces of metal components are a series of peaks and valleys on a macroscopic scale. When they are subjected to light pressure, the actual metal-to-metal contact will be only at the contacting peaks, a small percentage of the area. Contact resistance will be high. As the pressure is increased, the high spots are depressed and the actual metal-to-metal contact area is increased, thus decreasing the contact resistance. There is a limiting pressure above which the contact resistance remains uniform at room temperature. In most applications, the electrode material is softer than the workpieces. Consequently, the application of a suitable electrode force will produce better contact at the electrode-to-work interfaces than at the interface between the workpieces.

Influence of Electrodes. Electrodes play a vital role in the generation of heat because they conduct the welding current to the work. In the case of spot and seam welding, the electrode contact area largely controls the welding current density and the resulting weld size. Electrodes must have good electrical conductivity, but they must also have adequate strength and hardness to resist deformation caused by repeated applications of high electrode force. Deformation or "mushrooming" of the electrode face increases the contact area and decreases both current density and welding pressure. Weld quality will deteriorate as tip deformation proceeds. Consequently, the electrodes must be reshaped or replaced at intervals to maintain adequate heat generation for acceptable weld properties.

When the electrodes are slow in following a sudden decrease in total work thickness, a momentary reduction in pressure will occur. If this happens while the welding current is flowing, interface contact resistance and the rate of heat generation will increase. An excessive heating rate at the contacting surfaces tends to cause overheating and violent expulsion of molten metal. Molten metal is retained by a ring of unfused metal surrounding the weld nugget at the interfaces. A momentary reduction in electrode force permits the internal metal pressure to rupture this surrounding ring of unfused metal. Internal voids or excessive electrode indentation may occur. Weld properties may fall below acceptable levels and electrode wear will be greater than normal.

Influence of Surface Condition. The surface condition of the parts influences heat generation because contact resistance is affected by oxides, dirt, oil, and other foreign matter on the surfaces. The most uniform weld properties are obtained when the surfaces are clean.

The welding of parts with a nonuniform coating of oxides, scale, or other foreign matter on the surfaces causes variations in contact re-

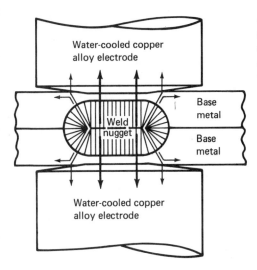

Fig. 1.5—Heat flow from spot, seam, and projection welds

sistance. This produces inconsistencies in heat generation. Heavy scale on the work surfaces may also become embedded in the electrode faces, causing rapid electrode deterioration. Oil and grease will pick up dirt which also will contribute to electrode deterioration.

Influence of Metal Composition. The electrical resistivity of a metal directly influences resistance heating during welding. In high-conductivity metals such as silver and copper, little heat is developed even under high current densities. The small amount of heat generated is rapidly transmitted into the surrounding work and the electrodes.

The composition of a metal determines its specific heat, melting temperature, latent heat of fusion, and thermal conductivity. These properties are related to the amount of heat required to melt the metal and produce a weld. The amounts of heat necessary to raise unit masses of most commercial metals to the fusion temperature are very nearly the same. For example, stainless steel and aluminum require the same Btu's per pound to reach fusion temperature, even though they differ widely in spot welding characteristics. The electrical and thermal conductivities of aluminum are about ten times greater than those of stainless steel. Conse-

quently, the heat loss into the electrodes and surrounding metal is greater with aluminum. Because of these factors, the welding current for aluminum must be considerably greater than that for stainless steel.

Heat Balance

Heat balance occurs when the depths of fusion (penetration) in the workpieces are approximately the same. The majority of spot and seam welding applications are confined to the welding of two equal thicknesses of the same metal with electrodes of the same alloy, shape, and size. Heat balance in these cases is automatic. However, in many applications, the heat generated in the parts is unbalanced.

Heat balance may be affected by:

(1) Relative electrical and thermal conductivities of the metals to be joined

(2) Relative geometry of the parts at the joint

(3) Thermal and electrical conductivities of the electrodes

(4) Geometry of the electrodes

Heating will be unbalanced when welding pieces of significantly different compositions, different thicknesses, or both. The unbalance can be minimized in many cases by part design, electrode material and design, or projection location in the case of projection welding. Heat balance can also be improved by using the shortest weld time and lowest current that will produce acceptable welds.

Heat Dissipation

During welding, heat is lost by conduction into the adjacent base metal and the electrodes as shown in Fig. 1.5. This heat dissipation continues at varying rates during current application and afterward until the weld has cooled to room temperature. It may be divided into two phases: (1) during the time of current application and (2) after the cessation of current flow. The extent of the first phase depends upon the composition and mass of the workpieces, the welding time, and the external cooling means. The composition and mass of the workpieces are determined by the design. External cooling depends upon the welding setup and the welding cycle.

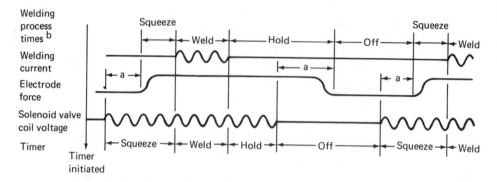

a. Machine operating time

b. Welding process "times" differ from timer "times" due to machine operating time

Fig. 1.6 — Basic welding cycle for spot and projection welding

The heat generated by a given amperage is inversely proportional to the electrical conductivity of the base metal. The thermal conductivity and temperature of the base metal determine the rate at which heat is dissipated or conducted from the weld zone.[2] In most cases, the thermal and electrical conductivities of a metal are similar. In a high-conductivity metal, such as copper or silver, high amperage is needed to generate the heat that is dissipated rapidly into the adjacent base metal and the electrodes. Spot, seam, and projection welding of these metals are very difficult.

If the electrodes remain in contact with the work after current flow ceases, they rapidly cool the weld nugget. The rate of heat dissipation into the surrounding base metal decreases with longer welding times because a larger volume of base metal is heated. This reduces the temperature gradient between the base metal and the weld nugget. For thick sheets of metal where long welding times are generally employed, the cooling rates will be slower than with thin sheets and short weld times.

If the electrodes are removed from the weld too quickly after the welding current is turned off, problems may result. With thin sheets, this procedure may cause excessive warpage. With thick sheets, adequate time is needed to cool and

2. Heat flow in welding is discussed in the *Welding Handbook*, Vol. 1, 7th ed.: 80-98.

solidify the large weld nugget while under pressure. It is usually best, therefore, to have the electrodes in contact with the work until the weld cools to a temperature where it is strong enough to sustain any loading imposed when the pressure is released.

The cooling time for a seam weld nugget is short when the electrodes are rotated continuously. Therefore, welding is commonly done with water flowing over the workpieces to remove the heat as rapidly as possible.

It is not always good practice to cool the weld zone too rapidly. With quench-hardenable alloy steels, it is usually best to retract the electrodes as quickly as possible to minimize heat dissipation to the electrodes and, thus, the cooling rate of the weld.

WELDING CYCLE

The welding cycle for spot, seam, and projection welding consists basically of four phases:

(1) *Squeeze time* — the time interval between timer initiation and the first application of current. This time interval is to assure that the electrodes contact the work and apply a force before welding current is applied.

(2) *Weld time* — the time that welding current is applied to the work in making a weld in single-impulse welding.

(3) *Hold time* — the time during which force is maintained to the work after after the last im-

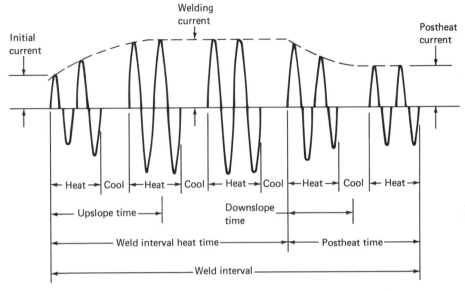

Fig. 1.7—Multiple-impulse welding with slope control and postheat

pulse of current ceases to flow. During this time, the weld nugget solidifies and is cooled until it has adequate strength.

(4) *Off time*—the time during which the electrodes are off the work and the work is moved to the next weld location. The term is generally applied where the welding cycle is repetitive.

Figure 1.6 shows a basic welding cycle. One or more of the following features may be added to this basic cycle to improve the physical and mechanical properties of the weld zone:

(1) Precompression force to seat the electrodes and workpieces together

(2) Preheat to reduce the thermal gradient in the metal at the start of weld time

(3) Forging force to consolidate the weld nugget

(4) Quench and temper to produce the desired weld strength properties in hardenable alloy steels

(5) Postheat to refine the weld grain size in steels

In some applications, the welding current is supplied intermittently during a *weld interval time;* it flows during *heat time* and ceases during *cool time,* as shown in Fig. 1.7.

WELDING CURRENT

Types

Both alternating current (ac) and direct current (dc) are used to produce spot, seam, and projection welds. The welding machine transforms line power to low voltage, high amperage welding power. Most applications use single-phase ac of the same frequency as the power line, usually 60 Hz. Direct current is used for applications that require high amperage because the load can be balanced on a 3-phase power line. Its use also reduces the power losses in secondary circuit. Direct current flow may be essentially constant for a timed period or in the form of a high-peaked pulse. The latter is normally produced from stored electrical energy.

Current Programming

With direct energy machines, the rate of current rise and fall can be programmed. The current rise period is commonly called *upslope time* and the current fall period is called *downslope time* (see Fig. 1.7). These features are available on machines equipped with electronic control systems.

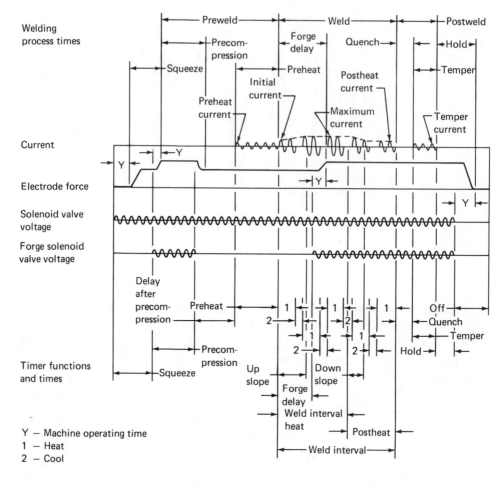

Note. When weld delay is required, it occurs before the start of weld time; forge delay time then will be initiated at the beginning of weld delay time. Both preheat and weld delay are not used in the same sequence.

Fig. 1.8 — A complex weld cycle — multiple impulse

Upslope is generally used to avoid overheating and expulsion of metal at the beginning of weld time when the base metal interface resistance is high. Downslope is used to control weld nugget solidification to avoid cracking in metals that are quench-hardenable or subject to hot tearing.

Prior to welding, the base metal can be preheated using a low current flow. Following the formation of the weld nugget, the current can be reduced to some lower value for postheating of the weld zone. This may be part of the weld interval, as shown in Fig. 1.7, or a separate application of current following a quench time period.

WELD TIME

The time of current application or weld time for other than stored energy power is controlled by electronic, mechanical, manual, or pneumatic means. Times commonly range from one half cycle (60 Hz) for very thin sheets to several seconds for thick plates. For the capacitor or magnetic type of stored energy machines, the weld time is determined by the electrical constant of the system.

Single-Impulse Welding

The use of one continuous application of current to make an individual weld is called single-impulse welding (see Fig. 1.6). Up or down current slope may be included in the time period.

Multiple-Impulse Welding

Multiple-impulse welding consists of two or more pulses of current separated by a preset cool time (see Fig. 1.7). The basic application of this sequence is for spot welding relatively thick steel sheet to control the rate of heating at the interface.

ELECTRODE FORCE

Completion of the electrical circuit through the electrodes and the work is assured by the application of electrode force. This force is produced by hydraulic, pneumatic, magnetic, or mechanical devices. The pressure developed at the interfaces depends upon the area of the electrode faces in contact with the workpieces. The functions of this force or pressure are to (1) bring the various interfaces into intimate contact, (2) reduce initial contact resistance at the interfaces, (3) suppress the expulsion of molten weld metal from the joint, and (4) consolidate the weld nugget.

Forces may be applied during the welding cycle as follows:

(1) A constant weld force.

(2) Precompression and weld forces—a high initial level to reduce initial contact resistance and bring the parts into intimate contact, followed by a lower level for welding.

(3) Precompression, weld, and forging forces—the first two levels as described in (2) followed by a forging force near the end of the weld time. Forging is used to reduce porosity and hot cracking in the weld nugget. This combination is shown in Fig. 1.8.

(4) Weld and forging forces.

EQUIPMENT

Spot, seam, and projection welding equipment consist of three basic elements: an electrical circuit, the control equipment, and a mechanical system.[3]

ELECTRICAL CIRCUIT

This consists of a welding transformer, a primary contactor, and a secondary circuit. The secondary circuit includes the electrodes that conduct the welding current to the work and the work itself. In some cases, a means of storing electrical energy is also used in the circuit. Both alternating current and direct current are used for resistance welding. The welding machine converts 60 Hz line power to low voltage, high amperage power in the secondary circuit of the welding machine.

Alternating Current

Most resistance welding machines produce single-phase alternating current (ac) of the same frequency as the power line, usually 60 Hz. The machine contains a single-phase transformer that provides the high welding currents required at a low voltage, normally in the range of 1 to 25 volts. Depending upon the thickness and type of material to be welded, currents may range from 1000 to 100 000 amperes. The typical electrical circuit design for this type of machine is shown in Fig. 1.9.

Direct Current

Welding machines may produce direct current of continuous polarity, pulses of current of alternating polarity, or high-peaked pulses of current. The latter type is produced by stored electrical energy.

---

3. Resistance welding equipment is covered in Ch. 13.

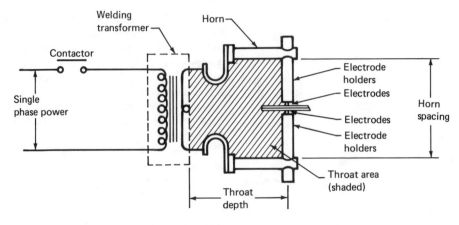

Fig. 1.9—Typical single-phase spot welding circuit

Rectifier Type Machines. These machines are direct energy type in that ac power from the plant distribution system passes through a welding transformer and is then rectified to dc power. Silicon diode rectifiers are widely used in the secondary circuit because of their inherent reliability and efficiency. The system can be single phase. However, one of the advantages of direct current systems is the ability to use a three-phase transformer to feed the rectifier system in the secondary circuit. This makes it possible to use balanced three-phase line power.

Frequency Converter Machines. This type of machine has a special welding transformer with a three-phase primary and a single-phase secondary. The primary current flow is controlled by ignitron tubes or silicon-controlled rectifiers (SCRs). Half cycles of three-phase power, either positive or negative, are conducted to the transformer for a timed period that depends upon the transformer design. The transformer output is a pulse of direct current. By switching the polarity of the primary half cycles, the polarity of the secondary current is reversed. A weld may be made with one or more dc pulses.

Stored Energy Machines. Stored energy machines are of electrostatic design. They draw power from a single-phase system, store it, and then suddenly discharge it to make the weld. These machines draw power from the supply over a relatively long time between welds, delivering it to the electrodes during a short weld time.

The equipment for electrostatic stored energy welding consists primarily of a bank of capacitors, a circuit for charging these capacitors to a predetermined voltage, and a system for discharging the capacitors through a suitable welding transformer. High voltage capacitors are generally used, the most common varying from 1500 to 3000 volts.

Electrodes

Resistance welding electrodes[4] perform four functions:

(1) Conduct the welding current to the work and, with spot and seam welding, determine the current density in the weld zone. In projection welding, the current density is determined by the size, shape, and number of projections.

(2) Transmit a force to the workpieces.

(3) Dissipate part of the heat from the weld zone.

(4) Maintain relative alignment and position of the workpieces in projection welding.

If the application of pressure were not involved, electrode material selection could be

---

4. Resistance welding electrodes are discussed more fully in Ch. 3.

made almost entirely on the basis of electrical and thermal conductivity. Since the electrodes are subjected to forces that are often of considerable magnitude, they must be capable of withstanding the imposed stresses at elevated temperatures without excessive deformation. Proper electrode shape is important because the current must be confined to a fixed area to achieve needed current density.

When only one spot or seam weld is to be made at a time, only one pair of electrodes is required. In this case, the force and current are applied to each weld by shaped electrodes. Several closely spaced projection welds can be made with one pair of welding dies (electrodes).

Electrodes of several copper alloys with satisfactory physical and mechanical properties are available commercially. Generally speaking, the harder the alloy, the lower its electrical and thermal conductivities. The choice of a suitable alloy for any application is based on compromise of its electrical and thermal properties with its mechanical qualities. Electrodes selected for aluminum welding, for instance, should have high conductivity at the expense of high compressive strength to minimize electrode sticking to the work. Electrodes for welding stainless steel, on the other hand, should sacrifice high conductivity to obtain good compressive strength to withstand the required electrode force.

Resistance to deformation or mushrooming depends upon the proportional limit and the hardness of the electrode alloy. The proportional limit is largely established by heat treatment. The temperature of the electrode face is the governing factor because this is where softening takes place.

The size and shape of electrodes are usually determined by the sheet thickness and the metal to be welded. Spot welding electrodes for steel should have flat faces of the proper diameter for the sheet thickness and should be of sufficient size to carry electrical, thermal, and mechanical loads.

CONTROL EQUIPMENT

Welding controls may provide one or more of the following principal functions:

(1) Initiate and terminate the flow of current to the welding transformer

(2) Control the magnitude of the current

(3) Actuate and release the electrode force mechanism at the proper times

They may be divided into three groups based on their functions: welding contactors, timing and sequencing controls, other current controls and regulators.

The welding contactor connects and disconnects the primary power to the welding transformer. It may be mechanical, magnetic, or electronic in operation. Mechanical contactors are operated with a foot pedal or a motor-driven cam. Magnetic contactors are actuated by an electromagnet operating against a spring and gravity that open the contactor when the electromagnet is de-energized. Electronic contactors use silicon-controlled rectifiers (SCRs) or ignitron or thyratron tubes to control the flow of primary current to the welding transformer.

The timing and sequence control establishes the welding sequence and the duration of each function of the sequence. This includes application of electrode force and current as well as the time intervals following each function.

The welding current output of a machine is controlled by transformer taps or an electronic heat control, or both. An electronic heat control is used in conjunction with ignitron tubes or SCRs. It controls current by delaying the firing of the ignitron tubes during each half cycle (60 Hz). Varying the firing delay time can be used to gradually increase or decrease the primary rms amperage. This provides upslope and downslope control of welding current.

Transformer taps are used to change the number of primary turns connected across the ac power line. This changes the turns ratio of the transformer with an increase or decrease in open circuit secondary voltage. Decreasing the turns ratio will increase the open circuit secondary voltage, the primary current, and the welding current.

MECHANICAL SYSTEMS

Spot, seam, and projection welding machines have essentially the same types of mechanical operation. The electrodes approach and retract from the work at controlled times and rates. Electrode force is applied by hydraulic, pneumatic, magnetic, or mechanical means. The rate of electrode approach must be rapid but controlled so that the electrode faces are not deformed from repeated blows. The locally heated weld metal expands and contracts rapidly during the welding cycle and the electrodes must follow this movement to maintain welding pressure and electrical contact. The ability of the machine to follow motion is influenced by the weight of moving parts, or their inertia, and by friction between the moving parts and the machine frame.

If the pressure between the electrodes and work drops rapidly during weld time, excessive surface heating may occur and an arc may form between the electrodes and the work. The arc will burn or pit the electrode faces and cause them to stick to the work. In some cases, the parts being welded may vaporize from the very high arc energy.

The electrode force used during the melting of the weld nugget may not be adequate to consolidate the weld metal and prevent internal porosity or cracking. Multilevel force machines may be employed to provide a high forging pressure during weld solidification. The magnitude of this pressure varies with the composition and thickness of the metal and the geometry of the parts. The forging pressure is often two to three times the welding pressure. Since the weld cools from the periphery inward, the forging pressure must be applied at or close to the current termination point.

SURFACE PREPARATION

For all types of resistance welding, the condition of the surfaces of the parts to be welded is of prime importance if consistent weld quality is required. The contact resistance of the faying surfaces has a significant influence on the amount of welding heat generated; hence, the electrical resistance of these surfaces must be nearly uniform for consistent results. They must be free of high resistance materials such as paint, scale, thick oxides, and heavy oil and grease. If it is necessary to prime paint the faying surfaces prior to welding, as is sometimes done, the welding operation must be performed immediately after applying the primer or special conducting primers must be used. For best results, the prime coat should be as thin as possible so that the electrode force will displace it and give metal-to-metal contact.

Paint should never be applied to outside surfaces before welding, because it will reduce electrode life and produce poor surface appearance. Heavy scale should be removed by mechanical or chemical methods. Light oil on steel is not harmful unless it has picked up dust or grit. Drawing compounds containing mineral fillers should be removed before welding.

The methods used for preparing surfaces for resistance welding differ for various metals and alloys.[5] A brief description of surface conditions and methods of cleaning follows.

Aluminum

The chemical affinity of aluminum for oxygen causes it to become coated with a thin, invisible film of oxide whenever it is exposed to air. The thin oxide film that forms on a freshly cleaned aluminum surface does not cause sufficient resistance to be troublesome for resistance welding. The permissible holding period, or elapsed time between cleaning and welding, may vary up to 48 hours or more, depending upon the cleaning process used, cleanliness of the shop, the particular alloy, and the application.

An aluminum surface may be mechanically cleaned for resistance welding with a fine grade

---

5. Methods for cleaning some metals and alloys are described in the appropriate chapters in Sec. 4, *Welding Handbook*, 6th ed. (This information may be revised in Vol. 4, 7th ed.)

of abrasive cloth, fine steel wool, or a fine, motor-driven scratch brush. Clad aluminum may also be cleaned by mechanical means but care must be taken not to damage the cladding. Numerous commercial chemical cleaners are available for aluminum. Chemical cleaning is usually preferred in large volume production for reasons of economy as well as uniformity and control.

Magnesium

Cleaning is particularly important since magnesium alloys readily alloy with copper at elevated temperatures. The contact resistance between the electrode and the work must be kept as low a possible. The alloys are supplied with either an oil or a chrome-pickle coating to protect the metal from oxidation during shipment and storage. For resistance welding, the protective coating must be removed to facilitate the oxide removal operation if sound and consistent welds are to be obtained.

Copper

Cleaning of copper alloys is important. The beryllium-coppers and aluminum-bronzes are particularly difficult to clean by chemical means. Mechanical means are preferred. In some instances, a flash coating of tin is employed to produce a uniformly higher surface resistance than pure copper would have.

Nickel

The maintenance of high standards of material cleanliness is of major importance in the resistance welding of nickel and its alloys. The presence of grease, dirt, oil, and paint increases the probability of sulfur embrittlement during welding and will result in defective welds. Oxide removal is necessary if heavy oxides are present from prior thermal treatments. Machining, grinding, blasting, or pickling may be employed. Wire brushing is not stisfactory.

Steels

Plain carbon and low alloy steels have relatively low resistance to corrosion in ordinary atmosphere. Hence, these metals are usually protected by a slushing oil during shipment, storage, and processing. This oil film has no harmful effects on the weld, provided the oily surfaces are not contaminated with shop dirt or other poorly conductant or dielectric materials.

Steels are supplied with various surface finishes. Some of the more common are (1) hot-rolled, unpickled; (2) hot-rolled, pickled, and oiled; and (3) cold-rolled with or without an anneal. The unpickled, hot-rolled steel must be pickled or mechanically cleaned. The hot-rolled pickled steel is weldable in the as-received condition except for possible wiping to remove loose dirt. The cold-rolled steel presents the best welding surface and, if properly protected by oil, requires no cleaning prior to welding other than wiping to remove any loose dirt.

The high alloy and stainless steels are non-corrosive and usually require no involved cleaning before welding. When exposed to elevated temperatures, stainless steels will acquire an oxide film; the thickness depends upon the temperature and time of exposure. The scale is an oxide of chromium which is effectively removed by pickling. Oil and grease should be removed by solvent or vapor degreasing prior to welding.

Coated Steels

The coatings and platings applied to carbon steel for the purpose of corrosion resistance or decoration lend themselves satisfactorily to resistance welding with few exceptions. In general, good results may be obtained without special cleaning processes. Welding of aluminized steel gives less difficulty with expulsion and pickup if the surfaces are wire brushed. Phosphate coatings increase the electrical resistance of the surfaces to a degree that welding current cannot pass through the sheets with low welding pressures. Higher pressures will produce welds, but slight variations in coating thickness may prevent welding.

Surface Preparation Control

Surface preparation control can be maintained by periodically measuring the room temperature contact resistance of the workpieces immediately following cleaning. The measurement is most readily taken from electrode tip to tip through two or more thicknesses of metal. Unit surface resistance varies inversely with pressure, temperature, and area of contact. The test conditions must be specified for the measurements to have significance in control of surface cleanliness.

SPOT WELDING

APPLICATIONS

Spot welding is used for the fabrication of sheet metal assemblies up to about 0.125-in. thickness when the design permits the use of lap joints and gastight seams are not required. The process is used in preference to mechanical fasteners, such as rivets or screws, when disassembly for maintenance is not required. Spot welding is much faster and more economical since a separate fastener is not needed for assembly.

The process is used extensively for joining low carbon steel sheet metal components for automobiles, cabinets, furniture, and similar products. Stainless steel, aluminum, and copper alloys are also spot welded commercially.

Occasionally the process is used to join steel plates up to 1/4-in. thick. However, loading of joints is limited and the joint overlap adds weight and cost to the assembly when compared to an arc welded butt joint.

ADVANTAGES AND LIMITATIONS

The major advantages of spot welding are high speed and adaptability for automation in the high production of sheet metal assemblies. It can be incorporated into assembly lines with other fabrication operations. Spot welding is also economical in many job shop operations using semiautomatic machines because it is faster than arc welding or brazing and requires less skill to perform.

The process also has limitations, some of which are:

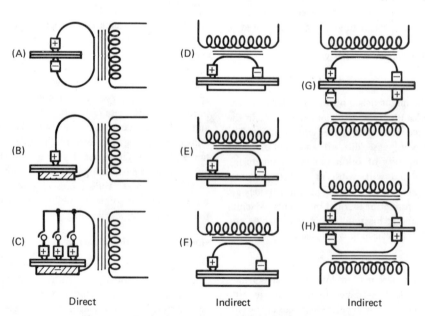

Direct   Indirect   Indirect

Fig. 1.10—Typical arrangements for single spot welds

(1) Disassembly for maintenance or repair is very difficult.

(2) A lap joint adds weight and material cost to the product when compared to a butt joint.

(3) The equipment costs are generally higher than for most arc welding processes.

(4) The short time, high current power requirement produces unfavorable line power demand, particularly with single-phase machines.

(5) Spot welds have low tensile and fatigue strengths because of the notch around the periphery of the nugget between the sheets.

(6) The full strength of the sheet cannot be utilized across a spot welded joint because fusion is intermittent and loading is eccentric due to the overlap.

PROCESS VARIATIONS

Direct and Indirect Welding

Spot welding is divided into two general categories: direct and indirect. For direct welding, both the welding current and the pressure are applied by the electrodes. With indirect welding, the welding current is introduced into one of the workpieces through a large contact adjacent to the electrode that applies the welding pressure. The spot weld is made at the electrode location.

Typical direct and indirect arrangements are shown in Figs. 1.10 and 1.11. Figure 1.10(A) shows the most common arrangement for spot welding that results in electrode indentations in both sheets. An electrode with a large contact area may be used on one side, as in Figs. 1.10(B) through (F), to reduce marking or to balance the heat at the interface. Figures 1.10(D) and (F) are similar arrangements except that the conducting lower anvil has been replaced by a nonconducting backup in the latter case. Figure 1.10(C) shows a setup for making a series of single spot welds in rapid succession using a single transformer connected to two or more welding stations. Only one set of electrodes is in contact with the work during weld time, producing a weld at that location. This arrangement is economical with respect to equipment costs.

Figures 1.10(G) and (H) show arrangements for welding with two transformers. The secondary circuits are connected in series and the primaries may be connected in series or parallel. Since the two secondary voltages are additive, this arrangement produces a relatively high voltage to overcome high resistance types of workpieces. This variation is sometimes called "push-pull" or "over-and-under" welding.

Parallel and Series Welding

Parallel and series welding variations are used for multiple spot welding. In parallel spot welding, two or more spot welds are made simultaneously with the welding currents from a single transformer flowing in parallel. This may be done using either a single secondary circuit with the electrodes connected in parallel as shown in Fig. 1.10(C), or multiple secondary circuits as shown in Figs. 1.11(A) and (B). The three-phase arrangement of Fig. 1.11(B) is lim-

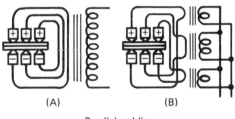

(A)   (B)

Parallel welding

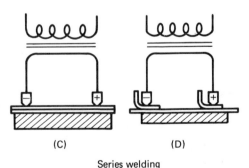

(C)   (D)

Series welding

Fig. 1.11—Typical arrangements for multiple spot welding (direct welding)

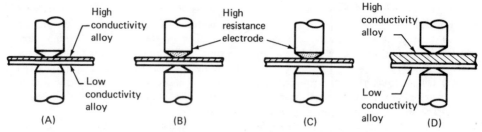

Fig. 1.12—Typical techniques for improving heat balance when spot welding dissimilar metals

ited to three welding stations only. With a single secondary circuit, the impedances of all current paths must be essentially equal for uniform current distribution to each spot weld. Surface condition, electrode design, electrode force, and material thicknesses must be the same at each welding station.

Series welding is illustrated in Figs. 1.11 (C) and (D). Welding current flows from one electrode through one spot weld into a third electrode or mandrel, then through a second spot weld to the other electrode. Welding amperage is affected by the total impedance of the series circuit. When two workpieces are spot welded in series, part of the welding current flows from one electrode through the adjacent piece to the other electrode. This shunted current does not contribute to welding. Its magnitude will depend upon the conductivity and thickness of the workpiece and the spot spacing.

HEAT BALANCE

Heat balance may present a problem when spot welding together unequal thicknesses of the same metal, equal thicknesses of two metals with a significant difference in electrical conductivities, or a combination of the two. Electrode configurations and compositions can be used to overcome unbalanced heating to some extent. If it is desired, for example, to spot weld equal thicknesses of a high conductivity alloy (electrical and thermal) to an alloy of low conductivity, improved heat balance could be obtained by one of the following techniques:

(1) Use an electrode with a smaller face area on the high conductivity alloy than that of the electrode on the lower conductivity alloy to

obtain approximately equal fusion. The electrode with the smaller face area will produce a higher current density in the higher conductivity alloy that will increase the heat generated and minimize the heat losses to the electrode across the contact area. This arrangement is shown in Fig. 1.12(A).

(2) Use an electrode material of higher resistance against the higher conductivity alloy to limit the heat losses to the electrode as shown in Fig. 1.12(B).

(3) Use a combination of (1) and (2), shown in Fig. 1.12(C).

(4) Increase the thickness of the high conductivity alloy as illustrated in Fig. 1.12(D).

The first three techniques for achieving heat balance in dissimilar metals can be applied in a general way to the spot welding of similar metals of unequal thickness. In this case, the thicker sheet has the higher resistance (low conductivity) and the nugget tends to penetrate deeper into it. Heat balance is improved by either decreasing the current density in the thicker sheet or the heat loss from the thinner sheet, or a combination of both.

In multiple layers of dissimilar thicknesses, a long weld time permits a more uniform distribution of heat in the asymmetrical resistance path between the electrodes. A correct heat balance may be obtained by using multiple-impulse (pulsation) welding or a single impulse of continuous current for an equivalent time.

JOINT DESIGN

In all cases, spot welds are made through two or more sections of equal or unequal thicknesses to produce a lap joint. One or more members may be flanges on parts or formed

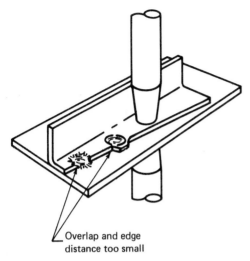

Overlap and edge
distance too small

**Fig. 1.13—Effect of improper overlap and
edge distances**

sections such as angles or channels. In all cases, the joint must be accessible from both sides to the spot welding electrodes or the dies in the case of indirect welding. In addition, the assembly must be designed so that the joints can be welded with standard machines, portable welding guns, or special purpose machines.

There are a number of factors that should be considered when designing for spot welding; several of these are edge distance, joint overlap, fit-up of parts, weld spacing, jont accessibility, marking or indentation, dissimilar thicknesses, and weld strength.

Edge Distance

This is the distance from the center of the weld nugget to the edge of the sheet. It must provide sufficient base metal to resist the internal pressure developed in the molten nugget during the welding cycle.

If spot welds are made too close to the edge of one or both members of a lap joint, the base metal at the edge of the member will overheat locally and upset outward as shown in Fig. 1.13. This reduces the restraint on the molten nugget and expulsion of molten metal may occur. The nugget may then be unsound, electrode indentation may be excessive, and weld strength

may be low. Minimum edge distance will be related to the base metal composition and strength, section thicknesses, electrode face contour, and the welding cycle.

Joint Overlap

The minimum permissible joint overlap is twice the minimum edge distance. However, other factors such as electrode clearance may require a larger overlap. If the overlap is too small, the edge distance will automatically be insufficient (see Fig. 1.13).

Fit-up

The mating parts should fit together along the joint with very little or no gap between them. Any force required to overcome joint gap will reduce the effective welding pressure. The force required to close the joint may vary as welding progresses and, thus, so will the welding pressure. The ultimate result may be significant variations in the strengths of individual welds.

Weld Spacing

When a number of spot welds are made successively along a joint, a portion of the secondary current will flow through the adjacent welds. This shunting effect must be considered when establishing the distance between adjacent spot welds and when establishing the welding machine settings.

The division of current will depend primarily upon the ratio of the resistances of the two paths, one through the adjacent welds and the other across the interface between the sheets. If the path length through the adjacent weld is long compared to the joint thickness, the shunting effect will be negligible. The minimum spacing between spot welds is related to the conductivity and thickness of the base metal, the diameter of the weld nugget, and the cleanliness of the faying surfaces. Minimum spacing for various sheet thicknesses is generally given in recommended practice tables. Welds can be made closer together using higher amperage, shorter weld time, and increased electrode force with a welding machine having a low inertia head. In some cases, an auxiliary weld timer or current control must be provided to produce the first weld with lower heat input.

Joint Accessibility

To accomplish spot welding, the design must consider the size and shape of commercially available electrodes and electrode holders as well as the type of spot welding equipment on which the welding will be done. Each joint must be accessible to the electrodes mounted on the welding machine.[6] For multiple spot welded joints, either the assembly or the welding machine (gun) must be moved unless a special multiple-station machine is provided.

Surface Marking

It is often desirable to produce welded assemblies with flush or invisible welds. When the welding current flows through the work between the electrodes, the work is resistance heated locally and tries to expand in all directions. Because of the pressure exerted by the electrodes, expansion transverse to the plane of the sheets is restricted. As the metal heats, it upsets radially in the plane of the sheets around the electrode face and usually produces a circular ridge as shown in Fig. 1.14. As the weld cools, contraction takes place almost entirely in the transverse direction and produces concave surfaces or marks at the electrode location. This is not to be confused with excessive electrode indentation into the work caused by improper welding procedures. Actual depth of a shrinkage depression seldom exceeds a few thousandths of an inch.

After some finishing operations such as painting, the marks may be very conspicuous. It is difficult to eliminate the marks completely, but they can be reduced materially by modifying the welding procedure. For example, the depth of fusion into the sheet can be minimized by welding in the shortest practical weld time.

Various techniques are used to minimize these markings. The common method is to use a large flat-faced electrode against the show side of the joint. This electrode should be made of a hard copper alloy to minimize wear. Another technique is to use an indirect variation of welding such as the arrangement shown in Fig. 1.15. It minimizes the current density and pressure in face sheet B.

Surface marking may occur when an electrode or its holder contacts the adjacent workpiece. Arcing between them may produce a small pit in the work that is objectionable from an appearance standpoint in some applications. Sometimes contact occurs when the electrode skids as it or the supporting machine components deflect under the load. In general, this will not be a problem if the proper joint overlap, electrodes, and equipment are used.

Dissimilar Thicknesses

There is a maximum section thickness ratio that can be effectively joined when spot welding two or more dissimilar thicknesses of the same metal. It is generally based on the thickness(es) of the outside sheet(s).

For carbon steels, the maximum ratio is 4 to 1 for two thicknesses. The welding of three metal thickness combinations with pointed type electrodes is limited to those where the thickness ratio of the two outside sheets is not greater than 2.5 to 1. For welding thickness combinations with higher ratios, the electrode face diameters must be selected to accommodate the different thicknesses.

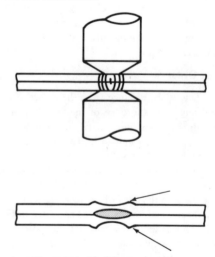

Fig. 1.14—Surface irregularity produced by spot welding

---

6. See Ch. 3 for information on electrodes and electrode holder designs.

The use of a large diameter flat electrode against one sheet will affect the size and shape of the weld nugget depending upon the thick-to-thin stacking arrangement of the sections. When the flat electrode contacts the heavier metal, a greater unbalance occurs as the thickness ratio increases. This results in less desirable welding conditions. Better welding conditions exist when the large flat electrode contacts the thinner section, although higher amperage and electrode force are necessary. The maximum ratio is then 4 to 1.

Minimum recommended spot spacing for a weld joining three thicknesses is 30 percent greater than the spacing required for welding two sections of the thicker outer sheet.

Weld Strength

The strength of a single spot weld in shear is determined by the cross-sectional area of the nugget in the plane of the interface between the two sections. In the case of a lap joint tested with the weld in shear, the eccentricity of load-ing will cause the joint to rotate as the load increases. The spot weld will fail either by shear through the nugget or by tearing the nugget out of the sheet. In general, the maximum shear strength is obtained when the nugget tears from the sheet. A nugget of some minimum diameter is required for this type of failure. Increasing the nugget diameter beyond this minimum will not significantly increase the weld strength.

Spot welds have relatively low strengths when stressed in tension by loading transverse to the plane of the sheets. This is due to the sharp notch between the sheets at the periphery of the weld nugget. Consequently, spot welded joints are not normally loaded in this manner.

The strength of multiple welded joints depends not only upon thickness but also spacing and pattern. A staggered pattern is preferred for multiple rows of welds rather than a rectangular pattern. The spacing between adjacent welds affects weld strength because of current shunting through previous welds. However, spacing

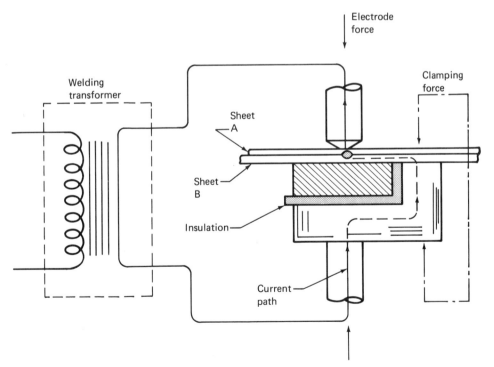

Fig. 1.15—Application of indirect welding to minimize marking on one side

also affects the number of welds that can be placed in a particular joint. Therefore, a compromise must be made between spacing and the number of welds in order to obtain the maximum joint strength.

ROLL SPOT WELDING

Roll spot welding consists of making a series of spaced spot welds in a row with a seam welding machine without retracting the electrode or removing the electrode force between welds. Electrode wheel rotation may or may not be stopped during the welding cycle. The radius of the wheel electrode, the contour of its face, and the weld time influence the shape of the nugget. The nugget is usually oval-shaped.

The weld spacing is obtained by adjustment of cool time with the wheel electrodes continuously rotating at a set speed. Hold time is effectively zero. Roll spot welding may also be done with interrupted electrode rotation when a hold time period is needed to consolidate the weld nugget as it cools.

When continuously moving electrodes are employed, as is commonly the case, weld time is usually shorter and welding amperage higher than those used for conventional spot welding. The higher amperage employed may sometimes require the use of a higher electrode force. Otherwise, recommended practices for spot welding apply.

ELECTRODE MAINTENANCE

Maintenance of electrodes is necessary for the production of consistent welds. An abnormal increase in the size of the electrode faces contacting the work is detrimental to weld strength and quality. For example, if a 1/4-in. diameter electrode face is allowed to increase to 5/16-in. diameter by mushrooming, the contact area will increase 50 percent with a corresponding decrease in current density and pressure. Depending somewhat upon the welding schedule, the result may be weak or defective welds. A danger sign is the production of poorly shaped spots which may be caused by:

(1) Noncircular electrode faces

(2) Too large a flat face on the electrode

(3) Concavity or convexity of the electrode face

(4) Misalignment of the electrodes with respect to the work

Correct electrode alignment is relatively easy to maintain with stationary welding machines and proper supporting fixtures. However, misalignment is common with portable gun type machines. The seriousness of this condition is dependent upon the ease with which the equipment can be manipulated and correctly positioned for welding. It is likely that the electrodes will have longer life between dressings on positioned work (stationary machines) than on nonpositioned work (portable welding guns).

WELDBONDING

Weldbonding is a combination of resistance spot or seam welding and adhesive bonding. The best technique is to apply the adhesive to the joint overlap and then weld through it, as is sometimes done with a primer coat. The adhesive is best applied in paste form, although films may be suitable for some applications. After welding, the adhesive is cured according to the manufacturer's recommendations.

Weldbonding improves the fatigue life and durability over that obtained with spot or seam welding alone. It may also improve buckling resistance with thin sheet, stress distribution, and rigidity of the joint. The presence of the adhesive in the joint tends to (1) dampen vibration and noise and (2) give some corrosion protection.

The adhesive, its application, and its cure add to manufacturing costs. The presence of the adhesive in the joint makes welding more difficult and may contribute to significant variations in weld quality.

The adhesive must flow from between the sheets when the electrode force is applied. Advanced curing of the adhesive may hamper flow and cause abnormally high resistance at the joint interface. High resistance may prevent current flow or cause excessive heating and metal expulsion. Application of a precompression electrode force prior to the welding cycle may help displace the adhesive (see Fig. 1.8). Regardless of conditions, not all of the adhesive will be displaced from between the sheets and, therefore, the contact resistance will be higher than with clean sheets.

The process has significant advantages for aluminum sheet metal fabrications. It may also be useful for joining thin steel sheets where improved joint strength and corrosion resistance may be realized.

SEAM WELDING

APPLICATIONS

Seam welding is primarily used to produce continuous gas- or liquid-tight joints in sheet metal tanks such as gasoline tanks for automobiles. Generally, two wheel electrodes are used. The process is sometimes used to weld the longitudinal seam in tubular sections with two wheel electrodes or one translating wheel and a stationary mandrel. A gastight seam is not normally required in structural tubing.

ADVANTAGES AND LIMITATIONS

This process has the same advantages and limitations as spot welding. An additional advantage is the ability to produce a continuous leaktight weld.

Seam welds must be made in a straight or uniformly curved path. Abrupt changes in welding direction or in joint contour along the path cannot be welded leaktight. This limits the design of the assembly.

Strength properties of seam welded lap joints are generally lower than those of fusion welded butt joints because of the eccentricity of loading on lap joints and the built-in notch along the nugget at the sheet interface.

PROCESS VARIATIONS

Lap Seam Welding

As with spot welding, lap joints can be seam welded using wheel electrodes, as shown in Fig. 1.16(A), or with one wheel and a mandrel electrode. The minimum joint overlap is the same as for spot welding.

Mash Seam Welding

Another type of weld commonly used is the mash seam weld. The overlap for this type of weld is usually considerably less than for the conventional lap joint as shown in Fig. 1.16(B).

The overlap is about 1 to 1.5 times the sheet thickness. The welded joint thickness will be in the range of 120 to 150 percent of the sheet thickness with proper welding procedures. Flat-faced wheel electrodes, wide enough to completely cover the overlap and control joint thickness, are normally used. High electrode force and continuous welding current are required. Electrode force, welding current, welding speed, overlap, and joint thickness are all interrelated and must be accurately controlled for consistent welding. The overlap must be accurately maintained and held to close tolerances. This is usually done by rigid clamping of the pieces to be welded or by tack welding them together in advance.

The show surface of a joint must be mashed as nearly flat as possible so that it will present a good appearance. In most cases, the electrode on the show surface is a mandrel that supports the piece parts to be joined. The welding wheel bears on the side that does not show. Proper positioning of the wheel with respect to the joint is required to obtain a smooth weld face acceptable for painting. The weld area may require some polishing before painting when the appearance of the finished product is important.

For consistent results in production, all parts must be accurately made within design tolerances. Locating and clamping jigs must consistently and properly position the parts for welding. In addition, all welding variables, including electrode force, weld current, and travel speed, must be accurately controlled.

Continuous seams with good appearance and free of crevices can be produced by this process variation. Crevice-free joints are necessary for applications, such as food containers or refrigerator liners, where cleanliness is important. Three problem areas that may be encountered with this method are:

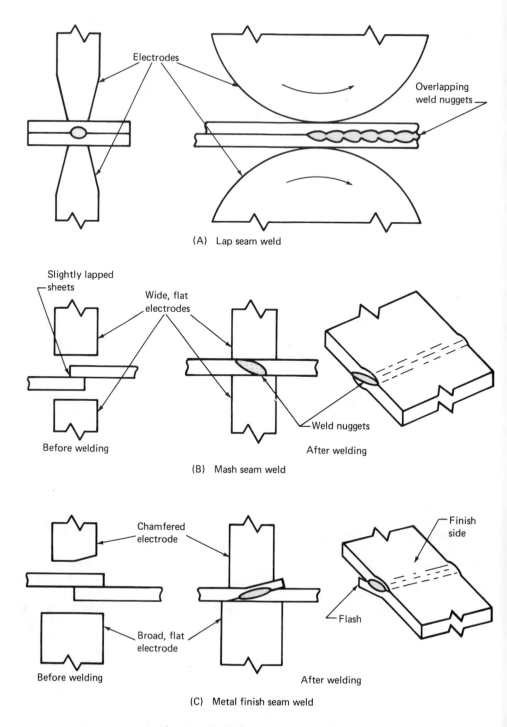

(A) Lap seam weld

(B) Mash seam weld

(C) Metal finish seam weld

Fig. 1.16—Variations of seam welding

(1) There is usually some offset at the joint because the seam cannot be mashed completely flat.

(2) The metal in the overlap must flow laterally as it is welded but this action is restrained by the fixturing or tack welds. Therefore, some distortion of the assembly is likely to take place.

(3) Very rigid fixturing is needed to resist weld distortion.

Low carbon steel and stainless steel may be mash seam welded for certain applications. However, metals that have a narrow plastic temperature range cannot be mash seam welded.

Metal Finish Seam Welding

This variation of seam welding is illustrated in Fig. 1.16(C). Lap and mash seam welds differ with respect to the amount of forging or, as the name implies, mash down. The lap weld has practically no mash down, while the thickness of a mash seam weld approaches that of one sheet thickness. In metal finish seam welding, mash down occurs on only one side of the joint. This variation of seam welding is a compromise between lap and mash seam welding.

The amount of deformation, or mash, is affected by the geometry of one electrode wheel face and the position of the joint with respect to that face. The wheel face is beveled on one side of the midpoint as shown in Fig. 1.17. This varies the amount of deformation across the joint. Good surface finish can be produced on the side of the joint against the flat wheel using proper welding procedures.

The location of the edge of the sheet contacting the flat-faced electrode, relative to the bevel on the other electrode, must be held within a close tolerance (see Fig. 1.17). With 0.031-in. thick low carbon steel sheet, for example, the edge must be within 0.016 in. of center. The overlap distance is not critical.

Higher amperage and electrode force are required than those for mash seam welding because of the greater overlap distance. The variation is applicable to the same metals and thicknesses as is mash seam welding.

Foil Butt Seam Welding

With this method, the edges of the sheets to be joined are butted together and a thin, narrow strip of foil is introduced on one or both sides of the joint as it is passed between conventional seam welding wheels. The 0.010-in. thick foil acts as a bridge to distribute welding current to the edges of both sheets, offers added electrical resistance, and helps contain the molten weld nugget as it grows and then cools. It also serves as filler metal to produce a flush or slightly reinforced weld joint. The foil must be guided accurately and centered on the joint for even current distribution to both edges. The foil may be roll spot welded to the sheets at low power before the joint is seam welded. The edges to be joined must fit tightly together and be clamped firmly in position.

For low carbon steel, welding speeds comparable to lap seam welding can be used, provided the joint has good fit-up. Very little forging is required and, thus, distortion is low.

Other

As with spot welding, two seam welds can be made in series using two welding heads. The two heads may be mounted side by side or in tandem. Two seams can be welded with the same welding current, and power demand will be only slightly greater than for a single weld.

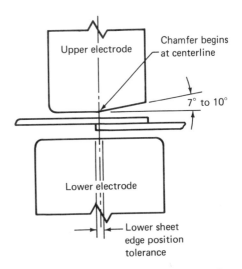

Fig. 1.17—Electrode face contour and joint position for metal finish seam welding

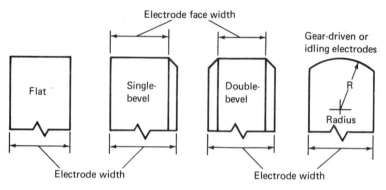

Fig. 1.18—Seam welding wheel face contours

A tandem wheel arrangement can reduce welding time by 50 percent since each half of a joint can be welded simultaneously. Thus, for a 72-in. long joint, two welding heads can be placed 36 in. apart with welding current flowing through the work from one wheel electrode to the other. A third continuous electrode is used on the other side of the joint. The full length of the joint can be welded with only 36 in. of travel.

HEAT BALANCE

Seam welding of dissimilar metals or unequal thicknesses presents the same heat balance problems as with spot welding. The techniques for improving heat balance are similar. The contact area between the work and electrode can be enlarged by increasing the wheel width or diameter, or both. One electrode wheel or mandrel may be made of an alloy of lower conductivity than the other.

WELDING CYCLE

Interrupted Current

Welding current is normally supplied in timed pulses (heat time) with a period of time (cool time) in between them. A nugget of weld metal is produced during each pulse of current.

Interrupted current is usually desirable for most seam welding operations for the following reasons:

(1) Good control of the heat is obtained.

(2) Each weld nugget in the seam is allowed to cool under pressure.

(3) Distortion of the workparts is minimized.

(4) It is easy to control spitting or burning.

(5) Sound welds with better surface appearance are possible.

To produce a gastight seam, the nuggets should overlap 15 to 20 percent of the nugget diameter. For maximum strength, the overlap should be 40 to 50 percent. The size of the nugget will depend upon heat time for a given welding speed and current. The amount of overlap will depend upon the cool time.

For a particular metal and sheet thickness, the number of welds (nuggets) per inch that can be produced economically will fall within a range. In general, as the sheet thickness decreases, the number of welds per inch must increase to obtain a strong, gastight seam weld. The ratio of welds per inch to welds per minute will establish the welding speed in inches per minute. The number of welds per minute is the number of cycles of ac per minute divided by the sum of the heat and cool times in cycles for a single weld.

To obtain the minimum number of welds per inch that will produce the required seam at a given welding speed, the heat time and welding currrent should be adjusted to give the required weld nugget geometry. The cool time should then be set to give the necessary nugget overlap. Since decreasing cool time may increase the heat buildup, nugget penetration may increase.

Continuous Current

With low carbon steel, welding current can be applied continuously along the length of the seam with high travel speeds if the current wave form available will produce the proper nugget size and spacing. In this case, weld quality is secondary to high production requirements. Continuous current can be used for sheet up to and including 0.040-in. thickness. Above this thickness, surface condition has a significant effect on welding and electrode life is short. Continuous current welds in a particular thickness can be made over a wide range of speeds. For example, two thicknesses of 0.040-in. steel stock can be welded at speeds ranging from 105 to 310 in./ min. The required amperage increases with speed.

WELDING SPEED

The speed of welding depends upon the metal being welded, stock thickness, and the weld strength and quality requirements. In general, permissible welding speeds are much lower with stainless steels and nonferrous metals because of restrictions on heating rate to avoid weld metal expulsion.

In some applications, it is necessary to stop the movement of the electrodes and work as each weld nugget is made. This is usually the case for sections over 0.188-in. thick or for metals that require postheating or forging cycles to produce the desired weld properties. Interrupted motion significantly reduces welding speed because of the relatively long time required for each weld.

With continuous motion, the welding current must be increased and heat time decreased as welding speed is increased to maintain weld quality and joint strength. There is some speed beyond which the required welding current may cause undesirable surface burning and electrode pickup. This will accelerate electrode wear.

ELECTRODES

Seam welding electrodes are normally wheels with diameters ranging from 2 to 24 inches. Common sizes have diameters of 7 to 12 in. and widths of 0.375 to 0.75 inches.

Various wheel contours are used, but the four basic types are: flat, single-bevel, double-bevel, and radius-faced. These are shown in Fig. 1.18. The face contour used is determined by the type of drive mechanism and the require-

ments for current and pressure distribution in the weld zone.

If the electrode is driven on the periphery by a steel roll, it is either knurl- or friction-driven; if by a train of gears, it is gear- or shaft-driven. Sometimes one electrode is driven and the other one idles. Gear type drives may be used to avoid interference with the work.

Radius-faced electrodes provide the best weld appearance and are the easiest to set up. Flat-faced electrodes are often used but are more difficult to set up with the two flat surfaces parallel and uniformly contacting the work. A knurl-driven electrode marks the weld seam. The width of the wheel is determined by the thickness and geometry of the part being welded. The diameter of the wheel is determined by the part configuration or the physical requirements of the welding machine.

Common practice is to cool both the electrodes and the work with flood cooling. If cooling water is detrimental to the work, the electrode shafts may be internally cooled.

The rate of electrode wear is determined by:

(1) The electrode material, face width, and operating temperature

(2) Abrasion by the drive rolls and the work

(3) The metal being welded and its surface condition

The width of the weld cross section at the interface between the two workpieces should range from 1.5 to 3 times the thickness of the thinner section. The ratio of weld width to sheet thickness normally decreases as the thickness increases. The weld width is always slightly less than the face width when commercial welding schedules are used.

EXTERNAL COOLING

Flood, immersion, or mist cooling is commonly used with seam welding. This is generally in addition to any internal cooling of the components in the secondary circuit of the welding machine. When external cooling is not used, electrode wear and warpage of the work may be excessive. For welding nonferrous metals and stainless steel, clean tap water is satisfactory. For ordinary steels, a 5 percent borax solution is commonly used to minimize rusting.

JOINT DESIGN

The various requirements that must be met in designing spot welded joints also apply, to seam welded joints. With seam welding, electrode design together with mounting and leak-tightness requirements place some limitations on design.

The wheel type electrodes are relatively large and require unobstructed access to the joint. Since the electrodes rotate during welding, they cannot be inserted into small recesses or internal corners. External flanges must change direction over large radii to be able to produce a strong, leaktight seam weld.

Curvature of the overlap plane is severely restricted by the diameters of the wheel electrodes. The smaller the radius of curvature, the more difficult the welding operation becomes. The diameter of the inside wheel must be less than twice the radius of curvature. The contact length between the outside wheel and the work is shorter and the contact length with the inside wheel is longer than with a flat flange. Welding speed must be slower because of this condition.

Figure 1.19 shows several common designs of seam welded joints that are generally similar to those used for spot welding applications. The most commonly used design, Fig. 1.19(A), is the simple lap joint in which the pieces or edges to be welded are overlapped sufficiently to prevent spitting of the weld metal from the edges of the stock. Common examples are the longitudinal seams in cans, buckets, water tanks, mufflers, and large diameter, thin-walled pipes.

Flange joints are forms of lap joints. The design in Fig. 1.19(B), in which one of the pieces is straight, is commonly used for welding flanged ends to containers of various types. In Fig. 1.19(C), both pieces are flanged. This design is used to join the two sections of automotive gasoline tanks. Often the flanged pieces are dished to obtain added strength, in which case it is necessary to mount one or both wheels at an angle to clear the work, as shown in Fig. 1.19(D). A practical limit is 6 degrees because greater angles cause excessive bearing thrust, lack of bearing clearance, and the use of large diameter electrodes.

PROJECTION WELDING

APPLICATIONS

Projection welding is primarily used to join a stamped, forged, or machined part to another part. One or more projections are produced on the parts during the forming operations. Fasteners or mounting devices, such as bolts, nuts, pins, brackets, and handles, can be projection welded to a sheet metal part. It is especially useful for producing several weld nuggets simultaneously between two parts. Marking of one part can be minimized by placing the projections in the other part.

The process is generally used for section thicknesses ranging from 0.02 to 0.125-in. thick. Thinner sections require special welding machines capable of following the rapid collapse of the projections. Various carbon and alloy steels and some nickel alloys can be projection welded.

ADVANTAGES AND LIMITATIONS

In general, projection welding can be used instead of spot welding to join small parts to each other or to larger parts. Economics of the two processes should be considered in the selection of one for a particular application. The chief advantages of projection welding include the following:

(1) A number of welds can be made simultaneously in one welding cycle of the machine. The limitation on the number of welds is the ability to apply uniform electrode force and welding current to each projection.

(2) Less overlap and closer weld spacing are possible because the current is concentrated by the projection, and shunting through adjacent welds is not a problem.

(3) Thickness ratios of at least 6 to 1 are possible because of the flexibility in projection

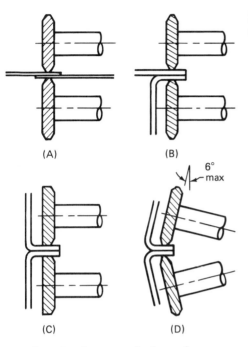

Fig. 1.19 — Common designs of seam-welded joints

size and location. The projections are normally placed on the thicker section.

(4) Projection welds can be located with greater accuracy and consistency than spot welds, and the welds are generally more consistent because of the uniformity of the projections. As a result, projection weld size can be smaller than spot welds.

(5) Marking of the show surface can be minimized by placing the projections in the other part. Any slight deformation on the show side can be sanded flush with the base metal.

(6) Large, flat-faced electrodes are used. Consequently, electrode wear is much lower than that with spot welding and this reduces maintenance costs. In some cases, the fixturing or part locators are combined with the welding dies or electrodes when joining small parts together.

(7) Oil, rust, scale, and coatings are less of a problem than with spot welding because the tip of the projection tends to break through the foreign material early in the welding cycle.

However, weld quality will be better with clean surfaces.

The process has several limitations in comparison to spot and seam welding; the most important ones are:

(1) The forming of projections may require an additional operation unless the parts are press-formed to design shape.

(2) With multiple welds, accurate control of projection height and precise alignment of the welding dies are necessary to equalize the electrode force and welding current.

(3) With sheet metal, the process is limited to thicknesses in which acceptable projections can be formed and for which suitable welding equipment is available.

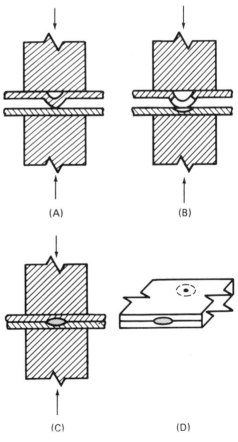

Fig. 1.20 — Formation of a projection weld

(4) Multiple welds must be made simultaneously and this requires higher capacity equipment than does spot welding. It also limits the practical size of the component containing the projections.

TYPES OF JOINTS

As with spot and seam welding, lap joints can be produced by projection welding. Generally, several domed or elongated projections are formed on one part. The number of projections used will depend upon the joint strength requirements, as with spot welding.

A forged or machined part can be welded to a sheet section using an annular projection on the part. Strong, leaktight joints can be produced. The projection is formed or machined on the end or shoulder of the part which, in turn, may be aligned with or pass through a hole in the sheet.

PROJECTION DESIGNS

Projections may be formed on a sheet of metal by embossing, on a solid piece of metal by machining or forging, or on an edge of stampings. The purpose of a projection is to localize the heat and pressure at a specific location on the joint. Welding current density is determined by the projection design.

The sequence of events during the formation of a projection weld is shown schematically in Fig. 1.20. In Fig. 1.20(A), the projection is shown in contact with the mating sheet. In Fig. 1.20(B), the current has started to flow through the projection, thereby heating it to welding temperature. The electrode force causes the heated projection to collapse rapidly and then fusion takes place as shown in Fig. 1.20(C). The completed weld is shown in Fig. 1.20(D).

Sheet Metal

A projection design for sheet metal should meet the following requirements:

(1) Be sufficiently rigid to support the initial electrode force before welding current is applied.

(2) Have adequate mass to heat a spot on

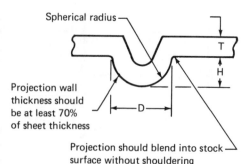

Fig. 1.21—General design of a
projection for steel sheet

the other surface to welding temperature. If too small, it will collapse before the other surface is adequately heated.

(3) Collapse without metal expulsion between the sheets or sheet separation after welding.

(4) Be easy to form and not be partially sheared from the sheet during the forming operation. Such projections may be weak and the resulting welds may be easily torn from the sheet on loading.

(5) Cause little distortion of the part during forming or welding.

The general design of a projection suitable for steel sheet is shown in Fig. 1.21. This design avoids the tendency to form the projection by

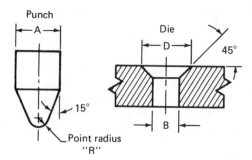

Material: tool steel hardened to 50-52 HRC

Fig. 1.22—Punch and die design for
projections in sheet metal

Table 1.1
Punch and die dimensions for spherical dome projections[a]

	Projection		Punch		Die	
				Point	Hole	Chamber
Thickness	Height,	Diameter,	Diameter	radius,	diameter,	diameter,
T	H, ±2%	D, ±5%	A	R, ±0.002	B, ±0.005	D
0.022-0.034	0.025	0.090	0.375	0.031	0.076	0.090
0.036-0.043	0.035	0.110	0.375	0.047	0.089	0.110
0.049-0.054	0.038	0.140	0.375	0.047	0.104	0.130
0.061-0.067	0.042	0.150	0.375	0.062	0.120	0.150
0.077	0.048	0.180	0.375	0.062	0.144	0.180
0.092	0.050	0.210	0.500	0.078	0.172	0.210
0.107	0.055	0.240	0.500	0.078	0.196	0.240
0.123	0.058	0.270	0.500	0.094	0.221	0.270
0.135	0.062	0.300	0.500	01.09	0.250	0.300
0.153	0.062	0.330	0.500	0.125	0.270	0.330
0.164	0.068	0.350	0.500	0.141	0.297	0.360
0.179	0.080	0.390	0.500	0.156	0.328	0.390
0.195	0.084	0.410	0.500	0.156	0.338	0.410
0.210	0.092	0.440	0.500	0.187	0.358	0.440
0.225	0.100	0.470	0.500	0.187	0.368	0.470
0.245	0.112	0.530	0.500	0.187	0.406	0.530

a. All dimensions are in inches.

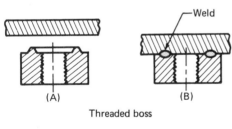

Threaded boss

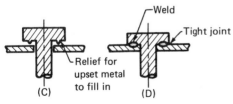

Shoulder stud

**Fig. 1.23—Examples of annular
projection welding**

shearing of the sheet or by significant thinning of the projection wall. The design of the punch and die to form this projection shape is illustrated in Fig. 1.22. The projection sizes recommended for various sheet thicknesses and the punch and die dimensions to produce the projections are given in Table 1.1.

Projections may be elongated to increase the nugget size and, thus, the strength of the weld. In this case, the contact between the projection and the mating section is linear. Elongated projections are generally used for the thicker sheet gages.

On thin sheet, an annular projection of small diameter may be used instead of a round projection. The annular projection has greater stiffness to resist collapse when electrode force is applied.

Machined or Forged Parts

Annular projections are frequently used on screw machine or forged parts to carry heavy loads and for applications that require a pressure-tight joint around a hole between two parts.

Such preparation also produces a high strength weld when a large stud or boss is welded to thin sheet metal. Figure 1.23 shows two applications of annular projections. Normally, the projection has a 90 degree included angle at the nose. The nose should be rounded, particularly to improve heat balance with heavy sections. Relief, as shown in Fig. 1.23(C), should be provided at the base of the projection for the upset metal to fill as the projection collapses. This will assure a tight joint without a gap, as shown in Fig. 1.23(D).

Various designs of weld fasteners are available commercially for projection welding applications. Typical examples are shown in Fig. 1.24. Projection design and number depend upon the application.

HEAT BALANCE

The factors that affect heat balance are:
(1) Projection design and location
(2) Thickness of the sections
(3) Thermal and electrical conductivities of the metals being welded
(4) Heating rate
(5) Electrode alloy

The distribution of heat in the two sections being projection welded together must be reasonably uniform to obtain strong welds, as in spot welding. Maintenance of heat balance may be a greater problem with projection welding because two or more welds are usually made simultaneously. Uniform division of welding current and electrode force is necessary to obtain even heating of all projections. However, the resistance across the projections will vary. Since the current paths are in parallel, the current will be distributed among them accordingly.

Projections must be designed to support the electrode force needed to obtain good electrical contact with the mating part and to collapse rapidly when heated. With multiple projections, slight variations in projection heights can affect heat balance. This can occur with wear of the projection-forming punches.

The major portion of the heat develops in the projections during the welding operation.

For this reason, the projections should be formed in the thicker of two pieces of the same metal or in the section with higher electrical conductivity when dissimilar metals are being joined. With high conductivity metals, heat dissipation is more rapid.

With unequal thicknesses, the projection design should be based on that required for the thinner section, even though the projection is formed in the thicker section.

As with spot welding, the conductivity of the electrode alloy can affect heat balance. Choice of electrode alloys can be used to improve heat balance.

WELDING CYCLE

Welding Current

The current for each projection is generally less than that required to produce a spot weld in the same metal and thickness combination. The projection will heat rapidly and excessive current will melt it and result in expulsion. However, the current must be high enough to create fusion before the projection has completely collapsed.

For multiple projections, the total welding current will approximately equal the current for one projection multiplied by the number of projections. Some adjustment may be required to account for normal projection tolerances, part designs, and the impedance of the secondary circuit.

Weld Time

Weld time is about the same for single or multiple projections for the same design. Although a short weld time may be desirable from a production standpoint, it will require correspondingly higher amperage. This may cause overheating and metal expulsion. In general, longer weld times and lower amperages are used for projection welding than those for spot welding.

In some cases, multiple impulse welding may be advantageous to control heating rate.

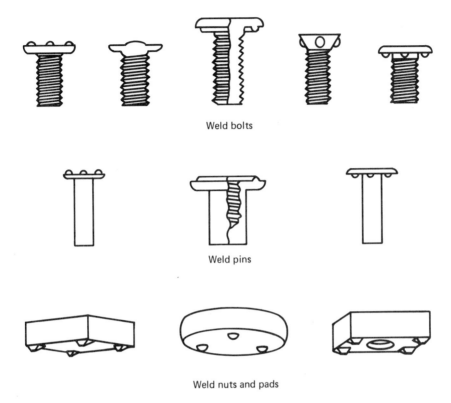

Weld bolts

Weld pins

Weld nuts and pads

Fig. 1.24—Typical commercial projection weld fasteners

This is helpful with thick sections and with metals of low thermal conductivity.

Electrode Force

The electrode force used for projection welding will depend upon the metal being welded, the projection design, and the number of projections in the joint. The force should be adequate to flatten the projections completely when they reach welding temperature, and to bring the workpieces in contact. Excessive force will prematurely collapse the projections and the weld nuggets will be ring-shaped with incomplete fusion in the center.

The welding machine must be capable of mechanically following the work with the electrodes as the projections collapse. Slow follow-up will permit metal expulsion before the workpieces are together.

ELECTRODES AND WELDING DIES

The areas of parts to be joined are frequently flat except for the projections. In such cases, large flat-faced electrodes are used. When the surfaces to be contacted are contoured, the electrodes are fitted to them. With such electrodes, the electrode force can be applied without distorting the parts, and the welding current can be introduced without overheating the contact areas.

For a single projection, the electrode face diameter should be at least twice the diameter of the projection. With multiple projections, the electrode face should extend a minimum of one projection diameter beyond the boundary of the projection pattern.

The best electrode material is one that is sufficiently hard (to minimize wear) but does not

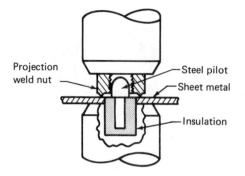

Fig. 1.25—Locating a weld nut with
an insulated pin

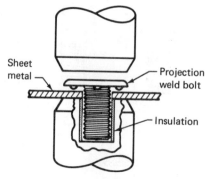

Fig. 1.26—Positioning a weld bolt
with an insulating sleeve

crack or cause surface burning on the part. If burning or cracking is encountered, a softer alloy of higher conductivity should be used. With multiple projections, electrode wear can upset the balance of welding currents and electrode forces on each projection. Then, the strength and quality of the welds may become unacceptable.

Electrodes for large production requirements often have inserts of Resistance Welder Manufacturers Association (RWMA) Group B material at the points of greatest wear. In some cases, it is more economical and equally satisfactory to use one-piece electrodes of RWMA Group A, Class 3 alloy.

Welding electrodes and locating dies for projection welding are usually combined. With the proper dies, it is possible to attain accuracy with projection welding equal to that of any other assembly process. The welding dies should meet the following requirements:

(1) Provide accurate positioning of the parts

(2) Permit rapid loading and unloading

(3) Have no alternative path for the welding current

(4) For ac welding, be made of nonmagnetic materials

(5) Be properly designed for operator safety

The dies must be mounted solidly on the welding machine. The parts are mated in one die and all the welds are made at once with one operation of the machine. One part may be located in relation to the other by punching holes in one with semipunchings in the other to match. The projections can usually be embossed or forged in the same operation.

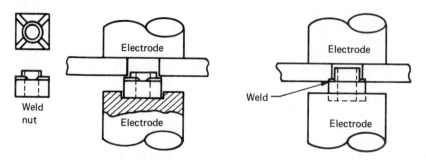

Fig. 1.27—Use of a recessed electrode to position a weld nut

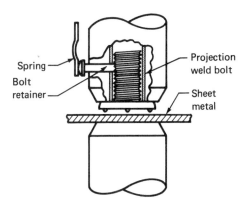

Fig. 1.28—Holding a bolt in the
upper electrode with a spring
retainer

In some designs, insulated pins or sleeves may be used in the electrode or dies to position and align the parts. Simple examples are shown in Figs. 1.25 and 1.26.

When the small part of an assembly can be placed on the bottom and the large part on top, it is a simple matter to hold the small part in a recessed lower electrode such as shown in Fig. 1.27. When it is desired to locate a small part on top of a larger part, a problem exists. Sometimes the small part can be located and held by a removable device and then welded with a flat upper electrode. Parts that nest into the upper electrode may be held by spring clips attached to the electrode. Figure 1.28 shows a spring-loaded retainer through the side of the electrode holding a bolt for welding. Vacuum may also be used to hold small parts in the upper electrode or die when part design permits.

The success of projection welding operations in production with respect to the electrodes depends largely upon the proper selection of materials, proper installation, and proper maintenance. If the dies are correctly designed and constructed, the installation is next in importance. First, the platens of the welding machine must be parallel to each other and perpendicular to the motion of the ram. The platens should also be smooth, clean, and free of nicks and pit marks. If they are not, the platens should be removed and machined smooth and flat before

installation of the dies. The check for parallelism of the platens should be made under intended operating forces. This can best be done by placing a steel block with smooth, parallel faces between the platens, applying the intended electrode force, and then checking for gap with feeler gages.

The next step is to check the bases of the die blocks. They must be clean, smooth, flat, and free from burrs and nicks. If not, they should be machined.

The dies are then installed on the machine. Most machines have tee slots at right angles to one another in the two platens to permit universal alignment of the dies. After the dies are properly lined up, they should be clamped securely to the platens. With the work in place in the dies, the position of the ram or knee of the machine should be adjusted for the proper stroke, including the necessary allowance for upset of the projections.

If the tips of the projections are in one plane and of uniform height, the setup is ready for trial welds. Non-uniformity of current or force on the projections may be caused by the following:

(1) Shunting of current through locators

(2) Unequal secondary circuit path lengths

(3) Excessive play in the welding head

(4) Too much deflection in the knee of the machine

The use of shims between die components or between dies and platens should be avoided. If shims must be used, they should only be clean, annealed, pure copper sheets of sufficient area to carry the secondary current.

If projections are located on curved or angled surfaces, accurate templates should be provided for checking the dies. Also, when curved parts are welded, or two or more parts are welded to others, the mill tolerances for the metal thicknesses involved may cause problems. These tolerances must be provided for in the design of the parts and arrangement of the projections.

JOINT DESIGN

Lap joint designs for projection welding are similar to those for spot welding. In general, joint overlap and edge distances for projection welding can be less than for spot welding. Most applications use multiple projections where the minimum distance between projections should be twice the projection diameter.

Part design at the joint location may be significantly limited because the welding electrodes normally contact several projections simultaneously. The electrodes must be mounted rigidly on the welding machine, and the supporting members must be strong enough to minimize deflection when electrode force is applied. Press type welding machines are commonly used for projection welding applications.

Fit-up is important with multiple projection welding. Each projection must be in contact with the mating surface to accomplish a weld. Uniformity of projection heights is a factor in good fit-up. The welding dies must be carefully designed and accurately manufactured to mate with the parts at the weld locations. They should not have to deform the parts to obtain good fit-up.

Where surface marking of one part must be minimized, the projections should be placed in the other part. A large, flat electrode on the show side of the joint should prevent electrode marking, although slight shrinkage may occur at each projection weld. This may be visible after some finishing operations.

When projection welds are used to attach other fasteners such as weld nuts and bolts, they must contain a sufficient number of projections to carry the design load. The design should be proven by applicable mechanical testing. Production quality control should be designed to ensure that weld quality does not drop below the design standards.

CROSS WIRE WELDING

General Principles

Resistance welding of crossed wires is, in effect, a form of projection welding. In practice, it usually consists of welding a number of parallel wires at right angles to one or more other wires or rods. There are many specific ways to perform the welding operation, depending upon production requirements, but the final finished prod-

uct is essentially the same regardless of the method used.

Crossed wire products include such items as stove and refrigerator racks, grills of all kinds, lamp shade frames, poultry equipment, wire baskets, fencing, grating, and concrete reinforcing mesh.

Wire racks may be welded in a press type projection welding machine or in special automatic indexing machines with hopper feed and a separate gun for each weld.

Concrete reinforcing mesh is made on continuous machines. The stay wires are fed from wire reels and the cross wires are fed either from wire reels on the side of the machine or from magazines of cut wire. The welded mesh is either rolled into coils like fencing or cut into mats and then stacked and bundled.

As in spot or projection welding, the wire or rod should be clean and free from scale or rust, dirt, paint, heavy grease, or other high resistance coatings. Plated or galvanized wire or rods may be used, but the coating at the weld will be destroyed.

Metals Welded

Low carbon steel wire is the most common metal welded. Typical machine settings for cross wire welding of this type of material are tabulated in Table 1.2. Next in importance are stainless steel and copper-nickel alloys. Compared to the values given in Table 1.2, stainless steel will require the same weld time, 60 percent of the amperage, and 2.5 times the electrode force. Copper-nickel alloys will require about the same weld time and amperage and about twice the electrode force.

Welding Technique

Normally, cross wire welds are not dressed after welding. Therefore, the major consideration may be appearance, with strength secondary in importance for some applications.

In setting up the welding machine, consideration must be given to the following:

(1) Design strength
(2) Appearance
(3) Welding electrodes

Table 1.2
Cross wire welding of low carbon steel wire

Wire dia., in.	Cold drawn wire				Hot drawn wire			
	Weld time, cycles	Electrode force, lbs.	Weld current, A	Weld strength, lbs.	Weld time, cycles	Electrode force, lbs.	Weld current, A	Weld strength, lbs.
	15% Setdown				15% Setdown			
1/16	5	100	600	450	5	100	600	350
1/8	10	125	1,800	975	10	125	1,850	750
3/16	17	360	3,300	2,000	17	360	3,500	1,500
1/4	23	580	4,500	3,700	23	580	4,900	2,800
5/16	30	825	6,200	5,100	30	825	6,600	4,600
3/8	40	1,100	7,400	6,700	40	1,100	7,700	6,200
7/16	50	1,400	9,300	9,600	50	1,400	10,000	8,800
1/2	60	1,700	10,300	12,200	60	1,700	11,000	11,500
	30% Setdown				30% Setdown			
1/16	5	150	800	500	5	150	800	400
1/8	10	260	2,650	1,125	10	260	2,770	850
3/16	17	600	5,000	2,400	17	600	5,100	1,700
1/4	23	850	6,700	4,200	23	850	7,100	3,000
5/16	30	1,450	9,300	6,100	30	1,450	9,600	5,000
3/8	40	2,060	11,300	8,350	40	2,060	11,800	6,800
7/16	50	2,900	13,800	11,300	50	2,900	14,000	9,600
1/2	60	3,400	15,800	13,600	60	3,400	16,500	12,400
	50% Setdown				50% Setdown			
1/16	5	200	1,000	550	5	200	1,000	450
1/8	10	350	3,400	1,250	10	350	3,500	900
3/16	17	750	6,000	2,500	17	750	6,300	1,800
1/4	23	1,240	8,600	4,400	23	1,240	9,000	3,100
5/16	30	2,000	11,400	6,500	30	2,000	12,000	5,300
3/8	40	3,000	14,400	8,800	40	3,000	14,900	7,200
7/16	50	4,450	17,400	11,900	50	4,450	18,000	10,200
1/2	60	5,300	21,000	14,600	60	5,300	22,000	13,000

a. Setdown, $\% = \dfrac{\text{Decrease in joint height}}{\text{Dia. of smaller wire}} \times 100$

(4) Electrode force

(5) Weld time

(6) Welding current (heat)

The particular application will determine which is most important, strength or appearance, when setting up for a particular crossed wire welding application. It is normally assumed that high strength welds with an acceptable appearance are desired.

The required electrode force, welding current, and weld time depend greatly upon the amount that the wires or rods are to be compressed together. This condition is called "setdown." It is the ratio of the decrease in joint height to the diameter of the smaller wire. Weld strength generally increases with "setdown" to a maximum.

The welding electrodes must be of the proper material and shape with provision for water cooling. RWMA Class II alloy electrodes usually have acceptable life, although electrode facings of harder alloys are sometimes used for special applications. Although flat electrodes are commonly used for cross wire welding, certain advantages are gained by shaping them to mate with the wires or rods being welded. Shaped electrodes provide better contact between the electrode and the work.

The electrode force depends upon the wire diameter, the specified setdown, the desired appearance, and the weld design strength. The electrode force will affect the appearance of the weld. The values given in Table 1.2 will produce welds with good appearance. Lower weld strengths than those shown in the table will result if higher forces are used without decreasing the weld time and increasing the welding current.

The weld time needed will depend upon the diameter of the wire to be welded. For best results, the values shown in the table should be used.

The welding current depends upon the diameter and the specified setdown. It should be slightly less than that which will result in spitting or expulsion of hot metal.

METALS WELDED

PROPERTIES INFLUENCING WELDABILITY

The following properties of metals have a bearing on their resistance weldability:

(1) Electrical resistivity

(2) Thermal conductivity

(3) Thermal expansion

(4) Hardness and strength

(5) Oxide-forming characteristics

(6) Plastic temperature range

(7) Metallurgical characteristics

Electrical Resistivity

This is probably the most important property from a resistance welding standpoint, since the heat generated by the welding current is directly proportional to resistance. More current is required to generate the heat for welding a metal of low resistivity than one of high resistivity. High currents require large transformers and power lines which increase equipment costs; therefore, metals of high electrical resistivity are considered to be more weldable than those of low resistivity.

Thermal Conductivity

The thermal conductivity is important because part of the generated heat is lost through conduction into the metal. This loss must be overcome by greater power input. Therefore, metals of high heat conductivity are less weldable than those of low conductivity. Thermal conductivity and electrical conductivity (reciprocal of resistivity) of the various metals closely parallel one another. Aluminum, for instance, is a good conductor of both heat and current, while stainless steel is a poor conductor of both.

Thermal Expansion

The coefficient of thermal expansion is a measure of the change in dimensions that takes place with a temperature change. When the co-efficient of thermal expansion is large, warping and buckling of welded assemblies can be expected.

Hardness and Strength

The hardness and strength of metals are important to resistance welding. Soft metals can be marked easily by the electrodes. Hard, strong metals require high electrode forces. Metals that retain their strength at elevated temperatures may require the use of welding machines capable of applying a forging force to the weld.

Oxide-Forming Characteristics

All commonly used metals oxidize in air, some more readily than others. The surface oxide normally has high electrical resistance. If this oxide film is very thick, the resistance may be so high that current can not flow through the pieces at normal secondary voltages. If the oxide film is extremely thin, it may have very little effect. For oxide film thicknesses in between these two extremes, welding current may flow, but very localized heating may occur at the interfaces. In the case of spot welding, over-heating between the electrodes and the work may cause surface flashing, pickup of metal on the electrode, and poor surface appearance. Furthermore, if the oxide film thickness varies from one part to another, the weld strength may not be consistent. The oxide-forming characteristics of the metal should be thoroughly understood and suitable precautions be observed. In welding aluminum alloys, for instance, the method of cleaning prior to welding and the time interval between cleaning and welding are important variables.

Aluminum alloys form oxides rapidly and, therefore, welding must be done within a short time after cleaning to avoid significant variations in surface contact resistance. Stainless steels, on the other hand, are oxidation-resistant at room temperature and precleaning is not usually necessary. The cleaning operation at the mill prior to packaging and shipping will usually be adequate for welding. Other metals have oxide-forming characteristics be-tween these two extremes and the cleaning requirements will depend upon the oxidation behavior of the particular metal. Whether a cleaning operation is necessary depends upon the amount of oxide present and how it will affect weld properties. Surface resistance measurements may be used to control cleanliness. In any case, all mill scale, heavy oxide from prior heat treatment, and extraneous material, such as paint, drawing compounds, or grease, must be removed prior to resistance welding.

Plastic Temperature Range

If the metal melts and flows in a narrow temperature range, the welding variables must be more closely controlled than with a metal having a wide plastic temperature range. This property may have considerable bearing on the welding procedure and equipment. Aluminum alloys have narrow plastic ranges and require precise control of welding current, electrode force, and electrode follow-up during welding. Therefore, projection welding of aluminum is not done commercially. Low carbon steel has a wide plastic range and is easily resistance welded.

Metallurgical Characteristics

With resistance welding, a small volume of metal is heated to its forging or melting temperature in a short time. The heated metal is then cooled rapidly by the electrodes and surrounding metal. Cold-worked metal will be annealed in the areas exposed to this thermal cycle. In addition, the rapid cooling may have a considerable hardening effect on some steels. High carbon steel, for instance, may harden so rapidly that the welds crack. A postweld tempering operation is necessary to avoid this cracking. Figure 1.8 shows a postweld quench and temper cycle following the weld interval.

LOW CARBON STEEL

Low carbon steels generally contain less than 0.25 percent carbon. The overall weldability of these steels is good. Their electrical resistivity is average. Hardenability is low. Welds with good strength can be obtained over a wide range of current, electrode force, and weld time settings.

HARDENABLE STEELS

Medium carbon steels may contain from 0.25 to 0.55 percent carbon; high carbon steels may contain 0.55 to 1.0 percent carbon. Low alloy steels contain up to 5.5 percent total alloying elements including cobalt, nickel, molybdenum, chromium, vanadium, tungsten, aluminum, and copper. Alloying additions produce certain desirable properties in steels. The steels may respond to heat treatment and the welds may be hard and brittle unless a postweld tempering cycle is employed. With seam welding, travel must stop during weld and postheat times. Special controls are available to do this on standard machines. In general, the weldability of these steels is poorer, due to their hardenability, than that of mild steel.

STAINLESS STEELS

Stainless steels contain relatively large amounts of chromium or chromium and nickel as alloying elements. They are divided into three groups, namely, martensitic, ferritic, and austenitic types. Whether the steel is hardenable or not depends upon the amounts of carbon, chromium, and nickel present.

Ferritic and Martensitic Types

These steels may be hardenable (martensitic types) or nonhardenable (ferritic types). When resistance welding the hardenable types, the precautions given for high carbon and low alloy steels must be followed. Hence, these steels have poor weldability characteristics. The nonhardenable types also have poor weldability characteristics because the weld zone will have a characteristic coarse-grained structure and low ductility. These steels are generally not suitable for applications where a ductile weld is required. With the martensitic types, a postweld heat treatment improves weld ductility. However, postweld heat treatment of the ferritic types is not beneficial.

Austenitic Type

There are a number of these steels, each having suitable properties for particular uses. The most common ones contain 18 percent chromium, 8 percent nickel, and approximately 0.10 percent carbon. Some are susceptible to carbide precipitation if heated for an appreciable time in a temperature range of 800° to 1600° F. They can be resistance welded without producing harmful carbide precipitation with short weld times. Less current is required than for low carbon steels since their electrical resistances are approximately seven times greater. Relatively high electrode forces are needed because of their high strengths at elevated temperatures. Austenitic stainless steels have higher coefficients of thermal expansion than carbon steels. As a result, seam welded assemblies may warp excessively if the design and welding procedures do not provide for this. It is advantageous to use short weld times to minimize the total heat input.

NICKEL ALLOYS

In general, nickel alloys are readily joined by resistance welding. The general requirements for welding are high electrode force, accurate timing, and cleanliness. These alloys are subject to embrittlement by sulfur, lead, and other low-melting-point metals when exposed to them at high temperatures. Oils, grease, lubricants, marking materials, and other foreign material, which might contain sulfur or lead, must be removed from the parts prior to welding or weld cracking may occur. Pickling prior to welding will only be necessary if a significant amount of oxide is present as indicated by surface discoloration.

Pure nickel can be welded rather easily. Some mechanical sticking of electrodes may be experienced because of the high electrical conductivity of nickel. A restricted dome electrode with a 170-degree cone angle is recommended for spot welding.

Monel 400[7] is an alloy of approximately two-thirds nickel and one-third copper. It has higher

---

7. A trademark

electrical resistivity than mild steel and is considerably stronger, particularly at elevated temperatures. Therefore, somewhat lower welding current and higher electrode force are required for spot and seam welding this alloy. Projection welding is accomplished without difficulty.

Monel K-500[7], which can be age-hardened by a subsequent 1100°F thermal treatment, has higher thermal and electrical resistivities and is stronger than Monel 400. Therefore, slightly lower welding current but higher electrode force is required for Monel K-500 than for Monel 400. Monel K-500 will crack in the age-hardened condition if subject to appreciable tensile stress at 1100°F. Therefore, spot, seam, and projection welding should be done on annealed material.

Inconel 600[7] contains approximately 78 percent nickel, 15 percent chromium, and 7 percent iron. It has even higher thermal and electrical resistivities than Monel 400, and is quite strong at elevated temperatures. Lower welding currents and higher electrode forces than for Monel 400 are required for this alloy. Recommended practices for spot welding are available. Projection and seam welding are quite readily accomplished using procedures similar to those for stainless steels.

Inconel X-750[7] and Inconel 722[7] are age-hardening variations of Inconel 600. They possess high strengths at elevated temperatures, and have very high electrical resistances. Relatively low welding currents and high electrode forces are used for these two alloys. Projection welding can be accomplished readily with machines having adequate force capacity. These alloys should be welded in the solution annealed condition only.

COPPER ALLOYS

Copper alloys have a wide range of weldability that varies almost directly with their electrical resistance. When the resistance is low, they are difficult to weld; when resistance is high, they are rather easy to weld. Machines having adequate current capacity and moderate forces are necessary. Because of the narrow plastic range of these alloys, it is very desirable to employ machines with low inertia heads.

Electronic control of current and time is also very desirable. Pure copper can not be spot, seam, or projection welded.

Copper-zinc alloys (brasses) become easier to weld as the zinc content increases because the electrical resistivity increases. The red brasses are difficult to weld while the brasses with high zinc content can be welded with a range of welding conditions, even though the required energy input is high compared to carbon steel. Relatively short welding times are recommended to prevent expulsion and sticking of the electrode to the work.

Copper-tin alloys (phosphor bronze), copper-silicon alloys (silicon bronze) and copper-aluminum alloys (aluminum bronze) are relatively easy to weld because of their relatively high electrical resistance. These alloys, particularly phosphor bronze, have a tendency to be hot short. This may result in cracking in the welds.

ALUMINUM AND MAGNESIUM ALLOYS

All the commercial aluminum and magnesium alloys that are produced in the form of sheet and extrusions may be spot and seam welded, provided the thicknesses involved are not too great. Properly designed equipment, correct surface preparation, and suitable welding procedures are necessary to produce satisfactory welds.

Aluminum and magnesium alloys have comparatively high thermal and electrical conductivities. To make spot and seam welds, high welding currents and relatively short welding times are necessary. One important factor influencing the choice of equipment for spot and seam welding is the rapid softening of the alloys at welding temperature. This necessitates fast movement of the welding electrode into the material as the nugget is formed. Although the distance is small, electrode movement must take place in a very short time, perhaps 0.002 to 0.005 second. Rapid acceleration of the welding machine head is necessary to maintain contact between the electrodes and the work. For this reason, low inertia equipment should be used. Projection welding of aluminum and magnesium is not done commercially because of their narrow plastic temperature range.

COATED AND PLATED STEELS

Many plated and coated steels can be spot, seam, or projection welded, but the weld quality usually is affected by the composition and the thickness of the coating. Coatings on steel are usually applied for corrosion resistance, decoration, or a combination of the two. Therefore, the welding procedures should assure a reasonable preservation of the coating function as well as produce welds of adequate strength. Strength requirements usually require machine settings similar to those for bare carbon steel. Adjustments to compensate for the coating will be determined by a number of factors including its effect on contact resistance, acceptable electrode marking, tendency of the coating to alloy with the base metal, and electrode sticking.

Coating thickness is the most important variable affecting weldability. It varies over relatively wide ranges in commercial practice. When thickness presents problems in welding, better welds can often be secured by decreasing the coating thickness. In some cases, it is possible to redesign parts for projection welding to obtain strong welds without impairing the corrosion resistance of the coating.

Satisfactory weld strength requires the development of a weld nugget between steel sheets which are initially separated by two layers of the coating metal. Coating thicknesses vary so widely in commerical practice that optimum machine settings for a particular application must be determined experimentally. A weld nugget of the desired size may be obtained without too much disturbance of the outside surfaces by using higher welding current, greater electrode force, and shorter weld time than for the same thickness of bare steel. Good electrode contact reduces the tendency of the electrode to alloy with the coating. However, it is extremely difficult to prevent alloying and pick-up around the periphery of the electrode face, particularly with low melting point coatings such as lead, tin, and zinc. Short welding times, good tip maintenance, and attention to electrode cooling

are the best preventive measures.

WELDING SCHEDULES

When setting up for welding a particular metal and joint design, a schedule must be established to produce welds that meet the design specifications. Previous experience can provide a starting point for the initial setup. If the application is a new one, reference to published information on the welding of the material by the designated process will serve as a guide for the initial setup.

Sample welds should be made and tested while changing one process variable at a time within a range to establish an acceptable value for that variable. It may be necessary to establish the effect of one variable at several levels of another. For example, weld or heat time and electrode force may be evaluated at several welding currents. Visual examination and destructive test results can be used to select an appropriate welding schedule. Finally, first production parts, or simulations thereof, should be welded and destructively tested. Final adjustments are then made to the welding schedule to meet design or specification requirements.

Starting schedules for many commercial alloys may be available from the equipment manufacturer. Some may also be found in the following publications:

(1) AWS C1.1, Recommended Practices for Resistance Welding

(2) AWS C1.2, Recommended Practices for Spot Welding Aluminum and Aluminum Alloys

(3) AWS C1.3, Recommended Practices for Resistance Welding Coated Low Carbon Steels

(4) AWS D8.5, Recommended Practices for Automotive Portable Gun Resistance Spot Welding

(5) AWS D8.7, Specification for Automotive Weld Quality, Resistance Spot Welding

(6) *Resistance Welding Manual*, Vols. 1 and 2, 3rd ed., Resistance Welder Manufacturers Association, 1956, 1961.

(7) *Metals Handbook*, Vol. 6, *Welding and Brazing,* 8th ed., American Society for Metals, 1971.

WELD QUALITY

The weld quality required depends primarily upon the application. It is affected by the composition and condition of the base metal, joint and part designs, electrode condition, and available equipment. In some applications, each weld must meet the minimum requirements of a rather stringent specification. This is true for aircraft and space vehicles. Other applications may have standards for satisfactory welds but permit some percentage of undersize or defective welds. Automotive components are examples.[8]

Design requirements may include surface appearance, minimum strength, and leaktightness with some seam welding applications. These should be monitored by a system of quality control that includes visual inspection and destructive examination of test samples or actual weldments. The most important factors that affect weld quality are as follows:

(1) Surface appearance
(2) Weld size
(3) Penetration
(4) Strength and ductility
(5) Internal discontinuities
(6) Sheet separation and expulsion
(7) Weld consistency

SURFACE APPEARANCE

The surface appearance of a resistance weld is no indication of its strength, size, or internal soundness. It is an indication of the conditions under which the weld was made, but it should not be used as the sole criterion for qualifying production welds. For example, a group of spot welds in a joint may have identical surface appearance, such as those shown in Fig. 1.29. Yet, all welds except the first may be greatly understrength because of the shunting of current through adjacent welds. Figure 1.30 shows an example of this. The diameter of the fused zone of the second weld is appreciably smaller than the first one due to shunting of current.

8. The requirements are given in AWS D8.7, Specification for Automotive Welding Quality—Resistance Spot Welding, latest edition.

However, they have identical surface appearance because all the welding current passes through the outside surface of both welds. This effect is more severe for closely spaced welds, welds in metal of low electrical resistivity, and welds in thick sheets (0.040-in. thick and greater).

Normally, the surface appearance of a spot, seam, or projection weld should be relatively smooth; round or oval in the case of contoured work; and free from surface fusion, electrode deposit, pits, cracks, deep electrode indentation, or any other condition that would indicate improper electrode maintenance or equipment operation. Table 1.3 lists some of the more common undesirable surface conditions, the causes, and the effects on weld quality.

WELD SIZE

The diameter or width of the fused zone must meet the requirements of the appropriate specifications or the design criteria. In the absence of such requirements, either accepted shop practices or the following general rules should be used.

(1) Spot welds that are reliably reproduced under normal production conditions should have a minimum nugget diameter of 3.5 to 4 times the thickness of the thinnest outside part of the joint. In cases of three or more dissimilar thicknesses, the nugget diameters between adjacent parts can be adjusted somewhat by the selection of the electrode design and materials.

(2) The individual nuggets in a pressure-tight seam weld should overlap a minimum of 25 percent. The width of the nugget should be at least 3.5 to 4 times the thickness of the thinnest outside part.

(3) Projection welds should have a nugget size equal to or larger than the diameter of the original projection.

There is a maximum limit to the nugget size of a spot, projection, or seam weld. This limitation is based more on the economics and practical limitations of producing a weld than on the laws of heat generation and dissipation that limit the minimum size of the weld. The maxi-

Fig. 1.29 — Surface appearance of two succeeding spot welds in 0.040 in. stainless steel sheet

mum useful nugget size cannot be specified in general terms. Each user should establish this limit in accordance with the design requirements and prevailing shop practices.

PENETRATION

Penetration is the depth to which the nugget extends into the pieces that are in contact with the electrodes. The minimum depth of penetration is generally accepted as 20 percent of the thickness of the outside piece. If penetration is less than 20 percent, the weld is said to be cold because the heat generated in the weld zone is too small, and normal variations in welding current, time, electrode force, etc., will cause undesirable changes in weld strength. In extreme cases, there may be no weld at all. The depth of penetration should not exceed 80 percent of the thickness. Greater penetration will result in expulsion, excessive indentation, and rapid electrode wear. Figure 1.31 shows normal, excessive, and insufficient penetration.

The depth of penetration into each outside piece should be approximately uniform for equal or nearly equal thicknesses. For dissimilar thickness ratios of three or more to one, the depth of penetration into the thicker piece need only be equal to that in the thinner piece.

STRENGTH AND DUCTILITY

Structures employing spot, seam, and projection welds are usually designed so that the

Fig. 1.30 — Cross section of spot welds in Fig. 1.29 showing the effect of current shunting

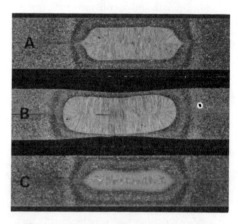

Fig. 1.31 — Penetration in spot welds; (A) — normal, (B) — excessive, (C) — insufficient

welds are loaded in shear when the parts are exposed to tension or compression loading. In some cases, the welds may be loaded in tension, where the direction of loading is normal to the plane of the joint, or a combination of tension and shear. In the case of flanged tank sections

Table 1.3
Undesirable surface conditions for spot welds

Type	Cause	Effect
1. Deep electrode indentation	Improperly dressed electrode face; lack of control of electrode force; excessively high rate of heat generation due to high contact resistance (low electrode force)	Loss of weld strength due to reduction of metal thickness at the periphery of the weld area; bad appearance
2. Surface fusion (usually accompanied by deep electrode indentation)	Scaly or dirty metal; low electrode force; misalignment of work; high welding current; electrodes improperly dressed; improper sequencing of pressure and current	Undersize welds due to heavy expulsion of molten metal; large cavity in weld zone extending through to surface; increased cost of removing burrs from outer surface of work; poor electrode life and loss of production time from more frequent electrode dressings
3. Irregular shaped weld	Misalignment of work; bad electrode wear or improper electrode dressing; badly fitting parts; electrode bearing on the radius of the flange; skidding; improper surface cleaning of electrodes	Reduced weld strength due to change in interface contact area and expulsion of molten metal
4. Electrode deposit on work (usually accompanied by surface fusion)	Scaly or dirty material; low electrode force or high welding current; improper maintenance of electrode contacting face; improper electrode material; improper sequencing of electrode force and weld current	Bad appearance; reduced corrosion resistance; reduced weld strength if molten metal is expelled; reduced electrode life
5. Cracks, deep cavities, or pin holes	Removing the electrode force before welds are cooled from liquidus; excessive heat generation resulting in heavy expulsion of molten metal; poorly fitting parts requiring most of the electrode force to bring the faying surfaces into contact	Reduction of fatigue strength if weld is in tension or if crack or imperfection extends into the periphery of weld area; increase in corrosion due to accumulation of corrosive substances in cavity or crack

that are seam welded along the flanges, the seam welds may be subjected to peeling action.

The strength requirements for spot and projection welds are normally specified in pounds per weld. For seam welds, the strength is usually specified in pounds per inch of joint length. It is good practice to specify weld strength that is greater than that of welds of minimum recommended nugget size, but not more than 150 percent.

The strength of spot and projection welds increases as the diameter becomes larger, even though the average unit stress decreases. The unit stress decreases because of the increasing tendency for failure to occur at the edge of the nugget as its size increases. In low carbon steel, for example, the calculated average shear stress in good welds at rupture will vary from 10 to 60 ksi. Low values apply to relatively large welds and high values to relatively small welds. In both instances, the actual tensile stress in the sheet at the weld periphery is at or near the ultimate tensile strength of the base metal. This factor tends to cause the shear strength of circular welds to vary linearly with diameter.

Single spot and projection welds are not strong in torsion where the axis of rotation is perpendicular to the plane of the welded parts. This strength tends to vary with the cube of the diameter. Little torsional deformation is obtained with brittle welds prior to failure. Angular displacements may vary from 5 to 180 degrees depending upon weld metal ductility. Torsion is normally used to shear welds across the interface to measure the cross section.

The ductility of a resistance weld is determined by the composition of the base metal and the effect of high temperatures and subsequent rapid cooling on that composition. Unfortunately, the standard methods of measuring ductility are not adaptable to spot, seam, and projection welds. The nearest thing to ductility measurement is the hardness test, since the hardness of a metal is usually an indication of its ductility. For a given alloy, ductility decreases with increasing hardness, but different alloys of the same hardness do not necessarily possess the same ductility.

Another method of indicating the ductility of a spot or projection weld is to determine the ratio of its direct tension strength to the tension-shear strength.[9] A weld with good ductility has a high ratio; a weld with poor ductility has a low ratio.

There are various methods which can be used in production welding to minimize the hardening effect of rapid cooling. Some of these are:

(1) Use long weld times to put heat into the work.

(2) Preheat the weld area with a preheat current.

(3) Temper the weld and heat-affected zones using a temper current at some interval after the weld time.

(4) Furnace anneal or temper the welded assembly.

9. These tests are described in Vol. 1, *Welding Handbook*, 7th ed.: 161-3.

These methods are not always practical. For instance, the first will produce greater distortion of the assembly and reduce production rates; the second and third methods require welding machine controls that provide these features; the fourth method involves an additional operation that may reduce the strength of cold worked base metal. If the assembly is quenched from the annealing temperature, it may cause excessive distortion.

INTERNAL DISCONTINUITIES

Internal discontinuities include cracks, porosity or spongy metal, large cavities, and, in the case of some coated metals, metallic inclusions. Generally speaking, these discontinuities will have no detrimental effect on the static or fatigue strength of a weld if they are located entirely in the central portion of the weld nugget. This is true because the stresses are essentially zero in the central portion of the weld nugget. On the other hand, it is extremely important that no defects occur at the periphery of a weld where the load stresses are highly concentrated.

Spot, seam, and projection welds in metal thicknesses of approximately 0.040 in. and greater may have small shrinkage cavities in the center of the weld nugget, as illustrated in Fig. 1.32(A). These cavities are less pronounced in some metals than in others due to the difference in forging action of the electrodes on the hot metal. Such shrinkage cavities are generally not detrimental in the usual applications. However, the cavity that results from heavy expulsion of molten metal, as shown in Fig. 1.32(B), may take up a very large part of the fused area and must be considered a defect. A certain number of expulsion cavities are to be expected in the production welding of most commercial steels. Heavy expulsion of molten metal is a result of improper welding conditions, and the number of such welds that can be accepted should be limited by specifications.

Internal defects in spot, seam, and projection welds are generally caused by low electrode force, high welding current, or any other conditions that produce excessive weld heat.

They are also caused by removing the electrode force too soon after welding current stops. When this occurs, the weld nugget is not properly forged during cooling. Such action may result with high-speed seam and roll spot welding.

SHEET SEPARATION

Sheet separation occurs at the faying sur-

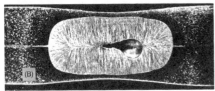

Fig. 1.32—Shrinkage cavities in spot welds: (A)—small, (B)—large

faces due to the expansion and contraction of the weld metal and the forging effect of the electrodes on the hot nugget. The amount of separation varies with the thickness of the stock, increasing with greater thickness. Normal separation is shown in Fig. 1.31(A).

Excessive sheet separation results from the same causes as surface indentation, to which it is related. Improperly dressed electrode faces act as punches under high electrode force. This tends to decrease the joint thickness, upset the weld metal radially, and force the sheets up around the electrodes. Excessive sheet separation is illustrated in Fig. 1.33 (note that one sheet is laminated).

Fig. 1.33—Excessive sheet separation

QUALITY CONTROL

Consistent weld quality can be maintained with proper control of the factors that tend to produce variations in the final product. These factors include:

(1) Joint design and fit-up
(2) Material thickness tolerance
(3) Composition, temper, and surface condition of the base metal
(4) Electrode material and shape
(5) Electrode and weldment cooling
(6) Welding cycle variables
(7) Postweld thermal treatments

The importance of joint design and fit-up is discussed elsewhere in this chapter. Large variations in part thickness, particularly with three or more thicknesses, may produce inconsistent fit-up which, in turn, can affect weld

quality. Changes in base metal alloy, temper, or surface conditions may require revision of the welding cycle to produce acceptable welds.

WELDING VARIABLES

When uniform parts are fed to a suitable welding machine operated to a qualified welding procedure specification, production of consistently acceptable welds is routine. Uniform production welding is achieved if essentially constant conditions are maintained. When the electrodes are properly dressed and the welding cycle is consistent, there should be very little variation in weld quality. Since it is very difficult to achieve absolute uniformity in production operations, such operations always should be set up to allow for some variations without causing significant differences in weld properties.

Electrode force is usually determined from graphs relating air or hydraulic pressure to force produced by the cylinder. Force gages are commercially available to verify settings for critical applications. In general, only static forces are measured. Proper maintenance and lubrication of moving parts help to assure uniformity of applied forces. Periodic electrode dressing assures the application of uniform pressures and current densities. Consistent secondary currents can be assured by the use of current or voltage regulators, maintenance of proper line voltage, consistent fit-up of parts, proper spacing of welds, and allowance for magnetic materials in the throat of the machine.

CONTROL ACCESSORIES

Controls for resistance welding have adequate life, reliability, timing accuracy, and firing precision for most applications. They utilize semiconductor components for resistance-capacitance timing as well as accurate cycle counting. Maintenance of the control functions alone will not compensate for all welding variables. Changes in work and electrode conditions, line voltage fluctuations, current shunting, cable wear, and electrode force variations can gradually or suddenly affect weld quality. A current regulator can compensate for some of these variations. Improved timing accuracy and current regulation will help to maintain weld quality indirectly. Various control accessories are available that can improve weld quality by automatic adjustment of control functions or by monitoring of the welding variables and indicating a malfunction.

Programmed Devices

Specialized accessories are available to expand the capability of standard NEMA[10] type weld controls. With these accessories, it is possible to increase overall weld quality. On occasion, some dual or multi-function accessories can be of extreme value. Some examples of this

----
10. National Electrical Manufacturers Association

type accessory are dual, triple, or multi-function of the following: weld time, phase shift heat control, or complete timing and heat control schedule.

Any timing or level adjustment can be set up on a multiple basis. The only requirement is that the control manufacturer be properly informed of the functional variables desired.

Monitors

Monitoring devices are electronic units that receive input signals from various points of the welding process, compare these input signals to a preset standard, and then display the result as acceptable or rejectable. Rejection can also be expanded to over or under an acceptable range. Various types of monitors are:

(1) Current detector, over and under

(2) Voltage (line) detector, over and under

(3) Tip voltage monitor

(4) Weld energy monitor (can be used as monitor only or as feedback unit to control weld time)

(5) Electrode movement monitor

(6) Electrode movement and current detector

Types 1 and 2 are single-variable readout devices; there are many specialized units that fall into this category. This style unit, usually simple in nature, will tend to prevent weld quality deterioration from excessive variations of this single variable. An "out of tolerance" signal from these units is in the form of an isolated relay which may be used to turn on an alarm or lock out the main welding control.

Type 3 is a more sophisticated monitoring device. To some extent, it reads the combination of secondary voltage, welding current, and electrode force using the voltage across the tips as its quality standard. If the input voltage, input current, metal conditions, or electrode force vary, they may produce a change in the voltage across the electrodes. If the voltage change is large enough to exceed a preset high or low limit, the unit will indicate the weld as a "reject." As long as the tip voltage remains within set limits, the unit will indicate weld quality as "acceptable."

Type 4, the weld energy monitor, can be used either as a monitor only or as a monitor with feedback control of the welding machine control. As a monitor, the unit integrates secondary voltage or current signals, or both, with an internal time base and calculates an energy input level for each weld.

The weld energy monitor is set up by running a group of sample welds at various machine settings and recording the upper and lower readings of acceptable welds. The user can then decide at what level to set the high and low limits on the unit. After this is accomplished, every weld reading that falls between these settings is indicated as an acceptable weld. Each weld that falls outside of these limits is indicated by a "reject" or "out of limits" signal. The "accept" and "out of limits" signals, plus a number of other signals, can be interfaced out of the weld energy monitor for use elsewhere, such as a computer or programmable control input.

Adaptive Controls

Adaptive or "feedback" control devices are electronic units that operate on a closed-loop or semi-loop principle. In general, this type system depends on signals generated by the welding process which are "fed back" into the circuitry and compared to a standard reference previously set up in the unit. As the weld time progresses, the circuitry attempts to make adjustments internally so that the feedback signal matches the standard reference.

Generally, the reference standard is set up by making sample welds with the weld controller set in a manual mode. In these systems, there are usually adjustable high and low limits to set, and any weld exceeding these preset values will be indicated as a questionable weld.

Current Regulator. In this unit, the welding transformer primary current is monitored by a current transformer. The current transformer output is fed to a special electronic control circuit. The output level of the control circuit is related to the primary current demand of the welding machine. Changes in this level are sensed by additional electronic circuitry that automatically adjusts the current control devices to increase or decrease the primary current to maintain a preset power level in the welding transformer.

Tip Voltage Regulator. This regulator receives a feedback signal from terminals at the welding electrodes. The unit compares the feedback signal to a previously set reference level and automatically adjusts the current control circuit to maintain the correct power at welding electrodes. The reference signal is determined when acceptable welds are made in the manual mode. This information is used to adjust the circuitry to the proper operating range. Circuitry also incorporated in this unit indicates when the unit can no longer maintain the preset level. Lights or an alarm indicate when the preset tolerances are exceeded.

Resistance Feedback Accessory. The accessory combines the resistance drop principle as a weld quality indicator with adjustable limit timers and indicators to provide the operator with information on the condition of the welding process. Also, the control makes in-process adjustments to the weld time to produce a good weld.

The process variable monitored is the change in the electrical resistance across the weld that occurs during the weld time. The resistance of a spot weld varies according to a predictable pattern as the nugget is being formed. For the first few cycles of weld time, the resistance may be erratic due to metal fit and surface condition variables. However, the general trend is for the resistance to increase as the metal is heated by the welding current.

At some point, the resistance will peak and then begin to decrease. It is theorized that this decrease in weld resistance is due to metal fusion and the resulting destruction of the interface resistance of the parts to be welded. It is possible to roughly relate the percentage drop in weld resistance from the peak value to weld nugget size. As an example, in the welding of two pieces of clean 0.060-in. thick mild steel, a resistance drop of 15 percent in a reasonable weld time with reasonable electrode force will be found to produce a good weld.

Weld Energy Monitor. In addition to its

monitoring capability, the unit can be interfaced to the main welding control to control the weld time. In this mode of operation, the device uses the same input (secondary current squared or secondary current combined with secondary voltage) for information.

With its added capability, the unit is in active control of the process rather than merely a passive device. It displays an energy level on a readout. Assuming that the proper energy level for a good weld has been established, the unit can be placed in control of the weld time to produce the proper energy for each weld.

Acoustic Emission Monitor. Acoustic emission testing is based on the fact that solid materials emit stress waves when stimulated mechanically or thermally. These waves can be picked up by a transducer and electronically processed for monitoring and controlling. Monitoring spot welds is an excellent example of in-process testing where the stimulus is provided by the welding process itself. Starting with the electrode set-down and ending with the electrode lift-off, spot welding produces a complex acoustic emission signal. The distinguishable elements of this acoustic record are associated with a number of different physical changes that occur during the welding process. The signals associated with expulsion (spitting or flashing) are generally of large amplitude. They can be easily distinguished from the rest of the acoustic emission associated with nugget formation.

The expulsion can be visible when it takes place at the electrode tips and invisible when it takes place between the metal parts. Expulsion occurs when an oversize weld nugget is formed, and it is generally an indication of overwelding. It will occur early in the weld cycle when the electrodes are clean or the metal is thin. It will occur later when the welding conditions have deteriorated. The detection of expulsion by acoustical emission can be accomplished with available techniques. Since the appearance of the first expulsion signal is the dynamic time-mark for the completion of a good weld, it can be used to signal the control to terminate welding current. Thus, the unit forms a feedback path between the welding process and the power input to the weld.

INSPECTION AND TESTING

A satisfactory resistance weld is the result of using correct welding settings and techniques and maintaining them for the duration of a particular production run. Factors such as current, time, electrode force, and material must be properly controlled. This is best done by periodic testing of workpieces or test samples for quality control.

The number of workpieces inspected as well as the inspection method may vary. The test pieces may be examined nondestructively or a certain number tested destructively. Statistical methods are then used to predict the quality of the whole lot. In any case, an inspector must be able to (1) recognize conditions that may cause variations in results, (2) make certain that the test pieces are representative samples, and (3) ascertain that production pieces are made under the same conditions as the test samples.

The inspector has many inspection means at his disposal, including visual, x-ray, magnetic particle, fluorescent penetrant, and destructive tests. For many types of weldments, the basis for acceptance is visual inspection. This may be true even for some weldments that are later to be tested by x-ray or other methods. It starts with the material prior to fabrication. After the parts are assembled in position for welding, any incorrect alignment of faying surfaces and other features of joint preparation that might affect the quality of the welded joint should be noted.

A high-quality, consistent weld cannot be judged entirely by appearance, but good welds should present a uniform and consistent appearance. The electrode indentation should not fade out because this would indicate electrode deterioration or cold welds.

No economical or practical nondestructive method of inspecting spot welds is available. Therefore, the practice is to periodically make a number of sample specimens for destructive testing. These pieces must be welded under the same conditions as the production parts. The material should be the same both in composition and thickness. Surface cleaning and all machine settings should be indentical. Care must be taken to prevent misleading effects from current shunting through adjacent welds.

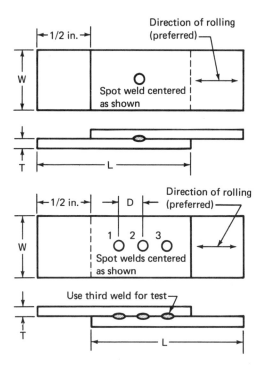

T, in.	W, in.	L, in.	D
up to 0.029	5/8	2	Minimum acceptable
0.030 to 0.058	1	3	weld spacing
0.059 to 0.125	1-1/2	4	

Fig. 1.34—Peel test specimens

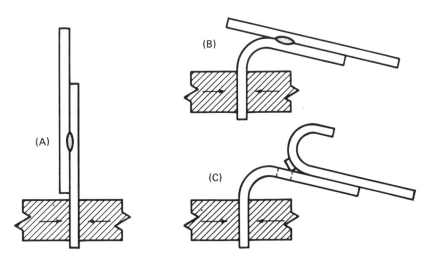

Fig. 1.35—Peel test

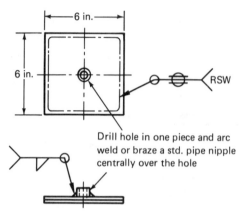

Fig. 1.36—Pillow test for seam welds

test for seam welds is the pillow test, a specimen for which is shown in Fig. 1.36. The specimen is tested for leaks with internal hydraulic pressure. An acceptable weld will not leak before the specimen ruptures. Standard practice in aircraft quality welding is to test welds for shear strength using the specimen shown in Fig. 1.37. This is done at intervals specified in the appropriate welding specification.

Sectioning of welded specimens is another method of weld evaluation. It consists of scribing the centerline of the weld, sawing to one side of this line, and smoothing the cut surface by grinding to the centerline. Etching allows macroscopic examination of the weld nugget for size, penetration, structure, and discontinuities.

The surfaces of a spot, seam, or projection weld should appear relatively smooth and free from surface melting, electrode deposits, pits, cracks, deep electrode indentations, or other conditions indicating that the proper electrode

Commercial quality welds are normally visually inspected and peel tested. Peel test specimen designs are shown in Fig. 1.34, and the method of testing in Fig. 1.35. A similar

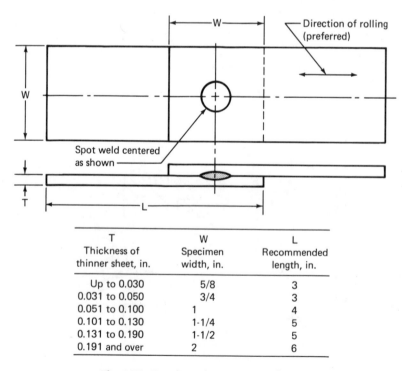

T Thickness of thinner sheet, in.	W Specimen width, in.	L Recommended length, in.
Up to 0.030	5/8	3
0.031 to 0.050	3/4	3
0.051 to 0.100	1	4
0.101 to 0.130	1-1/4	5
0.131 to 0.190	1-1/2	5
0.191 and over	2	6

Fig. 1.37—Tension-shear test specimen

face contour was not maintained or that the equipment was not functioning properly. The diameter of the fused area for resistance welds must meet the standards required by the specification or design drawings. Additional information on testing is found in AWS C1.1, Recommended Practices for Resistance Welding (latest edition).

STATISTICAL METHODS

The application of statistical control to production quality has three prime objectives:

(1) To reduce the number of rejections and machine shutdowns due to poor performance

(2) To assist in establishing the optimum specification limits of satisfactory quality

(3) To provide a reasonably reliable measure of actual production quality

If these objectives are achieved, the product will have high quality at low cost, and minimum scrap.

The basic principles of statistical control are widely used in industry. Briefly these principles are to:

(1) Select samples of actual production and test them for performance to specifications.

(2) Estimate the probable quality or con-formance of all production by analysis of samples.

(3) Predict the future quality by considering the trend of past and present quality.

The methods of sampling, extraction of data from the samples, and a decision as to whether to permit further use of the welding procedure constitute the quality control system.

In connection with the selection of the statistical control system, it will function faster and more economically if the number of variables to be kept under control is small and the charts and records that must be maintained are few in number. Obviously, the various machines that are used to weld the same or similar materials should be as nearly equal in adjustment and performance as possible.

A complicated system could be adopted that would cover a number of machines with different and varying characteristics. However, a more economical system can be used if all the machines are first stabilized and then similar machines are standardized before applying the statistical control.

The advantages, methods, and procedures for statistical quality control of resistance welding are described in AWS C1.1-66, Recommended Practices for Resistance Welding, 102-111.

SAFETY

MECHANICAL

Resistance welding equipment should be designed to avoid crushing of hands and other parts of the body. This may be achieved by:

(1) Arranging initiating controls, such as push buttons and foot switches, to prevent the operator from inadvertently operating them.

(2) Guarding the operator of multi-gun machines or large platen type machines that may require the hands to pass under the point of operation. Dual hand controls, physical barriers, electronic eye safety circuits, or similar devices should be used.

(3) Designing portable resistance welding guns so that the hands placed on the operating holders cannot be crushed, perhaps by suitably remote two-hand controls with appropriate guards.

PERSONAL EQUIPMENT

As with other welding operations, the personal protective equipment required is dependent upon the particular application. The following equipment is generally needed for resistance welding:

(1) Eye protection in the form of eye or face shields or hardened lens goggles is recommended. Face shields are preferred.

(2) Skin protection should be provided by non-flammable clothing with the minimum number of pockets and cuffs in which hot or molten particles can lodge.

(3) Protective footwear is advisable.

ELECTRICAL

Resistance welding equipment should be designed to avoid accidental contact with parts of the system that are electrically hazardous. All electrical equipment including control panels must be manufactured and installed in accordance with the safety requirements of appropriate local and national standards and codes. High voltage parts must be suitably insulated and protected by complete enclosures with access doors and panels interlocked to electrically isolate the machine. Doors and access panels of live electrical equipment at production floor level must be kept locked or interlocked to prevent access by unauthorized persons.

All electrical equipment must be suitably grounded and the transformer secondary may be grounded or provided with equivalent protection. External weld-initiating control circuits should operate at low voltage for portable equipment.

Additional information on safe practices for resistance welding is contained in ANSI Z49.1, Safety in Welding and Cutting (latest edition).

Metric Conversion Factors

$1 \text{ in.} = 25.4 \text{ mm}$

$1 \text{ lbf} = 4.45 \text{ N}$

$1 \text{ ksi} = 6.89 \text{ MPa}$

$1 \text{ in./min.} = 0.423 \text{ mm/s}$

$t_c = 0.56 (t_F - 32)$

SUPPLEMENTARY READING LIST

Adams, J. V., et al., Effect of projection geometry upon weld quality and strength. *Welding Journal,* 44(10): 466s-70s; 1965 Oct.

Andrews, D. R. and Broomhead, J., Quality assurance for resistance spot welding. *Welding Journal* 54(6); 431-5; 1975 June.

Buer, F. Y. and Begeman, M. L., Evaluation of resistance seam welds by a shear peel test. *Welding Journal,* 41(3): 120s-22s; 1962 Mar.

Chang, U. I., Mitchell, J. W., and Young, L. G., Evaluation techniques for electrode caps used in resistance spot welding. *Welding Journal,* 51(9): 617-25; 1972 Sept.

Chihoski, R. A., Variations in aluminum spot welds. *Welding Journal,* 51(12): 567s-78s; 1970 Dec.

Cunningham, A., et al., An analysis of the nugget formation in projection welding. *Welding Journal,* 45(7): 305s-13s; 1966 July.

Czohara, C. A. and Kilmer, T. H., Advancing the state of the art in resistance welding. *Welding Journal,* 55(4): 259-63; 1976 Apr.

Eichhorn, F. and Oppe, H. J., The projection welding of zinc coated sheet metal. *Welding Research Abroad,* 16(8): 20-31; 1970 Oct.

Johnson, K. I. and Needham, J. C., Design of resistance spot welding machine for quality control. *Welding Journal,* 51(3): 122s-32s; 1972 Mar.

Nadkarin, A. V. and Weber, E. P., A new dimension in resistance welding electrode materials. *Welding Journal,* 56(11): 331s-38s; 1977 Nov.

Riley, J. J. and Harris, J. F., Annual projection welding of tubular sections to low carbon steel sheet. *Welding Journal,* 45(7): 289s-304s; 1966 July.

Savage, W. F., Nippes, E. F., and Wassell, F. A., Static contact resistance of series spot welds. *Welding Journal,* 56(11): 365s-70s; 1977 Nov.

Savage, W. F., Nippes, E. F., and Wassell, F. A., Dynamic contact resistance of series spot welds. *Welding Journal,* 57(2): 43s-50s; 1978 Feb.

Ueda, T. and Kawataka, H., Fatigue strength of spot welded joints. *Welding Research Abroad,* 17(2): 39-52; 1971 Feb.

Wu, K. C., Electrode indentation criterion for resistance spot welding, *Welding Journal,* 47(10): 472s-78s; 1968 Oct.

Wu, K. C., Resistance spot welding of high contact resistance surfaces for weldbonding. *Welding Journal,* 54(12): 436s-43s; 1975 Dec.

2

Flash, Upset, and Percussion Welding

Chapter Committee

W. G. EMAUS, *Chairman*
  Anderson Automation, Inc.

R. A. GRENDA
  Weltronic Company

P. G. HARRIS
  Centerline (Windsor) Ltd.

F. R. HOCH
  Aluminum Company of America

A. E. LOHRBER
  Swift-Ohio Corporation

R. MATTESON
  Taylor-Winfield Corporation

F. PARKER
  Newcor, Inc.

D. ROTH
  *Kirkhof Transformer Division,
  FLX Corporation*

Welding Handbook Committee Members

A. F. MANZ
  Union Carbide Corporation

D. D. RAGER
  Reynolds Metals Company

2

Flash, Upset, and Percussion Welding

Flash, upset, and percussion welding are a family of welding processes that can be used to produce a butt joint between two parts of similar cross section by making a weld simultaneously across the entire joint area without the addition of filler metal. Upsetting force is applied at some point before, during, or after the heating cycle to bring the parts into intimate contact. It is the method of heating and time of force application that distinguish these three welding processes. Percussion welding may also be used to join the tip or end of a small part to a flat surface.

FLASH WELDING

FUNDAMENTALS OF THE PROCESS

Definition and General Description

Flash welding is a resistance welding process in which coalescence is produced simultaneously over the entire area of two abutting surfaces. Heat for welding is created by resistance at minute contact points to the flow of electric current and by arcs between the faying surfaces. Force is applied on the joint at the appropriate time to expel the molten metal from the joint and then upset the base metal.

Two parts to be joined are clamped in dies (electrodes) connected to the secondary of a resistance welding transformer. Voltage is applied as one part is advanced slowly toward the other. When contact occurs at surface irregularities, resistance heating occurs at these locations. High amperage causes rapid melting and explosion of the metal at the points of contact and then minute arcs form. This action is called "flashing." As the parts are moved together at a suitable rate, flashing is continued until the faying surfaces are covered with molten metal and a short length of each part reaches forging temperature. A weld is consummated by the application of an upsetting force to bring the molten faying surfaces in contact and then forge the parts together. Flashing voltage is terminated at the start of upsetting. The solidified metal expelled from the interface is called "flash."

Principles of Operation

Sequence of Welding. The basic steps in a flash welding sequence are as follows:
   (1) Position the parts in the machine
   (2) Clamp the parts in the dies (electrodes)
   (3) Apply the flashing voltage
   (4) Start platen motion to cause flashing
   (5) Flash at normal voltage

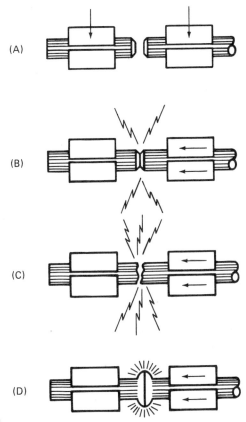

**Fig. 2.1—The basic steps in flash welding:
(A) position and clamp the parts;
(B) apply flashing voltage and start platen
motion; (C) flash; (D) upset and terminate
current**

(6) Terminate flashing
(7) Upset the weld zone
(8) Unclamp the weldment
(9) Return the platen and unload

Figure 2.1 illustrates these basic steps.
Additional steps such as preheat, dual-voltage
flashing, postheat, and trimming of the flash
may be added as the application dictates.

Flashing. This action takes place between the
faying surfaces as the movable part is advanced
toward the stationary part during welding. Heat
is generated at the joint and the temperatures of

the parts increase with time. Flashing action
(metal loss) increases with part temperature.

A graph relating part motion with time is
known as the flashing pattern. In most cases, a
flashing pattern should have an initial period of
constant velocity motion of one part toward the
other to facilitate the start of flashing. This
linear motion should then merge into an accelerating motion which should closely approximate a parabolic curve. This pattern of motion
is known as "parabolic flashing."

To produce a strong joint with uniform upset, the temperature distribution across the joint
should be uniform and the average temperature
of the faying surface should be the melting temperature of the metal. Once these conditions are
reached, further flashing is not necessary.

The steepness of the temperature gradient
corresponding to a stable temperature distribution
is a function of the part acceleration during parabolic flashing. In general, the higher the rate of
part acceleration, the steeper is the stable temperature gradient produced. Thus, the shape of
the temperature distribution curve in a particular
application can be controlled by appropriate
choice of flashing pattern. Since the compressive yield strength of a metal is temperature
sensitive, the behavior of the metal during the
upsetting portion of the welding cycle is markedly
dependent upon the flashing pattern. Therefore,
the choice of flashing pattern is extremely important for the production of sound flash welds.
The minimum flashing distance is the amount of
flashing required to produce a stable temperature
distribution. From a practical standpoint, the
flashing distance should be slightly greater than
the minimum acceptable amount to ensure that
a stable temperature distribution is always
achieved.

Upsetting. When a stable temperature distribution is achieved by flashing, the two parts
should be brought together rapidly and then upset together. The movable part should be accelerated rapidly so that the molten metal on the
flashing surfaces will be extruded before it can
solidify in the joint. Motion should continue
with sufficient force to upset the metal and weld
the two pieces together.

Upsetting current is sometimes applied as

the joint is being upset to maintain temperature by resistance heating. This permits upsetting of the joint with lower force than would be required without it. Upsetting current is normally adjusted by electronic heat control on the basis of either experience or welding tests.

Advantages and Limitations

Butt joints between parts with similar cross section can be made with friction[1] and upset welding as well as with flash welding. The major difference between friction welding and the other two processes is that the heat for friction welding is developed by rubbing friction between the faying surfaces rather than from electrical resistance. Upset welding is similar to flash welding except that the heat for welding is entirely from resistance heating at the joint; no flashing action occurs.

Some important advantages of flash welding are:

(1) Cross-sectional shapes other than circular can be flash welded; for example, angles, H-sections, and rectangles. Rotation of parts is not required.

(2) Parts of similar cross section can be welded with their axes aligned or at an angle to each other, within limits.

(3) The molten metal film on the faying surfaces and its ejection during upsetting acts to remove impurities from the interface.

(4) Preparation of the faying surfaces is not critical except for large parts that may require a bevel to initiate flashing.

(5) Rings of various cross sections can be welded.

(6) The heat-affected zones of flash welds are much narrower than those of upset welds.

Some limitations of the process are:

(1) The high single-phase power demand produces unbalance on the three-phase primary power lines.

(2) The molten metal particles ejected during flashing present a fire hazard, possible injury to the operator, and damage to shafts and bearings.

(3) Removal of flash and upset metal is

---

1. Friction welding is discussed in Chapter 7.

generally necessary and may require special equipment.

(4) Alignment of workpieces with small cross sections is sometimes difficult.

(5) The parts to be joined must have almost identical cross sections.

APPLICATIONS

Base Metals

Many ferrous and nonferrous alloys can be flash welded. Typical metals are carbon and low alloy steels, stainless steels, aluminum alloys, nickel alloys, and copper alloys. Titanium alloys can be flash welded, but an inert gas shield to displace air from around the joint is recommended to minimize embrittlement.

Dissimilar metals may be flash welded if their flashing and upsetting characteristics are similar. Some dissimilarity can be overcome with a difference in the initial extensions between the clamping dies, adjustment of flashing distance, and selections of welding variables. Typical examples are welding of aluminum to copper or a nickel alloy to steel.

Typical Products

The automotive industry uses wheel rims produced from flash welded rings formed from flat cold-rolled steel stock. The electrical industry uses motor and generator frames produced by flash welding plate and bar stock previously rolled into cylindrical form. Cylindrical transformer cases, circular flanges, and seals for power transformer cases are other examples. The aircraft industry utilizes flash welds in the manufacture of landing gear, control assemblies, and hollow propellor blades.

The petroleum industry uses oil drilling pipe with fittings attached by flash welding. Several major railroads are utilizing flash welding to join track of relatively high carbon steel. In most cases, welding is done in the field using welding machines and portable generating equipment mounted on railroad cars.

Miter joints are sometimes used in the production of rectangular frames for windows, doors, and other architectural trim. These products are commonly made of plain carbon and stainless steels, aluminum alloys, brasses, and bronzes. Usually the service loads are limited, but appearance requirements of the finished joints are extremely stringent.

EQUIPMENT

Typical Machine

A typical flash welding machine consists of six major parts:

(1) The machine bed to which is attached the fixed platen and a set of electrically insulated ways to support the movable platen

(2) The movable platen which is mounted on the electrically insulated ways

(3) Two clamping assemblies, one of which is rigidly attached to each platen to align and hold the parts to be welded

(4) A means for controlling the motion of the movable platen

(5) The welding transformer with adjustable taps

(6) The controls

Flash welding machines may be manual, semi-automatic, or fully automatic in their operation. However, most of them are either semi-automatic or fully automatic. With manual operation, the operator controls the speed of the platen from the time that flashing is initiated until the upset is completed. In semi-automatic operation, the operator usually initiates flashing manually and then energizes an automatic cycle that completes the weld. In fully automatic operation, the parts are loaded into the machine and the welding cycle is then completed automatically. The platen motion of many small flash welding machines is provided mechanically by a cam that is driven by an electric motor through a speed reducer. Large machines are usually hydraulically operated. Equipment for flash welding is discussed in Chapter 3.

Controls and Auxiliary Equipment

Electrical controls on flash welding machines are integral types designed to sequence the machine, control the welding current, and precisely control the platen position during flashing and upsetting. Electronic (SCR) contactors are widely used on machines drawing up to 1200 A from the power lines. Ignitron contactors are common on larger machines.

Preheat and postheat cycles are normally controlled with electronic timers and phase-shift heat controls. Timers for these functions may be initiated manually or automatically in the proper order during the welding period.

Dies

Flash welding dies are not in direct contact with the welding area as are spot and seam welding electrodes. Dies may be considered work holding and current conducting clamps. Since the current density in these dies is normally low, relatively hard materials with low electrical conductivity may be used. However, water cooling of the dies may be necessary in high production to avoid overheating.

There are no standardized designs for these dies since they must fit the contour of the parts to be welded. The size of the dies depends largely upon the geometry of the parts to be welded and the mechanical rigidity needed to maintain proper alignment of parts during upsetting. The dies are usually mechanically fastened to the welding machine platens.

Electrode contact area should be as large as practical to avoid local die burns. The contact surfaces may be incorporated in small inserts attached to the larger dies for low cost replacement and convenient detachment for redressing. A facing insert of RWMA[2] Group B material is frequently brazed to the die for maximum wear.

If the parts are backed up so that the clamping dies do not have to carry the upsetting force, clamping pressures need only be sufficient to provide good electrical contact. If the work cannot be backed up, it may be necessary to serrate the clamp inserts. In this case, they are usually made of hardened tool steel.

Flash welding dies tend to wear but do not mushroom. As wear takes place, the contact area may decrease and cause local hot spots (die burns). The dies should be kept clean. Flash and dirt will tend to embed in the die and cause hot spots and die burns. All bolts, nuts, and other die-holding devices should be tight.

---

2. Resistance Welder Manufacturers Association

Additional information on flash welding dies and materials is given in Chapter 3.

Fixtures and Backups

The function of fixtures for flash welding are (1) to locate rapidly and accurately two or more parts relative to each other, (2) to hold them in proper location while they are being welded, and (3) to permit easy release of the welded assembly. A fixture is either fastened to the machine or built into it. Parts are loaded directly into the fixture and welded.

Resistance welding processes are very rapid compared to other methods of joining. If maximum production is to be attained, fixtures must be easily loaded and unloaded. The following factors should be considered when designing a fixture:

(1) Quick-acting clamps, toggles, and other similar devices should be employed. Sometimes ejector pins are used to facilitate removal of the finished assembly.

(2) It must be designed so that welding current is not shunted through any locating device. This may require insulation of pins and locating strips.

(3) Usually, nonmagnetic materials are preferred because any magnetic material located in the throat of the machine will increase the electrical impedence and limit the maximum current which the machine can deliver.

(4) It should be made so that the operator can load and unload the parts with safety. This may require the use of swivel devices or slides so that the fixture can be moved out of the machine.

(5) A fixture must provide for movement of the parts as they are being clamped in the dies.

(6) All bearings, pins, slides, etc., should be protected from spatter and flash.

Backups are needed when the clamping dies cannot prevent slippage of the parts when the upsetting force is applied. Slippage usually occurs when the section of the part in the die is too short for effective clamping or the part is unable to withstand sufficient clamping force without damage.

A backup often consists of a steel bracket

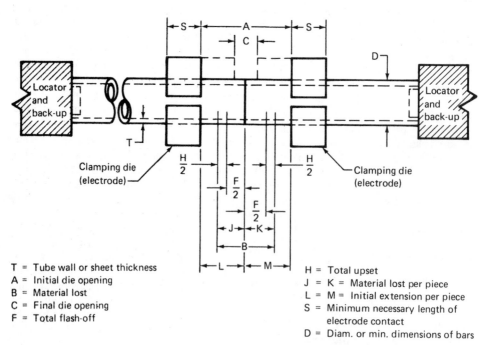

T = Tube wall or sheet thickness
A = Initial die opening
B = Material lost
C = Final die opening
F = Total flash-off

H = Total upset
J = K = Material lost per piece
L = M = Initial extension per piece
S = Minimum necessary length of electrode contact
D = Diam. or min. dimensions of bars

Fig. 2.2—Setup and material loss dimensions for flash welding of tube, sheet, and bar

that can be bolted in various positions to the platen. Brackets can have either fixed or adjustable stops against the parts.

WELDING PROCEDURES

Every welding operation involves numerous variables that affect the quality of the resulting weld. For this reason, a welding procedure should be developed that prescribes the settings for the welding variables to assure consistent weld quality. Flash welding involves dimensional, electrical, force, and time variables. The dimensional variables are shown in Figs. 2.2 and 2.3. The paths of the movable platen and the faying surfaces during flashing and upsetting are also shown on Fig. 2.3. The current, force, and time variables are shown on Fig. 2.4. Most operations do not utilize all of the variables shown. A simple flash welding cycle involves flashing at one voltage setting followed by upsetting.

Joint Design

Three common types of welds made by flash welding are shown in Fig. 2.5. Several basic design rules for flash welding are as follows:

(1) The design should provide for an even heat balance in the parts so that the ends to be welded will have nearly equal compressive strengths at the end of flashing time.

(2) The length of metal loss during flashing (flash-off) and upsetting must be included in the initial length when designing the part. With miter joints, the angle between the two members must be taken into account in the design.

(3) The parts must be designed so that they can be suitably clamped and held in accurate alignment during flashing and upsetting with the joint perpendicular to the upset force direction.

(4) The end preparation should be designed so that the flash material can escape from the joint, and that flashing starts at the center or the central area of the parts.

In general, the two parts to be welded should have the same cross section at the joint. Bosses may have to be machined, forged, or extruded on parts to meet this requirement.

In the flash welding of extruded or rolled shapes with different thicknesses within the cross-section, the temperature distribution during flashing will vary with section thickness. This tendency can often be counteracted by proper design of the clamping dies, provided the ratio of the thicknesses does not exceed about 4 to 1.

The recommended maximum joint lengths for several thicknesses of steel sheet are given in Table 2.1. The maximum diameters for steel tubing of various wall thicknesses are listed in Table 2.2. The limits can be exceeded in some cases using special procedures and equipment.

When flash welding rings, there is some ratio of circumference to cross-sectional area below which shunting of current becomes a problem. The power loss can be high. The minimum ratio will depend upon the electrical resistivity of the metal to be welded. With metals of high resistivity, such as stainless steel, the ratio can be lower than with low resistivity metals, such as aluminum.

Table 2.1
Recommended maximum joint lengths of flat steel sheet for flash welding

Sheet thickness, in.	Max. joint length, in.	Sheet thickness, in.	Max. joint length, in.
0.010	1.00	0.060	25.00
0.020	5.00	0.080	35.00
0.030	10.00	0.100	45.00
0.040	15.00	0.125	57.00
0.050	20.00	0.187	88.00

Table 2.2
Recommended maximum diameters of steel tubing for flash welding

Wall thicknesses, in.	Max. tubing diameter, in.	Wall thicknesses, in.	Max. tubing diameter, in.
0.020	0.50	0.125	4.00
0.030	0.75	0.187	6.00
0.050	1.25	0.250	9.00
0.062	1.50		
0.080	2.00		
0.100	3.00		

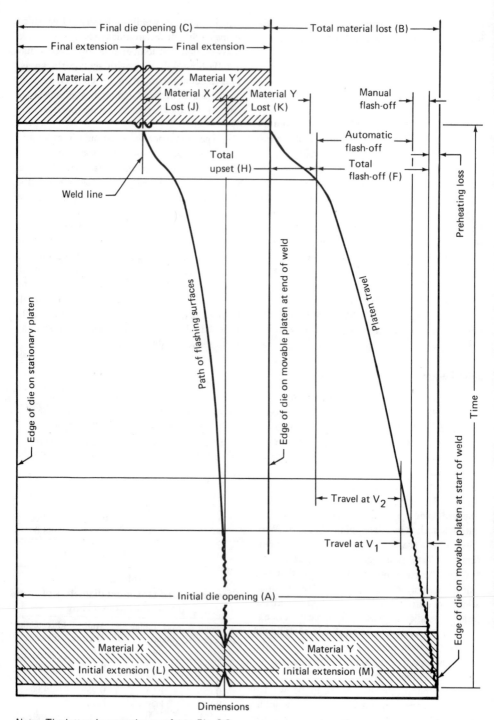

Note: The letters in parentheses refer to Fig. 2.2.

Fig. 2.3—Flash welding dimensional variables and motions

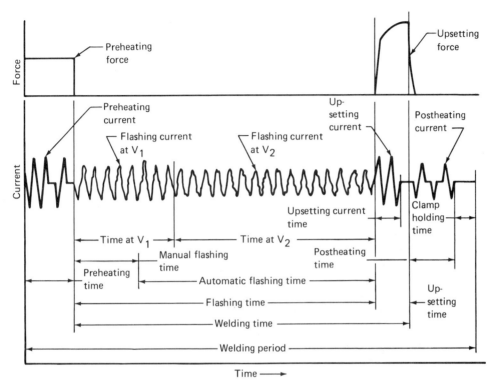

Fig. 2.4—Flash welding current, force, and time variables

When heavy sections are welded, it is often advisable to bevel the end of one part to facilitate the starting of flashing. Such beveling may eliminate the necessity for preheating or initially flashing at a voltage higher than normal. Suggested dimensions for beveling plate, rod, and tubing are shown in Fig. 2.6.

Heat Balance

Aligned Joints. When the two parts to be welded are of the same alloy and cross section, the heat generated in each of the parts during the weld cycle will be the same, provided the physical arrangement for welding is uniform. Flashoff and upsetting will also be equal in each part. In general, the heat balance between two parts of the same alloy will be adequate if their respective cross-sectional areas do not differ by more than normal manufacturing tolerances.

When flash welding two dissimilar metals, the metal loss during flashing may differ for each metal. Such behavior can be attributed to differences in electrical and thermal conductivities or melting temperatures, or both. To compensate for this, the extension from the clamping die of the more rapidly consumed part should be greater than that of the other part. In the case of aluminum and copper, the extension of the aluminum part should be twice that of the copper part.

Miter Joints. Flash welding of nonaligned sections (miter joints) may produce a joint with varying properties across it because of heat unbalance across the joint. Since the faying surfaces are not perpendicular to their respective part lengths, the volume of metal decreases across the joint to a minimum at the apex. Consequently, flashing and upsetting at the apex may vary significantly from that which occurs across the remainder of the joint.

Miter joints between round or square bars should have a minimum included angle of 150

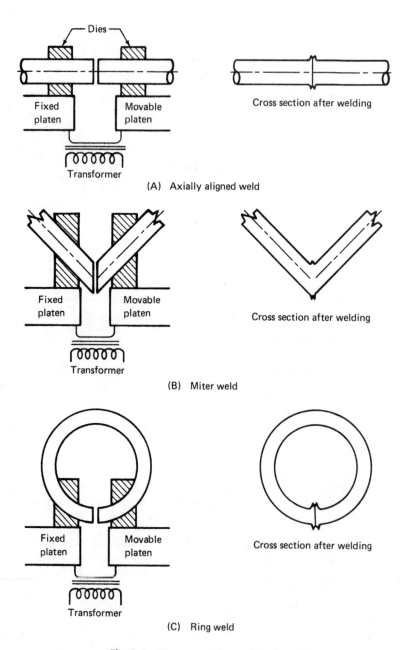

(A) Axially aligned weld

(B) Miter weld

(C) Ring weld

Fig. 2.5—Common types of flash welds

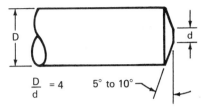

(A) Rods and bars of 0.25 in. and larger

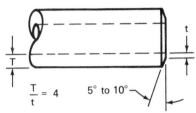

(B) Tubing of 0.188-in. wall and larger

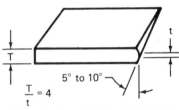

(C) Flat plate of 0.188-in. and thicker

Note: Bevel only one piece when D is 0.25 in. or larger and T is 0.188 in. or greater.

Fig. 2.6—End preparation for one part to facilitate the flashing of large sections

degrees. At smaller angles, the weld area at the apex will be poor quality because of the lack of adequate backup metal. Satisfactory miter joints may be made between thin rectangular sections in the same plane with an included angle as small as 90 degrees, provided the width of the

stock is greater than 20 times its thickness. If service loading produces a tensile stress at the apex, the outside corner should be trimmed to remove the poor quality joint area.

Surface Preparation

Surface preparation for flash welding is of minor importance and in most cases, none is required. Clamping surfaces usually require no special preparation unless excessive scale, rust, grease, or paint is present. The abutting surfaces should be reasonably clean to accomplish electrical contact. Once flashing starts, dirt or other foreign matter will not seriously interfere with the consummation of the weld.

Initial Die Opening

The initial die opening is the sum of the initial extensions of the two parts, as shown in Figs. 2.2 and 2.3. The initial extension for each part must provide for metal loss during flashing (flash-off) and upsetting as well as some undisturbed metal between the upset metal and the clamping die. Initial extensions for both parts are determined from available welding data or from welding tests. The initial die opening should not be too large. Otherwise, nonuniform upsetting and joint misalignment may occur.

Alignment

It is important that the parts to be welded are properly aligned in the welding machine so that flashing on the faying surfaces is uniform. If the parts are misaligned, flashing will occur only across opposing areas and heating will not be uniform. When upset, the parts will tend to slip past each other, as illustrated in Fig. 2.7. Alignment of parts should be given careful consideration in the design of the machine, the parts to be welded, and the tooling for welding them. This is especially true when the ratio of the width to thickness of sections is large.

Metal Loss

The final length of the welded assembly will be less than the sum of the initial part lengths because of flashing and upsetting losses. These losses must be established for each part of an assembly and then added to the part length so

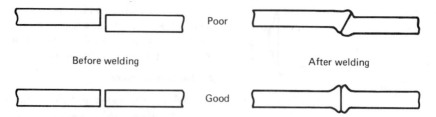

Fig. 2.7—Effect of part alignment on joint geometry

that the welded assembly will meet design requirements. Changes in welding procedures may require modification of part lengths.

Gas Shielding

In some applications, displacement of air from the joint area by an inert or reducing gas shield may improve joint quality by minimizing contamination by oxygen, nitrogen, or both. However, gas shielding cannot compensate for improper welding procedures, and it should only be used when required to produce joints of acceptable quality.

Argon or helium is particularly effective when flash welding reactive metals, such as titanium. At high temperatures, these metals are embrittled when they are exposed to air. Dry nitrogen may be effective with stainless and heat-resisting steels.

The value of a protective atmosphere depends upon the effectiveness of the shield design. The flash-off material may deposit on the gas shielding apparatus and interfere with its operation. Provisions for platen movement must be provided in the design.

Preheating

During preheating, the parts are brought into contact under light pressure and then the welding transformer is energized. The high current density resistance heats the metal between the dies. The temperature distribution across the joint during preheating approximates a sinusoidal waveform with the peak temperature point at the interface.

Three useful functions may be served by a preheating operation:

(1) It raises the temperature of the parts which, in turn, makes flashing easier to start and maintain.

(2) It produces a temperature distribution with a flatter gradient which persists throughout the flashing operations. This, in turn, distributes the upset over a longer length than is the case when no preheat is employed.

(3) It may extend the capacity of a machine and permit the joining of larger cross sections than would be otherwise possible.

Since preheating is usually a manual operation, even when the machine is capable of flash welding automatically, the reproducibility of the preheating operation is largely a function of operator skill.

Welding

Most commercial flash welding machines are operated automatically. The welding schedule is established for the particular operations by a series of test welds that are evaluated for quality. The machine is then set up to reproduce the qualified welding schedule for the particular application.

The operator may load and unload the machine and observe the welding cycle for consistency of operation. In some applications, automatic feed and ejection devices may be incorporated on the machine.

Postheating

When welding steels with extremely high alloy or carbon content, cracking is possible if the weld is cooled too rapidly to room temperature. In some cases, this condition may be avoided by preheating large parts which, in turn, will decrease the cooling rate subsequent to welding. When preheating the parts is not sufficient, cracking may be prevented either by postheating the joint in the welding machine by resistance heating or by immediately placing the weldment in a furnace operating at the desired temperature.

A postheat cycle may be incorporated in a flash welding machine using an electronic timer and phase-shift heat control. The postheat timer can be initiated at the end of upset or after a time delay. The desired temperature can be attained by adjustment of the heat control. Heat will be transferred from the weldment to the clamping dies during the postheat. This must be considered in the die design and material selection, and water cooling may be necessary.

Flash Removal

It is frequently necessary to remove the flash material from the welded joint. In some cases, this is done only for the sake of appearance. A joint is somewhat stronger in tension if the flash is not removed because of the larger cross section of the upset material. The notch effect at the weld line, however, may cause a reduction of fatigue strength. A portion of the upset material may be left in place when the design of parts indicates that reinforcement is beneficial.

It is generally easier to remove the flash immediately after welding while the metal is still hot. This can be done by a number of methods including machining, grinding, high-speed burring wheels, die trimming, oxyfuel gas cutting, high-speed sanding, and pinch-off type clamping dies. With some alloy steels, flash removal with cutting tools is often difficult because of their hardness. In these cases, either grinding or oxyfuel gas cutting is usually employed.

With soft metals such as aluminum and copper, the flash may be almost sheared off using pinch-off dies. These dies have sharp tapered faces which cut almost through the metal as upsetting takes place. The final die opening is small. The partially sheared flash is then easily removed by other means. The joint can then be smoothed by filing or grinding.

PROCESS VARIABLES

Flashing

Voltage. Flashing voltage is determined by the welding transformer tap setting. It should be selected to be as low as possible consistent with good flashing action. Electronic phase-shift heat control is not an effective means for reducing the flashing voltage. The secondary voltage wave form produced by this means is incompatible with good flashing action.

Changes in flashing voltage should be made only by changing the tap setting of the transformer. One system for providing two voltage ranges uses two primary contactors, each of which is connected to separate transformer taps. One contactor is energized to provide a high secondary voltage during the initial stages of flashing. The high voltage assists in starting the flashing action. The other contactor is energized after a predetermined time in the flashing operation to provide a normal secondary voltage. The first contactor is de-energized at the same time. The best flashing action is achieved with this arrangement.

Time. Flashing is carried out over a time interval to obtain the required flash-off of metal. The time required will be related to secondary voltage and the rate of metal loss as flashing progresses. Since a flashing pattern is generally parabolic, the variables are interrelated. In any case, smooth flashing action for some minimum flashing distance during some time interval is necessary to produce a sound, strong weld.

Upsetting

In the production of a satisfactory flash weld, the flashing and upsetting variables must be considered together since they are interrelated. The upset variables include the following:

(1) Flashing voltage cutoff
(2) Upsetting rate
(3) Upset distance
(4) Upset current magnitude and duration

Flashing Voltage Cutoff. In most cases, flashing voltage should be terminated at the same time that upsetting of the weld commences, but not beforehand. The point of voltage termination should be adjusted during actual welding tests to ensure that it does not take place before the faying surfaces make contact.

Upsetting Rate. Upsetting is initiated by increasing the acceleration of the moving part to bring the faying surfaces together quickly. The molten metal and oxides present on the surfaces are forced out of the joint as this occurs. Then the hot weld zone is upset. The upsetting rate must be sufficient to expel the molten metal before it solidifies and to produce the optimum upset while the metal has adequate plasticity.

The welding machine must apply a force to the movable platen to properly accelerate the part and overcome the resistance of the parts to plastic deformation. The force required depends upon the cross-sectional area of the joint, the pressure needed to upset the metal to be welded, and the mass of the movable platen. Table 2.3 gives the approximate minimum upsetting pressures for flash welding typical alloys. These values may be used for a first approximation of the welding machine size required to flash weld a particular joint area in one of these alloys.

Upset Distance. The magnitude of the upset distance must be sufficient to accomplish these two actions:

(1) The oxides and molten metal must be extruded to the surface from the center of the thickest section.

(2) The two faying surfaces must be brought into intimate metal-to-metal contact over the entire cross section.

The amount of upset required to obtain a sound flash weld is related in some complex way to the metal and the section thickness. Normally, an upset distance equal to one half of the part thickness should be adequate. If the flashing conditions produce relatively smooth flashed surfaces, a considerably smaller upset distance should be satisfactory for most metals. However, some heat-resisting alloys may require upset distances as large as 1 to 1.25 times the section thickness. Satisfactory welds are made in aluminum with upset distances about 50 percent greater than those employed with steels of similar thicknesses.

Upsetting Current. In some cases, the weld zone may tend to cool too rapidly after flashing is terminated. This may result in inadequate upset or cold cracking of the upset metal. The joint temperature can be maintained during upsetting by resistance heating with current supplied by the welding transformer. The current magnitude is commonly adjusted by electronic heat control.

Normally, upsetting current would be terminated at the end of upset. If the flash is to be mechanically trimmed immediately after welding, upsetting current may be maintained for an additional period to achieve the desired temperature for trimming.

WELD QUALITY

Effect of Welding Variables

Weld quality is significantly affected by the specific welding variables selected for the application. Table 2.4 indicates the effects of several variables on quality when they are excessive or insufficient in magnitude. Each variable is considered individually although more than one can produce the same result.

The discontinuities encountered in flash welding may be divided into two categories: metallurgical and mechanical.

Metallurgical Discontinuities

Base Metal Structure. Metallurgical discontin-

Table 2.3
Approximate minimum upset pressure requirements for flash welding various alloys

Material	Minimum upset pressure required, ksi
Plain carbon steels (0.4C max.)	10
Plain carbon steels (0.4C-1.2C), low and medium alloy steels	15-20
Stainless steels (400 series)	20
Titanium alloys	20-25
Stainless steels (300 series)	20-30
High temperature alloys	40-50
Aluminum alloys	10-50

Table 2.4
Effect of variables on flash weld quality

When variable is:	Welding variable				
	Flashing			Upsetting	
	Voltage	Rate	Time	Current	Distance or force
Excessive	Deep craters are formed that cause pockets and inclusions in weld; cast metal in weld	Tendency to freeze	Metal too plastic to upset properly	Burning of oxidation even to the extent of blowing out metal. Excessive deformation	Tendency to squeeze out too much plastic metal; flow lines bent parallel to weld line
Insufficient	Tendency to freeze; metal not plastic enough for proper upset	Intermittent flashing, which makes it difficult to develop sufficient heat in the metal for proper upset	Not plastic enough for proper upset	Longitudinal cracking through weld area. Inclusions and voids not properly forced out of weld	Failure to force molten metal from joint; cast metal retained in weld; oxides, inclusions and voids in weld

uities that often originate from conditions present in the base metal can usually be minimized by the application of appropriate materials acceptance specifications. The inherent fibrous structure of wrought mill products may cause anisotropic mechanical behavior. An out-turned fibrous structure at the weld line often results in some decrease in mechanical properties as compared to the base metal, particularly in ductility.

The decrease in ductility is not normally significant unless:

(1) The material is extremely inhomogeneous. Examples are severely banded steels, alloys with excessive stringer type inclusions, and mill products with seams and cold shuts produced during the fabrication process.

(2) The upset distance is excessive.

When excessive upset distance is employed, the fibrous structure may be completely reoriented transverse to the original structure.

Oxides. Another source of metallurgical discontinuities is the entrapment of oxides at the weld interface. Such defects are rare with good practice since proper upsetting should expel any oxides formed during the flashing operation.

Flat Spots. These metallurgical discontinuities are usually limited to ferrous alloys. Their exact cause is not certain. They appear in the form of smooth, irregular-shaped areas on the fracture surface through the weld interface.

There is excellent correlation between the location of flat spots and localized regions of carbon segregation in steels. In many cases, the cooling rates associated with flash welds are rapid enough to produce brittle, high carbon martensite at areas on the flashing interface where the carbon content happens to be greater than the nominal composition of the alloy. Microhardness tests and metallographic examination have confirmed the presence of high carbon martensite in the region surrounding a "flat spot" in almost every case, even in plain carbon steels. Furthermore, steels with banded microstructures appeared significantly more susceptible to this type of defect than unbanded steels.

Die Burns. These are a type of discontinuity produced by local overheating of the base metal at the interface between the clamping die and the part surface. They can usually be avoided completely when the parts are clean and mate properly with the dies.

Voids. Voids are usually the result of either insufficient upset or excessive flashing voltage.

Deep craters on the faying surfaces produced by excessive flashing voltage may not be completely eliminated with the correct upset. Such discontinuities are usually discovered during procedure qualification tests and are readily eliminated by decreasing the flashing voltage or increasing the upset distance. Figure 2.8 (A) and (B) indicate the appearance of flash welds with and without satisfactory upset.

Cracking. This type of discontinuity may be internal or external. It may be related to the metallurgical characteristics of the metal. Alloys that exhibit low ductility over some elevated temperature range may be susceptible to hot cracking. Such alloys, known as "hot-short" alloys, are somewhat difficult to flash weld, but usually can be welded with the proper conditions. Cold cracking may be found in hardenable steels. It can usually be eliminated by welding with conditions that provide a slow weld cooling rate, coupled with heat treatment as soon as possible after welding. Insufficient heating prior to or during upsetting may also cause cracking in the upset metal, as shown in Fig. 2.8(C). This can be eliminated by increasing the flashing distance or by resistance heating during upsetting.

Mechanical Discontinuities. These include misalignment of the faying surfaces prior to welding and nonuniform upsetting during welding. They are easily detected by visual inspection. Misalignment of the parts is corrected by adjustment of the clamping dies and fixtures. Nonuniform upsetting may be caused by part misalignment, insufficient clamping force, or excessive die opening at the start of upset. The latter can be corrected by decreasing the initial die opening and then adjusting the welding schedule, if necessary.

TESTING AND INSPECTION

No combination of presently known nondestructive tests can provide a conclusive evaluation of the quality of a flash welded joint. Fortunately, one of the major advantages of flash welding is that it can be highly mechanized and essentially automatic. Therefore, a consistent quality level is readily maintained after satisfactory welding conditions are established. The fact that no filler metal is employed means that the strength of the weld is primarily a function of the base metal composition and properties. Consequently, properly made flash welds should exhibit mechanical properties almost identical to those of the base metal.

In commercial practice, both destructive and nondestructive tests are employed to ensure maintenance of the quality level in critical flash welded products. The process control procedure usually includes the following:

(1) Material certification
(2) Qualification of welding procedure
(3) Visual inspection of the product

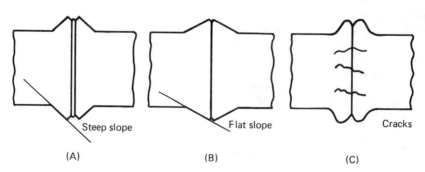

Steep slope   Flat slope   Cracks

(A)   (B)   (C)

Fig. 2.8—Visual indications of flash weld quality: (A) satisfactory heat and upset, (B) insufficient heat or upset or both, (C) cracks due to insufficient heat

(4) Destructive testing of random samples

When the product is used in a critical application, the above procedure is supplemented by other tests such as magnetic particle, fluorescent penetrant, and radiographic examination. When the welded joint is subsequently machined, routine measurement of the hardness of the weld area may also be specified. In addition, specifications may require proof testing of flash welded products.

Material Certification

Since material defects may cause flash weld discontinuities, each lot of raw material should be carefully inspected upon delivery to ensure that it meets specifications. Certified chemical analysis, mechanical property tests, macro-etch examination, and magnetic particle inspection may be applicable.

Procedure Qualification

Each new combination of material and section size to be flash welded normally requires qualification of the welding procedure to be employed. This usually consists of welding a number of test specimens that duplicate the material, section size, welding procedure, and heat treatment to be used in producing the product. All specimens should be visually inspected for cracks, die burns, misalignment, and other discontinuities. Where required, weld hardness should be measured. A tensile specimen should be machined from a test weld using the entire welded cross section where possible. The test results should be compared to the base metal properties and design requirements.

All pertinent welding conditions used in producing the qualification test should be recorded. The production run is then made using the same welding procedures.

Visual Inspection

Each completed weld in the production run should be visually examined for evidences of cracks, die burns, misalignment, or other external weld defects. Where specified, magnetic particle or fluorescent penetrant inspection is performed to assist in detecting flaws not visible

to the unaided eye. In critical applications, radiographic examination may also be specified.

Random Samples

Depending upon the size of the production run, a specified number of randomly chosen parts may be selected for destructive testing of the welds. The results of these destructive tests are normally subject to the same criteria specified in the welding procedure qualification test. A report of the results of all destructive tests is then prepared to certify the maintenance of the required average quality level for the lot.

Bend Tests

Notched bend tests may be used to force the fracture to occur along the interface for visual examination. A bend test may be useful as a qualitative means for establishing a welding schedule. However, such tests are not generally used for specification purposes.

WELDING OF STEEL

Typical data for the flash welding of steel tubing and flat sheets are given in Table 2.5. For welding solid, round, hexagonal, square, and rectangular steel bars, data are given in Table 2.6. Both tables are applicable to steels of low and medium forging strength. They give the recommended dimensions for setting up a flash welding machine to weld the various sections. Total flashing time is based on welding without preheating.

When setting up a schedule, the dimensional variables and flashing time are selected from the tables. The welding machine is adjusted to the lowest secondary voltage at which steady and consistent flashing can be obtained. The secondary voltages available are dependent upon the electrical design of the welding machine transformer.

The upsetting force used for a particular application depends upon the alloy and the cross-sectional area of the joint. The selection of equipment for steels should be based on the values of recommended upsetting pressures given in Table 2.7. Such values are based on welding without preheat.

Table 2.5
Data for flash welding steel tubing and flat sheets[ab]

T, in.	A, in.	B, in.	C, in.	F, in.	H, in.	J = K, in.	L = M, in.	Flash-ing time, s	D, in.	S, in. with loca-tor	S, in. with-out loca-tor
0.010	0.110	0.060	0.050	0.040	0.020	0.030	0.055	0.40	0.250	0.375	1.00
0.020	0.215	0.115	0.100	0.080	0.035	0.058	0.108	0.80	0.312	0.375	1.00
0.030	0.325	0.175	0.150	0.125	0.050	0.088	0.163	1.25	0.375	0.375	1.50
0.040	0.430	0.230	0.200	0.165	0.065	0.115	0.215	1.75	0.500	0.375	1.75
0.050	0.530	0.280	0.250	0.205	0.075	0.140	0.265	2.25	0.750	0.500	2.00
0.060	0.620	0.330	0.290	0.240	0.090	0.165	0.310	2.75	1.000	0.750	2.50
0.070	0.715	0.385	0.330	0.280	0.105	0.193	0.358	3.50	1.50	1.000	3.00
0.080	0.805	0.435	0.370	0.315	0.120	0.218	0.403	4.00	2.00	1.250	
0.090	0.885	0.475	0.410	0.345	0.130	0.238	0.443	4.50	2.50	1.750	
0.100	0.970	0.520	0.450	0.375	0.145	0.260	0.485	5.00	3.00	2.000	
0.110	1.060	0.570	0.490	0.410	0.160	0.285	0.530	5.75	3.50	2.25	
0.120	1.140	0.610	0.530	0.440	0.170	0.305	0.570	6.25	4.00	2.50	
0.130	1.225	0.650	0.575	0.470	0.180	0.325	0.613	7.00	4.50	2.75	
0.140	1.320	0.700	0.620	0.510	0.190	0.350	0.660	7.75	5.00	2.75	
0.150	1.390	0.730	0.660	0.530	0.200	0.365	0.695	8.50	5.50	3.00	
0.160	1.470	0.770	0.700	0.560	0.210	0.385	0.735	9.00	6.00	3.25	
0.170	1.540	0.800	0.740	0.580	0.220	0.400	0.770	9.75	6.50	3.50	
0.180	1.620	0.840	0.780	0.610	0.230	0.420	0.810	10.50	7.00	3.75	
0.190	1.690	0.870	0.820	0.630	0.240	0.435	0.845	11.25	7.50	4.00	
0.200	1.760	0.900	0.860	0.650	0.250	0.450	0.880	12.00	8.00	4.25	
0.250	2.010	1.010	1.000	0.730	0.280	0.505	1.005	16.00	8.50	4.50	
0.300	2.245	1.120	1.125	0.810	0.310	0.560	1.123	21.00	9.00	4.75	
0.350	2.460	1.210	1.250	0.880	0.330	0.605	1.230	27.00	9.50	5.00	
0.400	2.640	1.290	1.350	0.930	0.360	0.645	1.320	33.00			
0.450	2.780	1.350	1.430	0.970	0.380	0.675	1.390	38.00			
0.500	2.910	1.410	1.500	1.020	0.390	0.705	1.455	45.00			
0.550	3.040	1.465	1.575	1.055	0.410	0.733	1.520	50.00			
0.600	3.135	1.505	1.630	1.085	0.420	0.753	1.568	56.00			
0.650	3.245	1.555	1.690	1.125	0.430	0.778	1.623	63.00			
0.700	3.360	1.610	1.750	1.160	0.450	0.805	1.680	70.00			
0.800	3.525	1.675	1.850	1.210	0.465	0.838	1.763	83.00			
0.900	3.660	1.730	1.930	1.250	0.480	0.865	1.830	97.00			
1.000	3.800	1.800	2.000	1.300	0.500	0.900	1.900	110.00			

a. Date is based on welding two pieces with the same welding characteristics without preheat.
b. See Fig 2.2 to identify the dimensions given.

Table 2.6
Data for flash welding round, hexagonal, square, and rectangular steel bars[ab]

D, in.	A, in.	B, in.	C, in.	F, in.	H, in.	J = K, in.	L = M, in.	Flash-ing time, s	D, in.	S, in. with loca-tor	S, in. with-out loca-tor
0.050	0.100	0.050	0.050	0.040	0.010	0.025	0.050	0.40	0.250	0.375	1.00
0.100	0.182	0.082	0.100	0.062	0.020	0.041	0.091	0.75	0.312	0.375	1.00
0.150	0.270	0.120	0.150	0.090	0.030	0.060	0.135	1.15	0.375	0.375	1.50
0.200	0.350	0.150	0.200	0.110	0.040	0.075	0.175	1.50	0.500	0.375	1.75
0.250	0.430	0.180	0.250	0.130	0.050	0.090	0.215	1.90	0.750	0.500	2.00
0.300	0.510	0.210	0.300	0.150	0.060	0.105	0.255	2.25	1.000	0.750	2.50
0.350	0.600	0.250	0.350	0.180	0.070	0.125	0.300	2.75	1.50	1.000	3.00
0.400	0.685	0.285	0.400	0.205	0.080	0.143	0.343	3.25	2.00	1.25	
0.450	0.770	0.320	0.450	0.230	0.090	0.160	0.385	3.75			
0.500	0.850	0.350	0.500	0.250	0.100	0.175	0.425	4.25			
0.550	0.940	0.390	0.550	0.280	0.110	0.195	0.470	5.00			
0.600	1.025	0.425	0.600	0.305	0.120	0.213	0.513	5.50			
0.650	1.100	0.450	0.650	0.325	0.125	0.225	0.550	6.75			
0.700	1.180	0.480	0.700	0.350	0.130	0.240	0.590	7.50			
0.750	1.260	0.510	0.750	0.375	0.135	0.255	0.630	8.25			
0.800	1.340	0.540	0.800	0.400	0.140	0.270	0.670	9.00			
0.850	1.420	0.570	0.850	0.425	0.145	0.285	0.710	9.75			
0.900	1.500	0.600	0.900	0.450	0.150	0.300	0.750	10.50			
0.950	1.580	0.630	0.950	0.475	0.155	0.315	0.790	11.75			
1.000	1.660	0.660	1.000	0.500	0.160	0.330	0.830	13.00			
1.050	1.740	0.690	1.050	0.525	0.165	0.345	0.870	14.75			
1.100	1.820	0.720	1.100	0.550	0.170	0.360	0.910	16.50			
1.150	1.900	0.750	1.150	0.575	0.175	0.375	0.950	18.25			
1.200	1.980	0.780	1.200	0.600	0.180	0.390	0.990	20.00			
1.250	2.060	0.810	1.250	0.625	0.185	0.405	1.030	22.50			
1.300	2.140	0.840	1.300	0.650	0.190	0.420	1.070	25.00			
1.400	2.300	0.900	1.400	0.700	0.200	0.450	1.150	30.00			
1.500	2.460	0.960	1.500	0.750	0.210	0.480	1.230	38.00			
1.600	2.620	1.020	1.600	0.800	0.220	0.510	1.310	45.00			
1.700	2.780	1.080	1.700	0.850	0.230	0.540	1.390	54.00			
1.800	2.940	1.140	1.800	0.900	0.240	0.570	1.470	63.00			
1.900	3.100	1.200	1.900	0.950	0.250	0.600	1.550	75.00			
2.000	3.260	1.260	2.000	1.000	0.260	0.630	1.630	90.00			

a. Data is based on welding two pieces with the same welding characteristics without preheat.
b. See Fig. 2.2 to identify the dimensions given.

Table 2.7
Upsetting pressures for various classes of steels

Strength classification	Examples	Upsetting pressure, ksi
Low forging	SAE 1020, 1117, and high strength, low alloy steels	10
Medium forging	SAE 1045, 1065, 1335, 4130, 4140	15
High forging	SAE 4340, 8740, 3XX and 4XX stainless steels, high-speed tool steels	25
Extra high forging	Steels with high compressive strengths at elevated temperature	35

UPSET WELDING

FUNDAMENTALS OF THE PROCESS

Definition

Upset welding is a process which produces coalescence simultaneously over the entire area of two abutting surfaces or progressively along a joint. The heat for welding is obtained from the resistance to the flow of electric current through the metal at the joint. Force is applied to upset the joint and consummate a weld when the metal reaches welding temperature. In some cases, force is applied before heating starts to bring the faying surfaces in contact.

Principles of Operation

With this process, welding is essentially done in the solid state. The metal at the joint is resistance heated to a temperature where recrystallization can rapidly take place across the faying surfaces. A force is applied to the joint to bring the faying surfaces into intimate contact and then upset the metal. Upsetting hastens recrystallization at the interface and, at the same time, some metal is forced outward from this location. This tends to purge the joint of oxidized metal.

Process Variations

Upset welding has two variations:
(1) Joining two sections of the same cross section end-to-end (butt joint)
(2) Continuous welding of seams in roll-formed products such as pipe and tubing

The first variation can also be accomplished by flash welding and friction welding. The second variation is also done with high frequency welding.[3]

BUTT JOINTS

Metals Welded

A wide variety of metals in the form of wire, bar, strip, and tubing can be joined end-to-end by upset welding. These include:
(1) Carbon steels
(2) Stainless steels
(3) Aluminum alloys
(4) Brass
(5) Copper
(6) Nickel alloys
(7) Electrical resistance alloys
Carbon steels, stainless steels, copper alloys, and many aluminum alloys are easily welded with this process. Some aluminum alloys require precise control of the upsetting force.

Sequence of Operations

The essential operational steps to produce an upset welded butt joint are as follows:
(1) Load the machine with the parts aligned end-to-end

3. High frequency welding is discussed in Chapter 4.

(2) Clamp the parts securely

(3) Apply a welding force

(4) Initiate the welding current

(5) Apply an upsetting force

(6) Cut off the welding current

(7) Release the upsetting force

(8) Unclamp the weldment

(9) Return the movable platen and unload the weldment

The general arrangement for upset welding is shown in Fig. 2.9. One clamping die is stationary and the other is movable to accomplish upset. Upsetting force is applied through the clamping die or a mechanical backup, or both.

Joint Preparation

For uniform heating, the faying surfaces should be flat, comparatively smooth, and perpendicular to the direction of the upsetting force. Prior to welding, they should be cleaned to remove any dirt, oil, oxidation, or other materials that will impede welding across the interface.

The contact resistance between the faying surfaces is a function of the smoothness and cleanliness of the surfaces and the contact pressure. This resistance varies approximately with the reciprocal of the contact pressure, provided

the other factors are constant. As the temperature of the joint increases, the contact resistance changes, but it finally becomes zero when the weld is formed. Upset welding differs from flash welding in that no flashing takes place at any time during the welding cycle.

Generally, force and current are maintained throughout the entire welding cycle. Initially, the force is low to promote high initial contact resistance between the two parts. It is increased to a higher value to upset the joint when welding temperature is reached. After the prescribed upset is accomplished, the welding current is cut off and the force is removed.

Wire of small diameter is sometimes welded with chisel-shaped ends produced by shearing with a wire cutter. The wires to be joined are placed in the welding machine with the chisel edges in contact at 90 degrees to each other.

Equipment

Equipment for upset welding is generally designed to weld a particular family of alloys, such as steels, within a size range based on cross-sectional area. The mechanical capacity and electrical characteristics of the machine are matched to suit the application. Special designs may be required for certain aluminum alloys to provide adequate control of upsetting force.

Electric current for heating is provided by a resistance welding transformer. It converts line power to low voltage, high current power. No-load secondary voltages range from about 0.4 to 8 V, and the power line frequency may be 50 or 60 Hz. Secondary current is controlled by a transformer tap switch or by electronic phase-shift.

Basically, an upset welding machine has two platens, one of which is stationary and the other movable. The clamping dies are mounted on these platens. The clamps operate either in straight line motion or through an arc about an axis, depending upon the application. Force for upset butt welding is produced generally by a mechanical, pneumatic, or hydraulic system.

Heat Balance

Normally, the process is used to join together two pieces of the same alloy and cross-

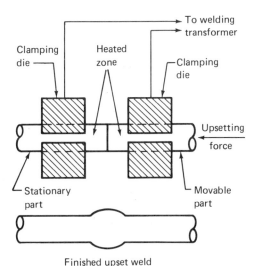

Finished upset weld

Fig. 2.9—General arrangement for upset welding of bars, rods, and pipes

sectional geometry. In this case, heat balance should be uniform across the joint. If the parts to be welded are similar in composition and cross section but of unequal mass, the part of larger mass should project from the clamping die somewhat farther than the other part. With dissimilar metals, the one with higher electrical conductivity should extend farther from the clamp than the other. When upset welding large parts that do not make good contact with each other, it is sometimes advantageous to interrupt the welding current periodically to allow the heat to distribute evenly into the parts.

Applications

Upset welding is used in wire mills and in the manufacture of products made from wire. In wire mill applications, the process is used for joining wire coils to each other to facilitate continuous processing. The process also is used for fabricating a rather wide variety of products from bar, strip, and tubing. Typical examples of the mill forms and products that have been upset welded are shown in Fig. 2.10. Wire and rod from 0.05 to 1.25-in. diameters can be upset welded.

Weld Quality

Butt joints can be made that have about the same properties as the base metal. With proper procedures, welds made in wires are difficult to locate after they have passed through the drawing processes. In many instances, the welds are considered part of the continuous wire.

Upset welds may be evaluated using tension tests. The tensile properties are then compared to those of the base metal. Metallographic and dye penetrant inspection techniques are also used. A common method for evaluating a butt weld in wire is a bend test. A welded sample is clamped in a vise with the weld interface located one wire diameter from the vise jaws. The sample then is bent back and forth until it breaks in two. If the fracture is through the weld interface and shows complete fusion or occurs outside the weld, the weld quality is considered satisfactory.

CONTINUOUS SEAM WELDING

General Description

In the manufacture of continuously welded pipe or tubing, coiled strip is fed into a set of forming rolls. These rolls progressively form the strip into cylindrical shape. The edges to be joined approach each other at an angle and culminate in a vee at the point of welding. A wheel electrode contacts each edge of the tube a short distance from the apex of the vee. Current from the power source travels from one electrode along the adjacent edge to the apex, where welding is taking place, and then back along the other edge to the second electrode. The edges are resistance heated by this current to welding temperature. The hot edges are then upset together by a set of pinch rolls to consummate a weld.

Equipment

Figure 2.11 shows a typical tube mill that utilizes upset welding for joining the longitudinal seam. Figure 2.11(A) shows the steel strip entering the strip guide assembly and the first stages of the forming section. The heat regulator, located behind the forming section, can be adjusted either manually or by phase-shift heat control. Figure 2.11(B) shows the rotary type of oil-cooled welding transformer. This welding equipment includes (1) a dressing tool assembly for dressing the welding electrodes without removing them from the welding machine and (2) a scarfing tool assembly that removes the upset metal after welding. In the third step, the welded tube enters the straightening and sizing section, shown in Fig. 2.11(C). Following this, the tubing is cut to the desired length.

Welding can be done using either ac or dc power. Alternating current machines may be operated on either 60 Hz single-phase power or on power of higher frequency produced by a single-phase alternator. Direct current machines are powered by a three-phase welding transformer-rectifier unit.

Welding Procedures

As the formed tube passes through the zone between the electrodes and the pinch rolls, there

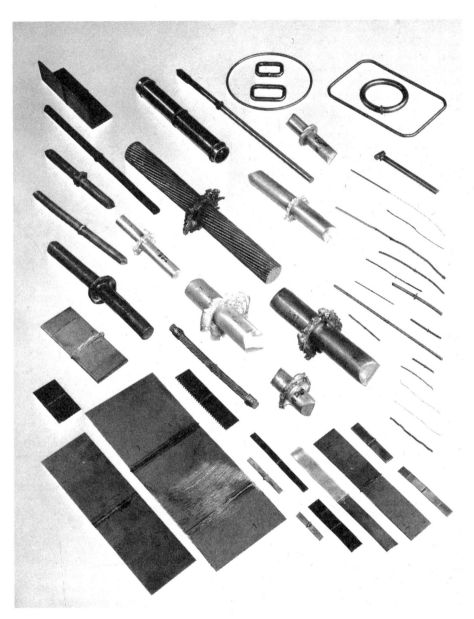

Fig. 2.10—Typical mill forms and products joined by upset welding

(A)

(B)

(C)

Fig. 2.11—Typical tube mill using welding for joining the longitudinal seam: (A) the strip guide assembly and first stages of the forming section; (B) the rotary type oil-cooled welding transformer; (C) the straightening and sizing section

is a variation in pressure across the joint. If no heat were generated along the edges, this pressure would be maximum at the center of the squeeze rolls. However, since heat is generated in the metal ahead of the squeeze roll center line, the metal gradually becomes plastic and the point of initial edge contact is slightly ahead of the squeeze roll axes. The point of maximum upsetting pressure is somewhat ahead of the squeeze roll center line.

The current flowing across the seam is distributed in inverse proportion to the resistance path between the two electrodes. This resistance, for the most part, is the contact resistance between the edges to be welded. Pressure is effective in reducing this contact resistance. As the temperature of the joint increases, the electrical resistance will increase and the pressure will decrease. A very sharp thermal gradient caused by the I^2R heating at the peaks of the ac cycle produce a "stitch effect." The stitch is normally of circular cross section, lying centrally in the weld area and parallel to the line of initial closure of the seam edges. It is the hottest portion of the weld. The stitch area is molten while the area between stitches is at a lower temperature. The patches of molten metal are relatively free to flow under the influence of the motor forces (current and magnetic flux) acting on them. Consequently, they are ejected from the stitch area. If the welding heat is excessive, too much metal is ejected and pinhole leaks may result. With too little heat, the individual stitches will not overlap sufficiently, resulting in an interrupted weld.

The longitudinal spacing of the stitches must have some limit. The spacing is a function of the power frequency and the travel speed of the tube being welded. With 60 Hz power, the speed of welding should be limited to approximately 90 ft/min. To weld tubing at higher speeds than this requires welding power of higher frequency. This is indicated in Table 2.8 which shows typical welding speeds using various sizes of 180 Hz power sources for steel tubing of several wall thicknesses.

It is desirable to have the outside corners of the edges close first as the formed tube moves through the machine so that the stitches will be inclined forward. This condition is known as an inverted vee. The advantages of using an inverted vee are twofold: (1) the angle deviation from the vertical reduces the forces tending to expel any molten metal in the joint, and (2) the major portion of the upset metal is extruded to the outside where it is easily removed. The tubing is normally formed so that the included angle of the vee is about 5 to 7 degrees.

Surface Burns

As in spot and seam welding, the current that provides the welding heat must enter the stock through electrode contacts. The resistance of these contacts must be kept to a minimum to avoid I^2R losses sufficient to result in surface burns on the tube. Burns are actually surface portions of the tube that are heated to their fusion or melting point. They tend to stick to or imbed themselves in the face of the wheel electrode. If

Table 2.8
Typical seam welding speeds for steel tubing using 180Hz power sources

Wall thickness, in.	Speed, ft/min.			
	125 kVA	200 kVA	300 kVA	500 kVA
0.050	150	200	—	—
0.065	110	100	200	—
0.083	72	105	145	—
0.095	—	85	115	—
0.109	—	66	90	—
0.125	—	50	70	140
0.134	—	—	60	125
0.156	—	—	—	85

large steel particles become embedded in the electrode face, the contact resistance will increase and cause more severe burning. This action continues to build up with each revolution of the electrode. To stop burning, the operation must be interrupted and the electrode cleaned or remachined.

To eliminate burns, the area of contact and the pressure between the electrode and the tube must be adequate. As a rule of thumb, each electrode should have sufficient contact area so that the current density will be less than 50 000 A/in.^{2} The relative shapes of the formed tube and the electrode should ensure that the maximum contact pressure occurs next to the seam.

Without the aid of some backup support, electrode contact pressure is limited by the ability of the tube to support the forces being applied. The maximum permissible pressure in the welding throat is a function of the yield strength of the metal and the ratio of tube diameter to wall thickness (D/t ratio). In extreme cases where the D/t ratio is high, a backup mandrel must be used to prevent distortion of the tube wall and misalignment of the joint.

INSPECTION AND TESTING

Upset welds can be inspected and tested in the same manner as flash welds. In general, the quality requirements for upset welds are not so stringent as those specified for flash welds. The process normally can not produce welds with the consistency available with flash welding.

PERCUSSION WELDING

FUNDAMENTALS OF THE PROCESS

Definition and General Description

Percussion welding is a joining process in which coalescence is produced simultaneously over the entire abutting surface by the heat from an arc. The arc is produced by a short pulse of electrical energy. Pressure is applied percussively during or immediately following the electrical pulse.

In general, "percussion welding" is the term used in the electronics industry for joining wires, contacts, leads, and similar items to a flat surface. On the other hand, if the item is a metal stud that is welded to a structure for attachment purposes, it is called "capacitor discharge stud welding"[4] even though the equipment and procedures are similar to percussion welding.

In the application of the process, the two parts are initially separated by a small projection on one part, or one part is moved toward the other. At the proper time, an arc is initiated between them. This arc heats the faying surfaces of both parts to welding temperature. Then, an impact force drives the parts together to produce a welded joint. There are basically two variations of percussion welding: magnetic force and capacitor discharge.

Although the steps may differ in certain applications because of process variations, the essential sequence of events in making a percussion weld are as follows:

(1) Load and clamp the parts into the machine.

(2) Apply a low force on the parts or release the driving system.

(3) Establish an arc between the faying surfaces with high voltage to ionize the gas between the parts or with high current to melt and

---

4. Capacitor discharge stud welding is discussed in Vol. 2, *Welding Handbook,* 7th Ed.; 262-3, 282-94.

vaporize a projection on one part.

(4) Move the parts together percussively with an applied force to extinguish the arc and consummate a weld.

(5) Release the force.

(6) Unclamp the welded assembly.

(7) Unload the machine.

Percussion welding is analogous to capacitor discharge stud welding. The differences between the two processes are the applications and the type of power source. Percussion welding may be used to join two parts of equal cross section end-to-end. These include wire, rod, and tubing. Welding current is supplied by a capacitor storage bank in these applications. The process may also be used to weld wires or contacts to large flat areas with power from a capacitor bank or a transformer.

Principles of Operation

Welding heat is generated by a high current arc between the two parts to be joined. The current density is very high, and this melts a thin layer of metal on the faying surfaces in a few milliseconds. Then, the molten surfaces are brought together in a percussive manner to consummate a weld.

The two process variations differ in the type of power supply, method of arc initiation, and work drive motion. With the capacitor discharge method, power is furnished by a capacitor storage bank. The arc is initiated by the voltage across the terminals of the capacitor bank (charging voltage) or a superimposed high voltage pulse. Motion may be imparted to the movable part by mechanical or pneumatic means.

For magnetic force welding, power is supplied by a welding transformer. The arc is initiated by vaporizing a small projection on one part with high current from the transformer. The vaporized metal provides an arc path. The percussive force is applied to the joint by an electromagnet that is synchronized with the welding current. Magnetic force percussion welds are made in less than one half cycle of 60 Hz. Consequently, the timing between the initiation of the arc and the application of magnetic force is critical.

Advantages and Limitations

Advantages. The extreme brevity of the arc in both versions of percussion welding limits melting to a very thin layer on the faying surfaces. Consequently, there is very little upset or flash on the periphery of the welded joint. Heat-treated or cold-worked metals can be welded without annealing them.

Filler metal is not used and there is no cast metal at the weld interface. A percussion welded joint usually possesses higher strength and conductivity than does a brazed joint. Also, unlike brazing, no special flux or atmosphere is required.

A particular advantage of the capacitor discharge method is that the capacitor charging rate is easily controlled and low compared to the discharge rate. The line power factor is better than with a single-phase ac machine. Both these factors give good operating efficiency and low power line demand.

Percussion welding can tolerate a slight amount of contamination on the faying surfaces because expulsion of the thin molten layer tends to carry any contaminants out of the joint.

Limitations. The percussion welding process is limited to butt joints between two like sections and to flat pads or contacts joined to flat surfaces. In addition, the total area that can be joined is limited since control of an arc path between two large surfaces is difficult.

Joints between two like sections can be accomplished more economically by other processes. Percussion welding is usually confined to the joining of dissimilar metals not normally considered weldable by other processes and to the production of joints where avoidance of upset is imperative. Another limitation of this process is that two separate pieces must be joined. It cannot be used to weld a ring of one piece.

APPLICATIONS

Weldable Metals

The magnetic force method is primarily used for joining electrical contacts to contactor arms. Combinations include copper to copper, silver-tungsten to copper, silver oxide to copper, and silver-cadmium oxide to brass. Areas from

0.040 to 1.27 in.² are being welded in production. Some metal loss occurs at the weld interface, and in most instances some flash must be removed from the periphery of the weld. Figure 2.12 shows several contact designs welded by this process. Figure 2.13 shows a section through a typical weld.

The capacitor discharge method is usually employed to produce the following types of joints:

(1) Butt joints between wires or rods

(2) Lead wire ends to flat conductors or terminals

(3) Contacts to relay arms

The wire is usually made of copper and may be solid or stranded, bare or tinned. The rods are usually copper, brass, or nickel-silver. Other alloys such as steel, alumel, chromel, aluminum, and tantalum may be welded to themselves or to other materials. The method is also applicable to reactive, refractory, and dissimilar metal welds, because the short weld time limits the contamination of the reactive metals and the formation of low-strength intermetallic zones in the joints.

Industrial Uses

Companies using percussion welding are mainly those in the electrical contact or component field. Large contact assemblies for relays and contactors are usually made on magnetic force percussion welding machines. Such machines can be automated for high production.

Hand-held capacitor discharge equipment may be used to weld wires to pins. This is particularly applicable to aerospace equipment that is subject to shock and vibration. The process is also used to weld electronic components to terminals.

EFFECT ON METALS WELDED

Heat Effect

A percussion weld is made in a very short time. It may take milliseconds when using magnetic force welding. Because of this short time, the heat-affected zones of percussion welds are shallow, usually less than 0.010 inch. There is little oxidation of mating surfaces and a minimum of alloying between dissimilar metals. Since the heat-affected depth is so small, heat-treated metals may be welded without softening them. The heat input is so concentrated and of such short duration that heat-sensitive components near the weld area are not affected by the welding cycle.

Fig. 2.12—Typical electrical contacts joined by magnetic force percussion welding

***Fig. 2.13—Photomacrograph of a section through a silver contact (top)
welded to a brass terminal (bottom)***

Heat balance between parts is usually not a factor of concern. Since percussion welding is essentially a dc process, polarity of the two parts involved may be important in some cases, as in arc welding.

Metal Loss

The metal loss that occurs during a percussion weld is not so great as in arc stud welding. The loss varies with the area of the weld and the type of welding machine. Metal loss can generally be ignored for parts to be joined by capacitor discharge percussion welding. However, it should be considered in magnetic force percussion welding.

Flash

This is the metal that is expelled at high velocity from the weld interface during a percussion weld. It can damage adjacent tooling and affect accuracy of assembly. Any flash attached to the weld joint should be removed so that it will not cause a problem in service.

MAGNETIC FORCE WELDING

Welding Machines

These machines use a low voltage power supply (20 to 35 volts from a transformer), a projection type arc starter, and an electromagnetic system to produce the weld force. A unit

generally consists of a modified press type resistance welding machine with specially designed transformer, controls, and tooling. Figure 2.14 shows a typical machine used to weld the type of parts shown in Fig. 2.12. An

***Fig. 2.14—Magnetic force percussion
welding machine***

air cylinder provides the initial force to bring the parts together.

Magnetic force percussion welding machines usually have an independent power source for the electromagnet so that the force magnitude and time of application can be varied with respect to the initiation of welding current. This is accomplished by using two separate transformers, one for welding power and one for the electromagnet power. The acceleration of the force member can be controlled by adjusting the magnitude of the electromagnet current, thereby providing a duration control for arc time.

Since welding is done during 1/2 cycle of 60 Hz, current flow is unidirectional. In some cases, the polarity of the two parts may have some effect on weld quality. In general, the same conditions that prevail in dc arc welding are also in effect in percussion welding with respect to polarity. The current is always passing through the transformer in the same direction and the core can become partially saturated. Consequently, the electrical controls should provide a low amplitude 1/2-cycle pulse in the opposite direction to deflux the transformer and electromagnet. This can be done during the loading time.

Joint Design

For welding two flat surfaces together, a projection similar to that for resistance welding must be formed on one piece as shown in Fig. 2.15. Its diameter and height must be developed for each application. The diameter must be large enough to support the initial force applied to the parts but too small to carry the welding current. The height determines the gap between the faying surfaces and, thus, the initial arc voltage. When large area contacts are welded, two projections may be required.

The surfaces to be joined must be flat and parallel during welding so that arcing will occur over the entire area. Areas that are not melted will probably not weld when impacted together.

Voltage and Current

It is necessary to establish and maintain the desired magnitude of voltage and current for the required weld area. These are determined by the projection design, the capacity of the welding transformer, and the impedance of the secondary circuit. The transformer should be of low impedance with secondary voltages higher than those commonly used in resistance welding.

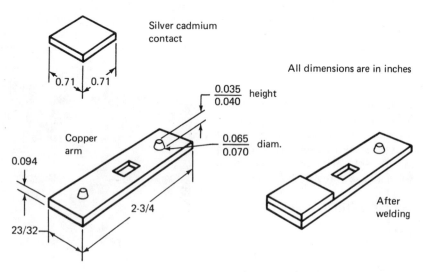

Fig. 2.15—Typical design of a magnetic force percussion welded contact assembly

Arc Time

Arc time can be considered as the time beginning with the explosion of the projection and ending when the two parts come together and the arc is quenched. The timing between the initiation of the arc and the application of magnetic force is very critical.

The arc time is a function of:

(1) Magnitude of magnetic force

(2) Timing of the magnetic force with relation to welding current

(3) Inertia or mass of the moving parts in the force system

(4) Height of the projection

(5) Magnitude of the welding current and the diameter of the projection

Acceleration of the movable head is directly proportional to the magnetic force applied and inversely proportional to the mass. The acceleration of the movable head with the two-transformer system can be controlled by adjusting the magnitude of the force current, which thereby provides a duration control for arc time, within limits.

CAPACITOR DISCHARGE WELDING

Two types of machines are presently used. One utilizes a high voltage, low capacitance system. Charging voltage ranges from 1 to 3 kV. With this system, wire end preparation is not critical since the applied potential is sufficient to ionize the air in the gap and start the arc.

The other system utilizes a low voltage, high capacitance energy source. This has the advantages of a safe working voltage (about 50V), a simple power supply, and low weld spatter. In some designs, the high voltage power is discharged through a transformer of low voltage output.

A low voltage system requires a 600V arc starting circuit and special wire end preparation. Once the air gap is ionized with the 600V (low amperage) circuit, the arc is sustained by the 50V circuit. The arc initiation circuit does no appreciable melting.

One type of low voltage machine consists of a hand-held gun and a portable power supply. The gun is designed to weld wires to terminals by holding a small flat or square terminal in one set of stationary jaws and the wire to be welded in a set of movable jaws. When the gun is triggered, springs move the wire toward the terminal at a high velocity. A feather edge on the end of the wire greatly improves arc starting. The arc is initiated at the point of contact of the wire and terminal. The welding current melts the feather edge on the wire faster than the wire is moving toward the terminal. The arc spreads over the wire area and melts a layer about 0.002 to 0.003-in. thick in each part. The arc is extinguished after about 150 to 600 microseconds as the two parts come in contact.

Another version of a portable, low voltage welding machine employs a high frequency pulse to initiate the arc. This feature eliminates the need for a special shape on the wire end.

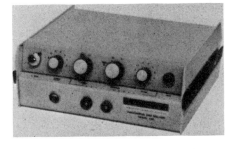 

Fig. 2.16—A portable capacitor discharge percussion welding power supply and hand-held gun

The machine uses an electromechanical actuator to accelerate the wire and to provide the necessary forging force. One version of this machine is shown in Fig. 2.16.

Low voltage semiautomatic and automatic machines are used to weld assemblies similar to the one shown in Fig. 2.17. Component leads are usually tinned annealed copper. Terminals may be brass, tinned brass, or nickel-silver alloys. Wires and leads of 0.006 to 0.102-in. diameters can be welded to terminals and plates of various thicknesses above 0.006 inch.

Controls for capacitor discharge equipment usually include those for welding voltage,

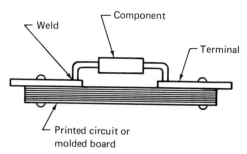

Fig. 2.17–A typical percussion welded electronic assembly

capacitance, and high frequency voltage when it is used. Control of the motion mechanism is also provided.

WELD QUALITY

The quality of percussion welds can be determined by metallographic examination and mechanical tests. Metallographic examination will show the weld interface and the widths of the heat-affected zones. In the case of dissimilar metals, it may reveal the degree of alloying at the interface. Microhardness tests on a metallographic section may indicate the effect of welding on the base metal.

Welded joints may be tested in tension, bending, or shear, depending upon the joint design. The effect of vibration may be important in some applications. The test method should be designed to qualify the welding procedures and weld joint properties for the intended applications.

Where the electrical integrity of the joint is important, appropriate resistance measurements should be made before and after a test and then the results compared. Some slight change in resistance may be expected.

SAFETY

MECHANICAL

The welding machine should be equipped with appropriate safety devices to prevent injury to the operator's hands or other parts of the body. Initiating devices, such as push buttons or foot switches, should be arranged and guarded to prevent them from being actuated inadvertently.

Machine guards, fixtures, or operating controls should prevent the hands of the operator from entering between the work-holding clamps or the parts. Dual hand controls, latches, presence-sensing devices, or any similar device may be used for this purpose.

ELECTRICAL

All doors and access panels on machines and controls should be kept locked or interlocked to prevent access by unauthorized personnel. When the equipment utilizes capacitors for energy storage, the interlocks should interrupt the power and short circuit all the capacitors through a suitable resistive load when the panel door is open. A manually operated switch or other positive device should also be provided in addition to the mechanical interlock or contacts. Use of this device will assure complete discharge of the capacitors.

PERSONAL

Flash guards of suitable fire resistant material should be provided to protect the operator and avoid fires. In addition, personal eye protection with suitable shaded lenses should be worn by the operator.

Additional information on safe practices for welding may be found in the American National Standard Z49.1, *Safety in Welding and Cutting* (latest edition), available from the *American Welding Society.*

Metric Conversion Factors

1 in. = 25.4 mm
1 in.² = 645 mm²
1 ft/min = 5.08 mm/s
1 psi = 6.89 kPa
1 ksi = 6.89 MPa

SUPPLEMENTARY READING LIST

Hind, R. H., Butt Welding in the tool industry. *Production Engineering*, 40(12): 785-793; 1961 Dec.

Holko, Kenneth H., Magnetic force upset welding dissimilar thickness stainless steel tee joints. *Welding Journal*, 49(9), 427-439s: 1970 Sept.

King, P. P. and Schnepf, Arc percussive butt welding of fine wire conductors. *Welding Journal*, 44(2): 100-105; 1965 Feb.

Kotechki, D. J., Cheever, D. L., and Howden, D. G., Capacitor discharge percussion welding: microtubes to tube sheets. *Welding Journal*, 53(9): 557-560; 1974 Sept.

Makara, A. M. and Sakhatski, G. P., The flash welding of high tensile steels. *Welding Research Abroad*: 44-48; 1969 Mar.

Manning, R. F. and Welch, J. B., Percussion welding using magnetic force – a production process. *Welding Journal*, 39(9): 903-907; 1960 Sept.

Moore, C. D., Flash welding aluminum to copper. *Machinery*, 99: 790-792; 1961, Oct. 4.

Petry, K. N., et al., Principles and practices in contact welding. *Welding Journal*, 49(2): 117-126; 1970 Feb.

Savage, W. F., Flash welding: the process and application. *Welding Journal*, 41(3): 227-237; 1962 Mar.

Savage, W. F., Flash welding: process variables and weld properties. *Welding Journal*, 41(3): 109s-119s; 1962 Mar.

Stieglitz, H. W., Flash welding copper to steel. *Metal Progress*, 80: 112; 1961 Nov.

Sullivan, J. F. and Savage, W. F., Effect of phase control during flashing on flash weld defects. *Welding Journal*, 50(5): 213s-221s; 1971 May.

3

Resistance Welding Equipment

Chapter Committee

W. G. EMAUS, *Chairman*
   Anderson Automation, Inc.

R. A. GRENDA
   Weltronic Company

P. G. HARRIS
   Centerline (Windsor) Ltd.

A. E. LOHRBER
   Swift-Ohio Corporation

R. MATTESON
   Taylor-Windfield Corporation

F. PARKER
   Newcor, Inc.

D. ROTH
   Kirkhof Transformer Division,
   FLX Corporation

Welding Handbook Committee Members

A. F. MANZ
   Union Carbide Corporation

D. D. RAGER
   Reynolds Metals Company

3

Resistance Welding Equipment

INTRODUCTION

The selection of resistance welding equipment is usually determined by the joint design, materials of construction, quality requirements, production schedules, and economic considerations. Standard resistance welding machines are designed to meet the requirements of Bulletin No. 16, Resistance Welding Equipment Standards, issued by the Resistance Welders Manufacturers Association (RWMA). These machines are capable of welding a wide variety of alloys and component sizes. Complex resistance welding equipment of special design may be necessary to meet the economic requirements of mass production.

A resistance welding machine has three principal elements:

(1) An electrical circuit consisting of a welding transformer and a secondary circuit including the electrodes which conduct the welding current to the work.

(2) A mechanical system consisting of a machine frame and associated mechanisms to hold the work and apply the welding force.

(3) The control equipment to initiate and time the duration of current flow. It also may control the current magnitude as well as the sequence and the time of other parts of the welding cycle.

Resistance welding machines are classified according to their electrical operation into two basic groups: direct energy and stored energy.

Machines in both groups may be designed to operate on either single-phase or three-phase power.

Most resistance welding machines are the single-phase direct energy type. This is the type of machine most commonly used because it is the simplest and least expensive in initial cost, installation, and maintenance. With properly designed controls, it will normally produce welds of commercial quality comparable to those made with other types of machines. Actually, the mechanical systems and secondary circuit designs are essentially the same for all types of welding machines, but the transformer design and control systems will differ considerably.

A single-phase welding machine will have a larger volt-ampere demand than a three-phase machine of equivalent rating. The demand of a single-phase machine causes unbalance on a three-phase power line. Also, the power factor is relatively low because of the inherent inductive reactance in the machine. Single-phase demand may not be a problem if the welding machine is a relatively small part of the total line load or a number of single-phase welding machines are connected to balance the load on the three phases of the power line.

A three-phase direct energy machine draws power from all three phases of the power line at a high power factor. The inductive reactance of the welding circuit is low because direct

92

current is used for welding. Consequently, the kVA demand of a three-phase machine is lower than that of an equivalent single-phase machine. This is a definite advantage where a large capacity machine is needed and the power line capacity is limited.

The principle of a stored energy machine is to accumulate and store electrical energy and then discharge it from storage to make the weld. The energy is normally stored electrostatically in a capacitor bank. Single-phase power is generally used for small bench models. The power demand is low because storage time is relatively long in comparison to the weld time.

SPOT AND PROJECTION WELDING MACHINES

ROCKER ARM TYPE

The simplest and most commonly used spot welding machine is the rocker arm design, so called because of the rocker movement of the upper horn. A horn is essentially a cylindrical arm or extension of an arm of a resistance welding machine which transmits the electrode force and, in most cases, the welding current. This type of machine is readily adaptable for the spot welding of most weldable metals. Three methods of operation are available, namely: (1) air, (2) foot, and (3) motor.

Air-operated machines, such as the one in Fig. 3.1, are the most popular type. With air operation, the welding cycle is generally controlled automatically with a combination control unit. They can operate very rapidly and are easily set up for welding.

Foot-operated machines are best suited for miscellaneous sheet metal fabrication, particularly for short production runs where consistent weld quality is not required. Motor-operated machines are normally used where compressed air is not readily available.

Standard rocker arm machines are generally available with throat depths of 12 to 36 in. and transformer capacity of 5 to 75 kVA. The general construction of these machines is the same with all three types of operation.

A rocker arm type machine, however operated, suffers a reduction of available welding force and welding current as the throat depth is increased. On the other hand, the throat depth can readily be changed by the substitution of different horns. Some machines have adjustable upper horns and, in some cases, adjustable lower horns. With extended horns, rather larger assemblies can be inserted into the machine throat, but machine performance may be reduced because of the increase in throat area.

Electrode Position

The travel path of the upper electrode is an arc about the fulcrum of the upper arm. The electrodes must be positioned so that both are in the plane of the horn axes. Also, the two horns should be parallel when the electrodes are in contact with the work. Even with parallel horns, electrode skidding can occur if the electrode holders or horns are not sufficiently rigid. Skidding can be reduced by changing to more rigid electrode holders, adjusting the position of the electrodes, or providing support to the lower horn. Because of the radial motion of the upper electrode, these machines are not recommended for projection welding.

Mechanical Design

The machine frame houses the transformer and tap switch and supports the mechanical and electrical components.

For air-operated machines, the stroke of the air cylinder must be proportioned to the required electrode spacing. Its diameter must be proportioned to the required electrode force and lever arm ratio Y/X in Fig. 3.1). For a given

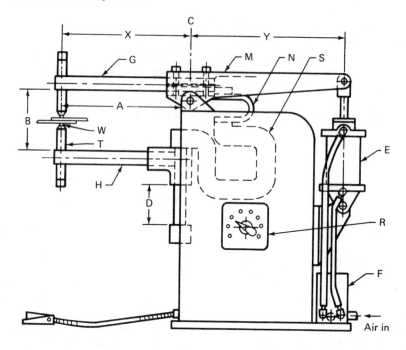

A - Throat depth
B - Horn spacing
C - Centerline of rocker arm
D - Lower arm adjustment
E - Air cylinder
F - Air valve
G - Upper horn

H - Lower horn
M - Rocker arm
N - Secondary flexible conductor
R - Current regulator (tap switch)
S - Transformer secondary
T - Electrode holder
W - Electrode

Fig. 3.1—Air-operated rocker arm spot welding machine

cylinder diameter, the available welding force will decrease as the throat depth is increased. Electrode spacing can be set by adjusting the position of the electrodes in the horns. In most cases, however, it is desirable to use a double-acting air cylinder with adjustable stroke.

The force exerted by a piston is equal to the product of its area and the air pressure. Most industrial air systems are operated at 80 psi pressure minimum, and cylinder size is determined on this basis.

Electrode force is the product of the piston force and the lever arm ratio Y/X. Consequently, it is in direct proportion to the air pressure as controlled by a pressure regulator. Air pressures below 20 psi should not be used because of possible erratic and inconsistent behavior of the air cylinder.

With foot and motor operated machines, the air cylinder is replaced by a stiff spring. The spring is compressed by a foot-operated lever arm or a motor-driven cam as it exerts a force on the end of the rocker arm. The amount of force is determined by the stiffness of the spring and the compression distance.

PRESS TYPE

Press type machines are recommended for all projection welding operations and many spot welding applications. With this type of machine, the movable welding head travels in a straight line in guide bearings or ways. These bearings must be of sufficient proportions to withstand any eccentric loading on the welding head. Standard press type welding machines are available with capacities of 5 to 500 kVA and with throat depths up to 36 inches. Nonstandard units with lower ratings, such as bench types, are widely used for radio, instrument, and jewelry work.

Press type machines are classified according to their use and method of force application. They may be designed for spot welding or projection welding, or both. Force may be applied by air or hydraulic cylinders, or manually with small bench units.

A few general guidelines for the selection of a machine of this type are as follows:

(1) Hydraulic operation is not normally used on machines rated below 200 kVA because of the higher cost compared to air operation.

(2) Air operation may be used on all sizes of machines. When high forces are required, however, air cylinders and valves will be quite large, operation will be slow, and air consumption will be high. When all factors are taken into consideration, most machines of 300 kVA and under are air operated, and machines of 500 kVA and above are hydraulically operated. In between, they may be operated by either method.

Fast electrode follow-up is particularly important when spot and projection welding relatively thin sections, particularly those of aluminum and other nonferrous metals. Air operation provides much faster follow-up than does hydraulic operation because of the compressibility of air. With hydraulic operation, follow-up must occur by fluid motion throughout the hydraulic system. A precharged air accumulator is sometimes used with this system to improve follow-up. Other systems use springs in combination with the electrode holders.

General Construction

Standard press type welding machines are designed and built on the unit principle for economy in manufacture. The same frame size is used with two or three transformers of different kVA ratings and with a range of throat depths. A typical press type welding machine is

shown in Fig. 3.2.

Projection welding machines have platens on which dies, fixtures, and other tooling are mounted. The platens have flat surfaces and usually have standard T-slots for bolting attachments. Machines designed for spot welding are equipped with horns and electrode holders. A combination unit will have both platens and horns, as shown in Fig. 3.2. Such a machine will have one throat depth as a projection welding machine and a greater throat depth as a spot welding machine. The platens, the ram, and the force cylinder are all on the same centerline. The distance from this centerline to the frame of the machine is the projection welding throat depth. On standard machines with horns, the spot welding electrodes are located 6 in. farther from the frame. This is true whether or not platens are used. The design throat depth should not be exceeded because the ram is not designed to take greater eccentric loading.

On projection and combination machines, the lower platen is mounted on a knee which may be adjusted vertically. The knee may be of cast iron, steel, copper, or bronze.

Mechanical Design

Air-Operated Machines. These machines are usually the direct-acting type where the electrode force is exerted by the air cylinder through the ram. Three general types of double-acting air cylinders are employed. These are illustrated in Fig. 3.3. In all cases, air for the pressure stroke enters at port A and exhausts at port B. For the return stroke, the air enters at port B and exhausts at port A.

Figure 3.3(A) shows a fixed-stroke cylinder with stroke adjustment. The stroke adjuster K limits the travel of piston P and the electrode opening.

An adjustable-stroke cylinder with a dummy piston is shown in Fig. 3.3(B). The dummy piston R is attached to the adjusting screw K which positions this piston. Chamber L is connected to port A through the hollow adjusting screw. The stroke of the force piston P is adjusted by the position of the dummy piston R above it. This cylinder design responds faster than a fixed-stroke cylinder because the volume

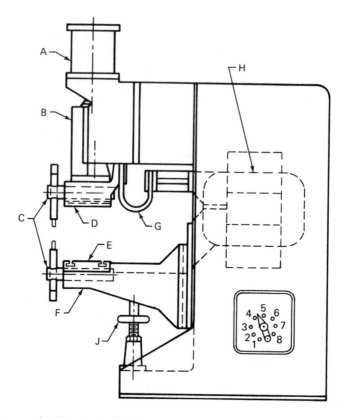

A - Air or hydraulic cylinder
B - Ram
C - Spot welding attachment
D - Upper platen
E - Lower platen

F - Knee
G - Flexible conductor
H - Transformer secondary
J - Knee support

Fig. 3.2—Press type combination spot and projection welding machine

(L) above piston P can be made smaller than that of a fixed-stroke cylinder of the same size.

A modification of the adjustable-stroke cylinder provides a retraction feature to provide additional electrode opening for loading and unloading of the machine or for electrode maintenance. This adjustable-retractable stroke cylinder is shown in Fig. 3.3(C). With this type, a third port C is connected to chamber H above the dummy piston R. If air is admitted to chamber H at a pressure slightly higher than the operating pressure in chamber L, piston R will move down to a position determined by the adjustable stop

X. This determines the up position for piston P and the electrode opening for welding. When the air from chamber H is exhausted to atmosphere, piston P will lift piston R with it until stop X contacts the cylinder head. This will increase the electrode opening for loading and unloading of the machine. Readmission of air to chamber H will return pistons P and R to welding position when the pressure in chamber H is slightly higher than that in chamber M. Flow control valves or cushions are usually used to control the operating speed of an air cylinder.

Several optional features are available on

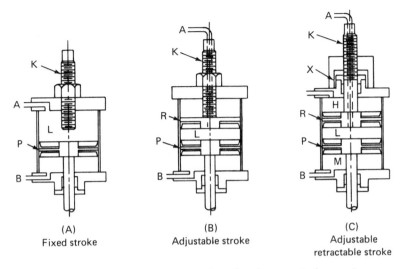

(A)
Fixed stroke

(B)
Adjustable stroke

(C)
Adjustable
retractable stroke

**Fig. 3.3—Typical air cylinder designs for air-operated press type
welding machines**

air-operated machines to provide either dual electrode force for forging action or fast follow-up of the electrodes as the weld nugget is formed. Such features usually incorporate a rubber diaphragm of some form to minimize friction in the force system.

Hydraulic Machines. With these machines, a hydraulic cylinder is used in place of an air cylinder. The designs for hydraulic cylinders are similar to those for air-operated cylinders (see Fig. 3.3). Hydraulic cylinders are generally smaller in diameter than air cylinders because higher pressures can be developed with a fluid system.

In the simplest type of hydraulic system, a constant speed motor drives a constant pressure, constant delivery pump. The output pressure of the pump is controlled by an adjustable relief valve. Fluid delivery is controlled with a four-way valve of similar design to that employed in an air system. Auxiliary devices include a sump, a filter, a heat exchanger, a gage, and, sometimes, an accumulator.

PORTABLE TYPE

A typical portable spot welding machine consists of four basic units:

(1) A portable welding gun or tool

(2) A welding transformer and, in some cases, a rectifier

(3) An electrical contactor and sequence timer

(4) A cable and hose unit to carry power and cooling water between the transformer and welding gun

A typical portable welding gun consists of a frame, an air or hydraulic actuating cylinder, hand grips, and an initiating switch. The unit may be suspended from an adjustable balancing unit. Some small units may use manual pressure to apply the electrode force.

There are two basic types of air or hydraulically operated guns. One is the scissor type which is analogous to a rocker arm spot welding machine. The other is the "C" type, so-called because of its shape. This type has action similar to a press type spot welding machine.

There are no standard designs for portable spot welding guns. They are used for applications where it is impractical to bring the work to the welding machine. The design of the gun usually is tailored to the configuration of the assembly to be welded. There may be some degree of standardization of gun sizes and shapes by a particular manufacturer or user that specializes in this type of equipment. These units may be applicable to a number of similar part designs.

The design of a gun is influenced by the electrode force required. To minimize the size and weight of a gun, a hydraulic cylinder is commonly used to provide forces greater than 750 lbs. However, air cylinders supplying up to 1500 lbs. are sometimes used for simplicity of the equipment.

Transformers for portable guns should produce open-circuit secondary voltages that are two to four times greater than those of transformers for stationary machines. The higher voltages are needed because of the added impedence of the cable between the transformer and the gun. The transformer is usually mounted some distance above the work to give more freedom and flexibility to the gun, but it can be close-coupled to minimize secondary impedance losses. A current-controlling tap switch, an air valve, and sometimes an air-hydraulic booster or a rectifier unit are mounted on the transformer.

An air-hydraulic booster is a piston device for transforming air pressure into high hydraulic pressure. The pressure increase is inversely proportional to the areas of the two pistons. The booster provides the necessary hydraulic pressure to the gun cylinder.

A combination control is required to operate a portable gun unit. It consists of a primary contactor and sequence timer. If an electronic tube contactor is used, the control is usually mounted separately but as close to the transformer as possible. If the contactor is a solid-state device, the compactness of this unit can permit mounting the control directly on the transformer.

The cable and hose assembly between the transformer and welding gun consists of electrical cables, air or hydraulic hoses, water-cooling hoses, and a cable to the initiating switch. The gun cable is normally a concentric, kickless type. The two conductors in the cable are intertwined in such a manner that the reactance is low. They are operated at high current density to reduce bulk. As a result, water cooling is necessary to keep them from overheating.

MULTIPLE SPOT WELDING TYPE

A multiple spot welding machine is a special purpose unit designed to weld a specific assembly. This type of machine should be considered when the production requirements and the number of spot welds on an assembly are so large that welding with a single spot machine is uneconomical. The principal advantages of these machines are:

(1) A number of welds can be made at the same time.

(2) Part dimensions and weld locations can be reasonably consistent.

(3) The equipment can be very reliable and easy to maintain.

Welding Station Design

Multiple spot welding machines utilize a number of transformers, usually of dual secondary design. Figure 3.4 shows typical standardized components that are used in designing a wide range of multiple spot welding machines. Force is applied directly to the electrode through a holder by an air or hydraulic cylinder. To make welds on close centers, the cylinder diameter must be small. This can be accomplished with tandem pistons on the same shaft, as shown in the illustration. A 2-in. diameter air cylinder of this design can develop an electrode force of 350 lb with normal shop air pressure. This force is adequate for spot welding two 0.025-in. thicknesses of cold rolled steel. Closer spot weld spacing may be obtained by using hydraulic cylinders of smaller diameter. The combination cylinder and electrode holder assemblies are commonly called welding guns.

A welding transformer with two insulated secondaries can power two separate welding circuits. This type of transformer is noted for its compact design and narrow width. If desired, only one of the dual secondaries need be used at one time. For higher secondary voltage, the two secondaries can be connected in series to feed one secondary circuit. To increase the welding current available to a single circuit, the secondaries can be connected in parallel. Welding guns and transformer units of this type can be designed to spot weld two sheets of cold

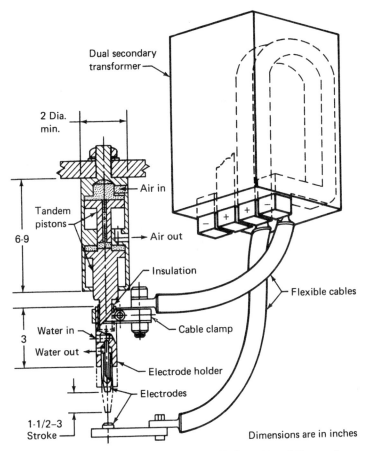

Dual secondary transformer

2 Dia. min.

Air in

Tandem pistons

6-9

Air out

Insulation

Flexible cables

Water in

Cable clamp

3

Water out

Electrode holder

Electrodes

1-1/2–3 Stroke

Dimensions are in inches

Fig. 3.4—Basic components of a multiple spot welding system

rolled carbon steel up to 0.125 in. thick.

For most applications, the lower electrode is made of a piece of solid copper alloy with one or more electrode alloy inserts that contact the part to be welded. It is normally water cooled to remove heat. The inserts generally are designed with large contact areas to resist wear. Pointed electrodes are not normally used against the show side to avoid marking. A typical machine is shown in Fig. 3.5.

Equalizing gun designs are often used where standard electrodes are needed on both sides of the weld to obtain good heat balance or where variations in parts will not permit consistent contact with a large, solid lower elec-

trode. The same basic welding gun is used for these designs but it is mounted on a special "C" frame similar to that for a portable spot welding gun. The entire assembly can move as electrode force is applied to the weld location.

Machine Designs

Multiple transformer machines are used extensively in the manufacture of formed sheet metal products. Because of their broad usage and requirements, many designs of multiple transformer machines are available. The machines may be designed as welding stations in large, high production, automated assembly lines, or they may be used independently.

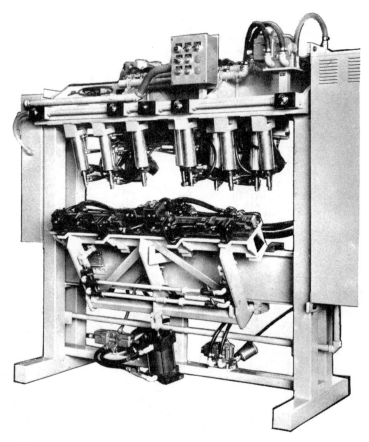

Fig. 3.5—A typical multiple transformer spot welding machine with a tilt table for loading parts

Independent machines may be loaded and unloaded either manually or automatically.

In designing a machine for a particular assembly, a number of factors must be considered. These include:

(1) Shape, size, and complexity of the part
(2) Metal composition and thickness
(3) Required weld appearance
(4) Production rate requirements
(5) Available equipment (presses, frames, and dial tables)
(6) Changeover time for different assemblies
(7) Cost of the equipment

ROLL SPOT AND SEAM WELDING MACHINES

A roll spot or seam welding machine is similar in principle to a spot welding machine except that wheel-shaped electrodes are substituted for the electrode tips used in spot welding. Both roll spot and seam welding can be performed on the same type of machine.

The essential elements of a standard seam welding machine are as follows:

(1) A main frame that houses the welding transformer and tap switch

(2) A welding head consisting of an air cylinder, a ram, and an upper electrode mounting and drive mechanism

(3) The lower electrode mounting and drive mechanism, if used

(4) The secondary circuit connections

(5) Electronic controls and contactor

(6) Wheel electrodes

The main frame, transformer, tap switch, ram, and air cylinder are essentially the same as those of a standard press type spot or projection welding machine. Hydraulic cylinders are seldom used on seam welding machines because the electrode force requirements are not normally high. To provide for electrode wear, either an adjustable connection is used between the ram and the piston rod or an adjustable-stroke air cylinder is employed. In addition, the position of the lower electrode and its mounting arrangement is sometimes adjustable. This adjustment is used to position the work at a proper height for convenient operation.

Most seam welding of thin gages is done using continuous drive systems. With thick gages, intermittent drive systems must be used to maintain electrode force on the weld nugget as it solidifies. The thickness range that can be welded with each drive system will depend upon the metal being joined.

The majority of continuous drive mechanisms use a constant speed, ac electric motor with a variable speed drive. The speed range depends upon the drive design and the electrode diameter. Good flexibility may also be obtained with a constant torque, variable speed dc drive.

TYPES

There are three general types of seam welding machines:

(1) *Circular,* where the axis of rotation of each electrode is perpendicular to the front of the machine. This type is used for long seams in flat work and for circumferential welds, such as welding the heads into containers. Such a machine is shown in Fig. 3.6.

(2) *Longitudinal,* where the axes of rotation of the electrodes are parallel to the front of the machine. This type is used for such applications as the welding of side seams in cylindrical containers and short seams in flat work.

(3) *Universal,* where the electrodes may be set in either the circular or longitudinal position. This is accomplished with a swivel type upper head in which the electrode and its bearing can be rotated 90 degrees about a vertical axis. Two interchangeable lower arms are used, one for circular operation and the other for longitudinal operation.

Figure 3.7 shows the electrode arrangements for standard types of machines as well as methods of electrode drive.

ELECTRODE DRIVE MECHANISMS

Knurl or Friction Roller

The knurl or friction roller drive has either the upper or the lower electrode, or both, driven by a friction wheel on the periphery of the electrode. When these friction rolls have knurled teeth, they are known as knurls or knurl drives. Knurl or friction roller drive will maintain a constant welding speed as the electrode diameter decreases from wear. Typical drive arrangements are shown in Figs. 3.7 (A), (B), (D), and (E).

A knurl drive is commonly used on machines for seam welding galvanized steel, terne plate, scaly stock, or other materials where the electrodes are likely to pick up surface material from the parts being welded. The knurl drive wheel tends to break up the material on the electrode face. Where the nature of the work permits, both electrodes should be knurl-driven to provide a more positive drive and lessen the possibility of skidding.

Gear Drive

With this method, the electrode shaft is driven by a gear train powered by a variable-speed drive, as shown in Figs. 3.7 (C) and (F). Only one electrode should be driven to avoid skidding. Otherwise, a differential gear box is necessary. This type of drive is generally less

Fig. 3.6—A standard circular seam welding machine with a special fixture

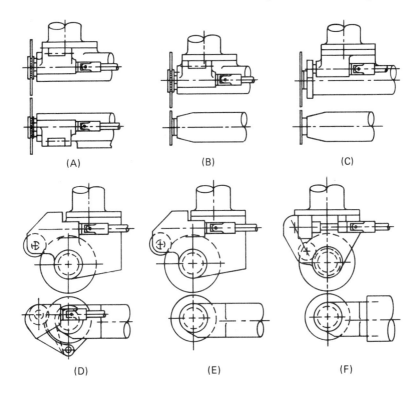

(A) Circular with double knurl or friction roller drive
(B) Circular with upper knurl or friction roller drive
(C) Circular with upper gear drive
(D) Longitudinal with double knurl or friction roller drive
(E) Longitudinal with upper knurl or friction roller drive
(F) Universal with upper gear drive

Fig. 3.7—Typical seam welding electrode and drive arrangements

desirable than a knurl drive because the welding speed decreases as the electrode wears. This can be compensated for by gradually increasing the drive speed.

The most important applications for a gear-driven machine are the welding of aluminum and magnesium and the fabrication of small diameter containers. Standard seam welding machines are designed with some minimum distance between electrode centers for each machine size. If one of the electrodes must be small to fit inside a container, the other must be correspondingly larger to account for the required center distance. If the ratio of the two

electrode diameters exceeds about 2 to 1, the smaller electrode should be driven and the large one should idle to minimize electrode skidding.

SPECIAL PURPOSE MACHINES

Special purpose machines are available for specific applications. Such machines can be generally grouped as traveling electrode type, traveling fixture type, and portable seam welding machines.

Traveling Electrode Type

With this type of machine, the seam to be welded is clamped or otherwise positioned on a fixed mandrel or shoe of some type and the ram and wheel electrode are moved along the seam. The mandrel or shoe is the lower electrode. The ram and electrode are moved by an air or hydraulic cylinder or by a motor-driven screw. Sometimes two upper electrodes operating in series are used side by side or in tandem. Figure 3.8 shows a typical traveling electrode machine.

Traveling Fixture Type

In the traveling fixture type, the upper electrode remains in a fixed position. The fixture and work are moved under the electrode by a suitable driving system. Multiple electrodes can also be used to advantage with this type of machine, such as the one shown in Fig. 3.9.

Portable Type

Portable seam welding guns may be used for work that is too large and bulky to be fed through a standard machine. The gun consists of a pair of motor-driven wheel electrodes and bearings, together with an air cylinder and associated mechanism for applying the electrode force. Welding current is supplied in the same manner as for portable spot welding. A variable-speed dc drive may be used where a wide range of welding speeds is desirable. The motor and speed reducer are mounted directly on the welding gun frame.

Fig. 3.8—Traveling electrode seam welding machine

Fig. 3.9—*Traveling fixture seam welding machine with two electrodes in tandem*

COOLING

One requirement in seam welding is the proper cooling of the machine, the electrodes, the current-carrying bearings, and other components of the secondary circuit. Cooling of the work is also important in most applications to minimize warpage from the local heating. Water jets spraying on both the work and the welding electrodes are usually satisfactory. Welding under water may be done in special cases.

Another method of cooling the weldment is a water mist that removes the heat by evaporation. A mist is produced by mixing air and water in proper proportions in a nozzle.

FLASH AND UPSET WELDING MACHINES

Flash and upset welding machines are similar in construction. The major difference is the motion of the movable platen during welding and the mechanisms used to impart the motion. Flash welding is generally preferred for joining components of equal cross section end-to-end. Upset welding is normally used to weld wire, rod, or bar of small cross section and to join continuously the seam in pipe or tubing. Flash welding machines are generally of much larger capacity than upset welding machines.

FLASH WELDING MACHINES

General Construction

A standard flash welding machine basically

consists of a main frame, a stationary platen, a movable platen, clamping mechanisms and fixtures, a transformer, a tap switch, electrical controls, and a flashing and upsetting mechanism. The stationary platen is generally fixed in position, although some designs provide a limited amount of adjustment for electrode and work alignment. The movable platen is mounted on ways on the frame and connected to the flashing and upsetting mechanism. Both platens are usually of cast or fabricated steel, although some small welding machines may have cast bronze, cast iron, or copper platens. The platens are connected to the transformer secondary. Electrodes that hold the parts and conduct the welding current to them are mounted on the platens. The transformer and tap switch are generally located within or immediately behind the frame with short, heavy duty copper leads to the platens.

The depth of the frame and, consequently, the width of the platens depends upon the size of the parts to be welded as well as the clamping mechanism design. Upsetting force should be aligned as nearly as possible to the geometric center of the parts to minimize machine deflection. Dual flashing and upsetting cylinders or cams are sometimes used with wide platens to provide uniform loading or clearance for long pieces to extend over the mechanism.

Trasformer and Controls

A flash welding transformer is essentially the same as those used for other types of single-phase resistance welding machines. A tap switch in the primary circuit is normally used to adjust flashing voltage. An autotransformer is sometimes used to extend the range of adjustment of secondary voltage. The primary power to the transformer is switched with an electronic contactor. Phase-shift heat control may be used with the contactor to provide low power for preheating or postweld heat treatment in the machine. With ignitron contactors, auxiliary load resistors must be connected in parallel with the transformer primary for proper operation of the ignitrons.

Programming of secondary current for preheating prior to flashing and postheating of the completed weld in the machine can be done with appropriate controls.

Flashing and Upsetting Mechanisms

In the operation of a flash welding machine, the parts are moved together using a predetermined travel pattern. This movement must be carefully controlled to consistently produce sound welds. After the appropriate flashing time, the pieces are rapidly brought into contact and upset. The upsetting action must be accurately synchronized with the termination of flashing.

The type of mechanism used for flashing and upsetting will depend upon the size of the welding machine and the application requirements. Some mechanisms permit the faying surfaces to be butted together under pressure and then preheated. After the appropriate temperature is reached, the pieces are separated and then the flashing and upsetting sequence is initiated. The movable platen may be actuated with a motor-driven cam or with an air or hydraulic cylinder.

Motor-operated machines normally use an ac motor with a variable speed drive which, in turn, drives a rotary cam. The cam is designed to produce a specific flashing pattern. It may contain an insert block to upset the joint at the end of flashing. The speed of the cam determines the flashing time. The platen may be moved directly by the cam or through a lever system. The motor may operate intermittently for each welding cycle or continuously. With continuous operation, the drive is engaged through a clutch on the output shaft of the speed reducer. A typical motor-operated flash welding machine is shown in Fig. 3.10.

A motor-driven flashing cam may be used in combination with an air or hydraulic upsetting mechanism, particularly on larger machines. Such a combination provides adjustment of upset speed, distance, and force independently of the flashing pattern. Current flow is synchronized with the mechanical motion of the platen by limit switches.

Medium and large flash welding machines use hydraulically operated flashing and upsetting mechanisms. These machines are capable of applying high upsetting forces for large sections. They are accurate in operation and are readily

Fig. 3.10—Automatic motor-operated flash welding machine

set up for a wide range of work requirements. A large hydraulic flash welding machine is shown in Fig. 3.11. A servo system is used to control the platen motion for flashing and upsetting. The servo system may be actuated by a pilot cam mechanism or by an electrical signal generated from the secondary voltage or the primary current. Choice of operating mode depends upon the application. The control may be programmed to include preheating and postheating. An accumulator is generally required to provide an adequate volume of hydraulic fluid from the pumping unit during upsetting.

Electro-hydraulic servo systems are generally of two designs. In one design, the servo valve meters the fluid directly to the hydraulic cylinder for position control. With the other design, the servo valve meters the fluid to a small control cylinder that operates a follower valve on a separate hydraulic system. The first design is simple and straightforward, but the second system has two distinct advantages. First, it has two separate hydraulic circuits for improved valve life. Second, the speed of response is fast and control of the platen position is accurate.

Clamping Mechanisms and Fixtures

Several designs of clamping mechanisms are available to accommodate different types of parts. These designs may be grouped generally as operating in either the vertical or the horizontal position. In special cases, the mechanisms may be mounted in other positions.

Vertical Clamping. The movement of the electrode may be in a plane perpendicular to the platen ways. The electrode may move either through a slight arc or in a straight line. If operating through an arc, a clamping arm pivots about a trunnion. This design is generally known as the "alligator" type. A machine with this type of clamping arrangement is shown in Fig. 3.12. Clamping force may be applied by an air or hydraulic cylinder operating directly or through a leverage or cam-operated mechanism. Vertical clamping is commonly used for bar stock and other compact sections.

Horizontal Clamping. With this design, the motion of the electrodes is parallel to the platen ways and generally in a straight line, as shown in Fig. 3.11. The major advantage of this type of

Fig. 3.11—Automatic hydraulically operated flash welding machine with horizontal clamping

Fig. 3.12—An automatic flash welding machine with vertical alligator type clamping

Fig. 3.13—An air-operated automatic upset welding machine

clamping mechanism is that the secondary of the welding transformer can be connected to both halves of the electrodes for uniform transfer of welding current into the work. This arrangement is highly desirable for welding parts with large cross sections. Clamping force can be applied with one of the mechanisms described for vertical clamping.

Fixtures. Fixtures may be used to support and align the parts for welding as well as to back up the parts to prevent slippage in the electrodes during upsetting. They are usually adjustable to accommodate the geometry and length of the parts. The design must be sturdy to withstand the upsetting force without deflecting. When the parts can be backed up, the clamping force on the electrodes can be limited to that needed to ensure good electrical contact and maintain satisfactory joint alignment.

UPSET WELDING MACHINES

Upset welding machines are quite similar to flash welding machines in principle except that no flashing mechanism is required. A typical upset welding machine, such as the one in Fig. 3.13, consists of a main frame that houses a transformer and tap switch, electrodes to hold the parts and conduct the welding current, and means to upset the joint. A primary contactor is used to control welding current.

The simplest type of upset welding machine is manually operated. In this machine, the pieces to be welded are clamped in position in the electrodes. A force is exerted on the movable platen with a hand-operated leverage system. Welding current is applied, and when the abutting parts reach welding temperature, they are upset together to accomplish the weld. The current is manually shut off at the proper time during the welding cycle. The work is then removed from the electrodes. A limit switch or a timing device may be used to terminate the welding current automatically after the weld has upset a predetermined amount.

Automatic machines may use springs or air cylinders to provide upset force (see Fig. 3.13). Either device can provide uniform force consistently. Spring or air operated machines are particularly adapted for welding nonferrous metals having narrow plastic ranges.

There are three standard sizes of upset welding machines rated at 2, 5, and 10 kVA.

Normal upset forces are 12, 70, and 120 lb, respectively. However, larger units are available.

Upset welding is used extensively for the welding of small wires, rods, and tubes in the manufacture of items such as chain links, refrigerator and stove racks, automotive seat frames, and for joining coils of wire for further processing. This process is often selected for applications where the upset is not objectionable for the design. It is best adapted for joints between parts with relatively small cross section where uniformity of welding current is not a problem.

RESISTANCE WELDING CONTROLS

The principal functions of resistance welding controls are to (1) provide signals to control machine actions, (2) start and stop the flow of current to the welding transformer, and (3) control the magnitude of the current. There are three general groups of controls: timing and sequencing controls, welding contactors, and auxiliary controls. If the machine is simple in construction, such as a foot-operated spot welding machine, its controls are simple. If the welding cycle is complex, such as that used in welding heat-resistant alloys to military specifications, the machine controls are complex. Many types of resistance welding machine controls are available. Most are produced to the requirements of the National Electrical Manufacturers Association (NEMA) standard[1].

TIMING AND SEQUENCE CONTROLS

Sequence Weld Timers

A sequence weld timer is a device to control the sequence and duration of the elements of a complete resistance welding cycle. It may

also control other mechanical movements of the machine such as driving or indexing mechanisms. Sequence weld timers are used on spot, seam, and projection welding machines.

The four basic steps in any spot, seam, or projection welding cycle are as follows:
(1) Squeeze time
(2) Weld time
(3) Hold time
(4) Off time

Squeeze time is the interval between the initial application of electrode force on the work and the first application of current. Weld time is the duration of welding current flow with single impulse welding. Hold time is the period during which the electrode force is maintained on the weld after the last impulse of current ceases. Off time is the period during which the electrodes are retracted from the work during repetitive welding. During off time, the work is moved to the next weld location.

Certain welding machines are manual or motor-operated. With these machines, a cam may be used to initiate a weld timer that controls only the duration of welding current.

A multiple impulse weld timer provides for a number of current pulses with an interval between them. It controls the duration of each pulse, called heat time, as well as the interval

1. Standard Publication No. ICS 5-1978, *Resistance Welding Control,* National Electrical Manufacturers Association, Washington, DC.

between them, or *cool time*. The sum of the heat and cool times is know as the *weld interval*.

Timers and combination controls are divided into two types: nonsynchronous and synchronous precision. The first type is less expensive than the second because timing accuracy requirements are not so stringent. A synchronous precision control can be used for applications where a nonsynchronous one is adequate. However, a nonsynchronous control may not be suitable for some applications where timing accuracy is important.

Single-phase and three-phase resistance welding controls are similar except for the firing sequence of the rectifier devices in the contactor and the techniques of electronic heat control. The timing and control functions are about the same except that terminology may vary between the two types of equipment. Three-phase controls are predominently synchronous precision types.

Synchronous Precision Controls

This type of control utilizes synchronous precision timers for accurate timing of all periods of current flow. The timer closes the primary circuit of the welding transformer at precisely the same point (electrical angle) with respect to the ac line voltage. Another distinction of a synchronous precision timer is that accuracy is absolute and equal to the set value. Thus, the current wave form and the energy delivered to the welding transformer are consistent for each weld. A synchronous precision control always contains an electronic heat control unit.

Control of the exact time when the primary circuit is closed is vital for precise results. If a highly inductive circuit, characteristic of a welding transformer, is connected to the primary lines at a point corresponding to zero current (if current were flowing), then no transient will occur. The current during the first half cycle will be practically the same as the steady-state current throughout the weld. On the other hand, when the primary circuit is closed at the time the current would be maximum (if current were flowing), a transient current occurs in the primary. Its peaks are greater in magnitude than those corresponding to normal or steady-state current. The transients in the secondary winding of the transformer are not so great, but they are sufficient to materially affect the energy delivered to the weld.

Nonsynchronous Controls

A nonsynchronous control utilizes a weld or weld interval timer that may initiate or terminate the flow of welding current at any time with respect to the line voltage wave form. Consequently, time as well as current input can vary. In addition, timing accuracy may not be equivalent to that of synchronous precision timers. However, this type of control is adequate for most commercial applications on steel products.

Classifications of Sequence Weld Timers

Sequence weld timers are classified by NEMA according to the functions they control and the timing precision. These are as follows:

(1) Types 1A, 1AS, and A1A that control weld time only.

(2) Types 1B and 1BS that control heat and cool times for multiple impulse welding operations.

(3) Types 3B and A3B which are sequence timers that control squeeze, weld, hold, and off times.

(4) Types 3C and A3C which are similar to Type 3B except that a squeeze delay or initial squeeze timer is provided to account for the electrode travel time to contact the work. These types of timers are used for high-speed repetitive welding.

(5) Types 5B and A5B, which are also similar to a Type 3B, are designed for multiple impulse welding applications. They control heat, cool, and weld interval times instead of weld time.

(6) Type 7B is a sequence timer used in conjunction with a synchronous precision weld timer (Type 1AS) to control squeeze, weld, hold, and off times.

(7) Type 9B is similar to a Type 7B except that a synchronous precision multiple impulse weld timer (Type 1BS) is used with it.

In the designations, S indicates a synchronous precision timer and the prefix A indicates an absolute cycle timer. The latter type is 100 percent accurate and synchronous.

Timing Mechanisms

Several types of timers are employed to control the duration of various functions during the welding cycle. The two most common types are RC timers and digital counters.

RC Timer. This type is based on the constant time interval required to either charge or discharge a condenser through a resistive circuit. The timing accuracy decreases with increasing time span. This timer is the one most often used in electronic welding controls.

Digital Counter. This type is accurate regardless of the set time. It employs semiconductor circuits to count the actual number of power supply frequency cycles and to initiate some action after a preselected number of cycles has elapsed.

Others. Some operations, such as postheating of flash or upset welds, are not critical with respect to timing accuracy. Pneumatic or motor-operated timers may be suitable for these applications. Timing ranges may vary from a few seconds to several minutes.

CONTACTORS

A contactor is used to close and open the primary power line to the welding transformer. Electronic contactors are normally used with resistance welding machines that are controlled by electronic timing controls. Magnetic contactors may be used with some low power machines where timing is not critical, such as manually-operated spot welding machines or upset welding machines.

Combination controls, both synchronous precision and nonsynchronous types, incorporate electronic contactors to switch the primary current to the machine transformer. Either ignitron tubes or silicon-controlled rectifiers (SCR) are used as switching devices. The choice will depend upon the performance requirements of the equipment.

Electronic contactors have two inversely connected ignitron tubes or SCRs in a power line to control the current flowing in that line. One device conducts the positive half cycles and the

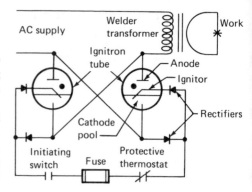

Fig. 3.14—A single-phase welding machine with an ignitron contactor and control circuit

other the negative half cycles of current. In the case of single-phase equipment, only one set is needed in one of the power lines, as shown in Fig. 3.14. With a three-phase frequency converter machine, one set is required in each leg of the transformer for a total three sets, as shown in Fig. 3.15.

An ignitron tube is constructed as shown in Fig. 3.16. It consists of a vacuum-tight stainless steel container with a pool of mercury in the bottom which serves as the cathode. A graphite anode is supported at the top by an insulating glass bushing to which the power lead is connected. An ignitor above the mercury pool starts the flow of current through the tube. The tube has a cooling water jacket to remove the heat generated by the conducting arc within the tube. Current flows in one direction only from the graphite anode to the mercury cathode.

The ignitor will initiate primary current conduction when some minimum voltage drop exists across the tube. This is done by sending a pulse of current from the ignitor to the mercury pool at some instant during the half cycle. The ignitron tube will continue conducting current until the applied voltage ceases at the end of the half cycle. The ignitor must start conduction for each half cycle during the period that primary current flow is desired.

The control circuit of an ignitron contactor with nonsynchronous timing is very simple, as

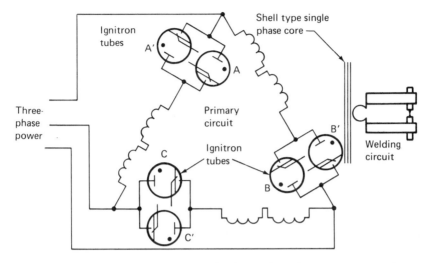

Fig. 3.15 — Electrical arrangement of a three-phase frequency converter spot welding machine

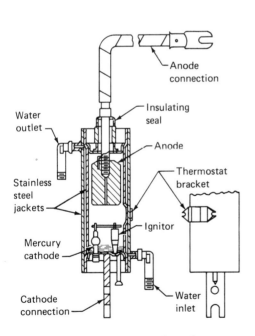

Fig. 3.16 — Cross section of an ignitron tube

Fig. 3.14 indicates. The rectifiers in the ignitor circuit allow current flow from the ignitor to the mercury pool only. The ignitron tubes will fire during the time that the initiating switch is closed to energize the ignitors. The switch is normally actuated by a timer.

The contactor energizes the welding circuit when the control circuit is closed. This control circuit requires no additional power since its power is obtained from the main circuit. Each ignitron tube is fired for the appropriate half cycle by its ignitor as long as the control circuit is energized.

A thermostatic switch is mounted on one or both of the ignitron tubes to stop operation if the tube temperature becomes too high.

With synchronous precision timing and electronic heat control, ignitron tube conduction must be initiated at a specific point on the half-cycle voltage wave. This is done by controlling the ignitor circuits with SCRs, as shown in Fig. 3.17, or with thyratron tubes in older units. The control signals the SCRs or thyratrons when to conduct current to the ignitors. With this arrangement, the control circuit need not carry ignitor current.

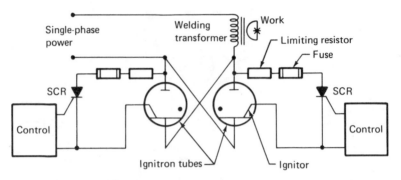

Fig. 3.17—An ignitron contactor with silicon-controlled rectifiers to initiate tube firing

The effect of duty cycle on the selection of ignitron tube size for a particular application is as important as the current rating of the tube. Duty cycle averaging time is a matter of a few seconds, depending upon the tube size. Tube manufacturers' rating charts should be consulted when selecting the tube size for a specific application.

Ignitron tubes require a minimum load current to ensure satisfactory operation. If the primary current demand will be less than 40 A, an auxiliary load resistor should be permanently connected in parallel with the welding transformer. This resistor will always provide the required minimum loading on the tubes.

If sufficient current for proper ignition is not furnished to the ignitors, the ignitor circuit will try to carry the welding transformer load. Obviously, this action could severely overload the ignitor circuit and damage it. Ignitor circuits are usually protected by fuses to avoid this problem.

Water-cooled, silicon-controlled rectifiers are available as functional replacements for ignitron tubes. They come in a variety of sizes. Their selection for resistance welding controls must be approached in the same manner as for ignitron tubes. SCR firing circuits are somewhat different from those for ignitron tubes and the designs vary among control manufacturers. One feature of SCR controls is the ability to mount them in any position on a gun welding trans-

former. This is not possible with ignitron tubes which must be mounted vertically.

AUXILIARY CONTROLS

Heat Control

The control of the heat or current ouput of the welding machine can be accomplished with adjustable taps on the welding transformer primary and with electronic heat control. A tap switch changes the ratio of transformer turns for major adjustment of welding current. When an intermediate setting is needed or for fine adjustment, electronic heat control is used.

In electronic heat control circuits, SCRs or thyratron tubes are used to control the ignitor circuits of the ignitron tubes (see Fig. 3.17). The ignitron conduction can be delayed from the start of the half cycle by withholding the signal to the ignitor. This delay is normally controlled by a heat adjustment dial. With an SCR contactor, the action is similar.

Referring to Fig. 3.17, if the ac voltage produced by the control is 180 degrees out of phase with the ac supply voltage, the SCRs cannot conduct. The ignitor circuits do not fire, and the ignitron tubes block the flow of primary current. As the out-of-phase or delay angle of the control voltage is decreased to approximately 130 degrees, the SCR will fire the ignitor of the associated ignitron tube late in the half cycle. Current will flow through the ignitron tube, but the rms value will be low. As the delay angle is decreased, the ignitor will fire the ignitron tube earlier, and it will conduct current for a greater part of the half cycle. The rms current will increase. When the delay angle equals the power factor angle of the load, 100 percent rms primary current will be conducted to the welding transformer. Figure 3.18 illustrates this concept for welding machine loads with four different power factors. The higher the power factor (lower angle), the wider is the range of heat control.

The reduction in heat or energy varies as the square of the current. Thus, if the rms current can be varied from 100 to 20 percent, the heat will vary from 100 to 4 percent.

Complete control from 100 percent to zero is not feasible. With ignitron tubes, the limiting minimum value is 40 percent rms current for 220 volt systems and 20 percent rms current for 440 volt systems. With SCR contactors, the limiting values are significantly lower.

Automatic heat control is normally the basis of all auxiliary controls that change the welding amperage during a welding sequence. These include current and voltage regulators as well as upslope, downslope, and temper controls.

To minimize variations in welding current, the heat control should be operated as near to full heat as possible. At low settings, a small change of the dial setting can significantly change the rms current. Line voltage disturbances, such as the operation of another welding machine, can distort the line voltage wave form sufficiently to produce such a change. Major changes in welding transformer output should be made by changing the tap switch.

The power demand is always greater when heat control is used to adjust the magnitude of

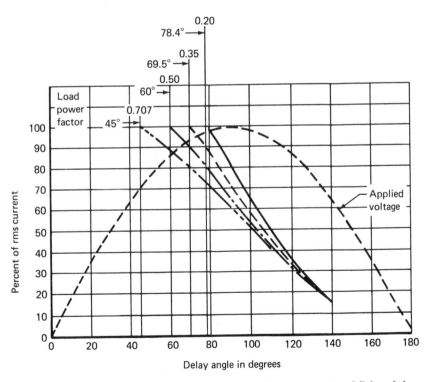

Fig. 3.18—The relationship between percent of rms current and firing delay angle for different power factors

the welding current. In general, the kVA demand with heat control follows a linear relationship with current. For example, if the welding current is adjusted by heat control to 80 percent of its maximum value, the kVA demand will be about 80 percent of maximum. However, if the welding current is reduced to 80 percent of its maximum value by changing the transformer tap switch, the kVA demand will be only about 64 percent of maximum.

Upslope and Downslope Controls

Upslope control is used to start the welding current at some low value and control its rate of rise to some maximum value during a period of several cycles. It is frequently used to minimize or prevent the expulsion of molten metal from between the faying surfaces when welding coated steels and some nonferrous metals, particularly aluminum.

Downslope control is used to decrease the welding current from maximum to a lower value called the *postheat current*. The gradual decrease in current reduces the cooling rate of the weld. It may be useful when welding hardenable steels

to minimize the cooling rate and the cracking tendency.

Upslope and downslope of welding current are illustrated in Fig. 3.19. The accepted nomenclature for the various parts of a welding current cycle are also shown.

Quench and Temper Control

The quench and temper control is a device that applies a temper cycle to the completed weld after a quench period during which no current flows. In each case, the time period is adjustable. Temper current magnitude is normally adjustable with heat control.

This control is frequently used when spot welding hardenable steels in thickness ranges of 0.016 to 0.125 inch. After the weld is made, it is then rapidly cooled during quench time to hasten formation of martensite. The quench time must be long enough for martensite to form. The minimum time will depend upon the weld thickness and steel composition. A current pulse is then applied to reheat the weld zone and thus temper (soften) the martensite. Although this cycle cannot duplicate furnace heat treatment,

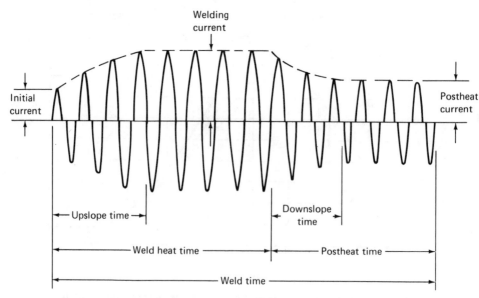

Fig. 3.19—Welding current with upslope and downslope features

it usually will prevent weld cracking.

Forge Delay Control

This control will initiate a forging force at a definite time interval after the start of weld time or weld interval. It is used to apply two levels of force to a weld, namely, a welding force and a forging force. Obviously, the welding machine must be designed to perform a dual force function.

Dual force is used for spot welding some aluminum alloys. The principle is to produce the weld and then apply a high force during cooling to avoid the formation of cracks. It is common to downslope the welding current to retard the cooling rate with the application of forging force.

Electronic Current Regulator

An electronic current regulator is designed to maintain a constant welding current under adverse conditions. This device will make corrections for either line voltage fluctuations or impedance changes generally caused by insertion of magnetic material into the throat of the welding machine.

It first compares the primary current, as measured by a current transformer or other device (feedback signal), to a previously adjusted satisfactory level (command signal); then it varies the phase-shift heat control network to make these signals equal and opposite.

Electronic Voltage Regulator

An electronic voltage regulator is designed to maintain a constant voltage at the welding machine transformer in the presence of line voltage variation. There are various forms of this device. Some designs make an arbitrary correction that is dependent upon voltage rise or drop, and others compare the line voltage (feedback) to a previously adjusted (command) operating voltage. Both types cause the phase-shift heat control network to respond and change the ignitron tube conduction point. This type of regulator is simpler than the electronic current regulator.

Load Distribution Control

A load distribution control is used with resistance welding machines having two or more transformers. This control distributes the electrical power demand by energizing the welding transformers in sequence on one or more phases. Reconnection is normally provided to energize the transformers simultaneously on two or more phases.

This control generally contains several single-function timers to control mechanical functions, such as squeeze and hold timers, and two or more weld periods. In addition, it has a contactor for each transformer. The weld timers are a function type but are weld safe; that is, the termination of weld time is not dependent upon conduction of a single electronic device. Accessories such as heat control and upslope control are occasionally added to this type of control.

A less expensive version of this control uses only one ignitron or SCR contactor and a series of magnetic contactors. The ignitron or SCR contactor switches the primary current on and off. The magnetic contactors connect the welding transformers in succession to the contactor circuit during a nonconductive period.

STANDARD CONTROLS

Resistance welding control functions can vary from control of only weld time, in the case of manually operated spot welding machines, to sequencing and timing a number of events with automatic spot and seam welding machines. A typical complex weld cycle is shown in Fig. 3.20. It is evident that a control unit for this type of operation requires a number of timing and sequencing features.

Standard NEMA single-phase resistance welding controls and the types of timers that are used with each control are given in Table 3.1. Timer and auxiliary control functions and the standard timing ranges for single-phase controls are given in Table 3.2. The functions performed by a particular control unit depend upon the intended applications of the welding machine.

Timer and auxiliary control functions and their timing ranges for three-phase controls are similar to those for single-phase controls. Timing ranges for specific functions may vary slightly because a three-phase machine is nor-

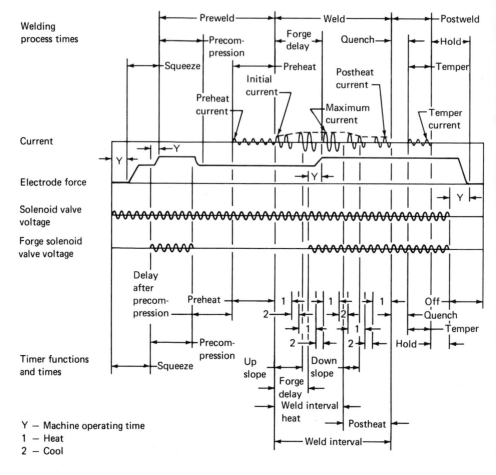

Welding process times

Current

Electrode force

Solenoid valve voltage

Forge solenoid valve voltage

Timer functions and times

Y — Machine operating time
1 — Heat
2 — Cool

Note. When weld delay is required, it occurs before the start of weld time; forge delay time then will be initiated at the beginning of weld delay time. Both preheat and weld delay are not used in the same sequence.

Fig. 3.20—A complex weld cycle with multiple impulses

mally used for welding nonferrous metals, primarily aluminum.

MONITORING AND ADAPTIVE CONTROLS

There are a number of factors that affect the consistency of resistance spot welds during a production run. These include line voltage variations, electrode deterioration, changes in surface resistance, shunt paths, and variations in the force system. There are several systems available to

monitor specific welding variables or actions that occur during the welding cycle. If the monitor detects a fault, it can do one or more of the following:

(1) Set off an alarm or signal light
(2) Document the information
(3) Reject or identify the faulty part
(4) Interrupt the process until the problem is corrected
(5) Alter time or current for the next weld
(6) Change a variable during the weld cycle to ensure a good weld

Table 3.1
Standard NEMA single-phase
resistance welding controls

Control type	Timer types	Applications
Nonsynchronous types		
N1A	1A	Single weld pulse
N1B	1B	Multiple weld pulse
N2	3B	Single-pulse weld sequence
N3	5B	Multiple-impulse weld sequence
N6	3C	Single-pulse weld sequence
Synchronous precision types		
S1H	1AS or 1CS	Single weld pulse
S2H	1AS and 7B	Single-pulse weld sequence
S3H, S4H	1BS and 9B	Multiple-impulse weld sequence
S5H	1BS	Multiple weld sequence

Variables that affect process stability and weld consistency include weld time, welding current, impedance, energy, and electrode force. Physical changes that take place in the weld zone are temperature, expansion and contraction, electrical resistance, and, in some cases, metal expulsion.

Monitoring devices can compute either weld energy or impedance by measuring welding voltage, current, resistance, or time. When the computed value falls outside acceptable limits, the unit can notify the operator or automatically adjust one or more of the variables prior to the next weld.

Another monitoring device uses an acoustic emission device to sense the first sound generated by metal expulsion between the faying surfaces. The device immediately terminates the welding current. The principle behind this device is that every weld reaches its maximum size, as limited by expulsion.

Table 3.2
Timing ranges of standard single-phase resistance welding timers

Function	Timing range, cycles (60 Hz)		
	Nonsynchronous	Synchronous	Absolute cycle
Timer functions			
Squeeze	3-120, 3-30		1-99, 1-59
Squeeze delay[1]	3-30		1-59
Initial squeeze[1]	3-60		
Weld	3-120, 2-30	1-60, 1-10	1-99, 1-59
Weld interval	9-360		(1-8 heat times)[2]
Heat	3-30	1-30	1-99
Cool	3-30	1-30	1-29
Hold	3-60, 2-30		1-99, 1-29
Off	3-60, 3-30		1-99, 1-29
Auxiliary control functions			
Upslope	3-30	1-10, 1-30	1-9, 1-29
Weld heat	3-60	1-60	1-99
Weld interval heat	3-180	3-180	(1-8 heat times)[2]
Downslope	3-30	1-10, 1-30	1-9, 1-29
Quench	9-360	9-360	1-99
Temper	3-120	1-120	1-99
Forge delay	3-360	1-360, 0.1-59.9	1-99, 0.1-99.9
Preheat	3-120	1-120	1-99
Precompression	3-10		1-9
Delay after precompression	3-10		1-9

1. Only one of these functions is used.
2. The timer counts the number of heat times rather than the overall time.

ELECTRICAL CHARACTERISTICS

SINGLE-PHASE EQUIPMENT

The typical electrical system of a single-phase resistance welding machine consists of (1) a transformer, (2) a tap switch, and (3) a secondary circuit including the electrodes.

The welding transformer, in principle, is the same as any other iron-core transformer. The chief difference is that its secondary circuit has only one or two turns. Stationary machines have only one turn. Portable gun welding transformers may have two turns that can be connected in series or parallel, depending upon the output requirements.

Transformer Rating

Resistance welding transformers are normally rated on the basis of temperature rise limitations on the components. The standard rating in kVA is based on the ability of a transformer to produce that power at a 50 percent duty cycle without exceeding design limitations. This means that a transformer can produce its rated power for a total time of 30 seconds during each minute of operation without exceeding temperature limitations if it is being properly cooled.

Duty cycle is the percentage of time that the transformer is actually producing power during a one minute integrating period. It is expressed by the formula:

$$\text{Duty cycle, percent} = \frac{\text{Weld time, cycles (60 Hz)} \times (\text{welds/min})}{36}$$

For example, if a machine is producing 36 welds per minute with a weld time of 12 cycles (60 Hz), its operating duty cycle is 12 percent.

If a welding transformer is operated at less than 50 percent duty cycle, it can be operated at a higher power level than its thermal rating. The maximum permissible kVA input for a standard resistance welding transformer at a particular duty cycle can be determined using the following equation:

$$kVA_i = 7.07\, kVA_r/(DC)^{1/2}$$

where: kVA_i = maximum input power

kVA_r = standard power rating at 50 percent duty cycle

DC = operating duty cycle, percent

For example, a welding transformer rated at 100 kVA may be operated at 141 kVA at 25 percent duty cycle without overheating.

Tap Switches

Tap switches are devices for connecting various primary taps on the transformer to the supply lines. They are usually rotary type and designed for flush mounting in an opening in the machine frame or, in some cases, directly on the transformer. The switches are designed to accommodate the arrangements of the transformer taps. Straight rotary designs are normally used with 4, 6, or 8 tap transformers. In addition, there may be a series-parallel switch that connects two sections of the primary in series or in parallel. This provides a wider range of secondary voltages.

Most switch handles have locking buttons so that the contacts are centered in each operating position. In addition, some switches have an "off" position which acts as a disconnect. A tap switch should not be operated when the transformer is energized. Otherwise, arc-over between points will damage the contact surfaces of the tap switch.

AC Secondary Circuit

The geometry of the secondary circuit or loop, the size of the conducting components, and the presence of magnetic material in the loop will affect the electrical characteristics of an ac welding machine. Available welding current and kVA demand will be influenced by the impedance of the secondary circuit.

The electrical impedance of an ac welding machine should be minimized to permit the delivery of the required welding current at minimum kVA demand. The electrical impedance will be smaller when:

(1) The throat area of the welding machine is decreased.

(2) The electrical resistance of the secondary circuit is decreased.

(3) The sizes of the secondary conductors are increased.

(4) The amount of magnetic material in the throat of the machine is decreased.

Power Factor Correction

Series Capacitors. Welding machines of good design effectively minimize the impedance of the secondary circuit. However, the size of the work to be welded and associated fixturing may require a large throat depth or throat height. This requirement may add considerable inductance to the secondary circuit. The increased inductance causes a reactive voltage drop which in turn decreases the power factor. To compensate for this, a higher secondary voltage will be required and the necessary electrical kVA demand will be increased.

Low power factor and intermittent, high electrical demand are not desirable to the electric utility, which must maintain a stable power supply to other customers. One method of reducing line kVA demand and improving power factor is the use of series capacitors in the primary circuit. A specific amount of capacitance can be connected in series with the transformer of a welding machine to neutralize the inductance of the machine and improve the power factor. This, in turn, will reduce the demand from the power line.

This power factor correction method will increase the voltage applied to the welding machine transformer. High voltage insulation is therefore required. A transformer tap switch is not used because it changes the series resonant condition. The welding current is changed with phase-shift heat control or a tapped autotransformer.

The resistance of the secondary circuit limits the current in any high power factor system. Since the metal being welded has resistance, the welding current may vary significantly with slight changes in metal thickness or cleanliness. This may affect weld consistency and quality, particularly with alloys of high resistance.

Voltages appearing across the welding machine transformer and the series capacitors are higher than the electrical supply voltage. There-fore, special high voltage electrical control panels are normally required. A protective overvoltage device, a discharge resistor, and a contact to ground are generally provided for safe operation and maintenance.

Three-phase welding systems have largely replaced single-phase series capacitor installations. A welding machine with an inherently good power factor is generally less troublesome than a series capacitor installation.

Shunt Capacitors. Shunt capacitors are seldom used with resistance welding equipment. The initial high inrush of current may actually increase the line demand. However, shunt capacitors may be preferred to series capacitors if the welding time is comparatively long, as in non-interrupted resistance seam welding.

DC Secondary Circuit

One method of decreasing impedance losses in the secondary circuit is to rectify the secondary power to dc. Single-phase dc resistance welding machines utilize a center-tapped secondary and a full wave silicon diode rectifier bank. With this system, kVA rating of a machine does not have to be increased to provide for a larger throat area. For a given size and application, the kVA demand of a dc machine will be significantly lower than that of an ac machine. The reason for this is the high power factor of about 90 percent for dc machines, compared to 25 to 30 percent for ac machines.

This type of power is particularly useful for portable gun welding applications. The impedance loss in the cable connecting the gun and transformer is much lower with dc than with ac. This, in turn, decreases the kVA demand and the required size of the welding transformer. It is also advantageous for spot and seam welding operations where the amount of magnetic material in the machine throat increases or decreases as welding proceeds.

DIRECT ENERGY THREE-PHASE EQUIPMENT

Frequency Converter Type

This type of machine has a specially designed transformer with three primary windings, each of which is connected across one of the three input phases. There is one secondary winding which is interleaved among the primary windings and connected to the secondary conductors.

Referring to Fig. 3.15, the transformer primary windings are connected to the power lines by three electronic contactors. Ignitron tubes or SCRs may be used as contactors. A welding control causes ignitron tubes A, B, and C to conduct in sequence. With the correct sequence and conduction time, current is passed through the three primary windings in the same direction. This causes unidirectional current to flow in the secondary circuit. Ignitron tubes A, B, and C are then shut off at the end of a preselected time. Ignitron tubes A', B', and C' are caused to conduct next with the correct sequence and conduction time, and current will flow in the opposite direction through the primary windings and the secondary circuit. This action effectively applies a reversing "dc" voltage to the primary windings.

The maximum duration of unidirectional primary current flow is governed primarily by the size of the transformer and its saturation characteristics. It is common practice to have two maximum dc pulse lengths. One is a short time of about 5 cycles (60 Hz) for high current applications and the other is usually 10 cycles with weld-ing current limited to about 50 percent of maximum. Specially designed massive transformers may permit the use of high current for the longer time period.

Figure 3.21 shows a typical current-force diagram for this type of machine. Programming may be provided for other functions such as preheat current, precompression force, and temper current. Single- or multiple-impulse welds may be made.

DC Rectifier Type

A three-phase dc rectifier type of welding machine is similar to the single-phase type in that a welding transformer powers a rectifier bank. The output of the rectifiers is fed into the welding circuit. Some machines use half-wave rectification as shown in Fig. 3.22(A). In this case, the transformer secondary is wye connected. Other machines, particularly earlier versions, have full-wave rectification with the transformer secondary connected in delta arrangement as shown in Fig. 3.22(B).

Welding current is controlled by electronic heat control, sometimes in conjunction with a transformer top switch. The design of the primary circuit and control varies among equipment manufacturers. The secondary current output of a three-phase machine is much smoother

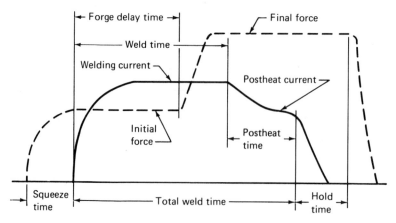

Fig. 3.21—Typical current-force diagram for frequency converter or dc rectifier types of three-phase spot welding machines

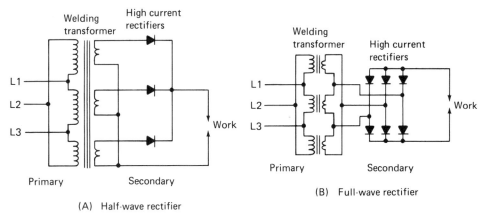

Fig. 3.22 —Electrical arrangements for three-phase dc rectifier welding machines

than that of a single-phase machine. In addition, power demand is balanced on the power line.

The three-phase rectifier consists of silicon diodes mounted on water-cooled conductors. The arrangement of conductors and diodes is electrically symmetrical. The impedance of each diode circuit must be similar so that each diode will share the load (current) equally. The diodes themselves must have similar electrical characteristics. Each diode has long life if it is properly applied and used. Welding current may be provided continuously as long as the thermal rating of the machine is not exceeded.

A typical current-force diagram for this type of machine is similar to that of Fig. 3.21. In addition, programming may be provided for other functions such as preheating, upslope, downslope, and tempering. Single- or multiple-impulse welds may be made.

STORED ENERGY EQUIPMENT

Equipment of this type is usually confined to small units suitable for bench mounting. They are powered from a single-phase line. Many designs of welding heads or portable tongs that are connected to their power units with cables are available. They are used for a wide variety of applications including assembly of small

Fig. 3.23 —A bench-mounted stored energy spot welding machine

electrical components of nonferrous alloys and the spot welding of foils.

Electrode force may range from a few ounces to several pounds. Calibrated springs are used in a manual force system to apply the electrode force. Stored energy is used to produce the welding current pulse. The welding current amplitude, duration, and wave shape are determined by the electrical characteristics of the power source including capacitance, reactance, resistance, and capacitor voltage. Welding times are often substantially shorter than one half cycle of 60 Hz.

Figure 3.23 shows a typical foot-operated bench type welding machine of this type with a maximum electrode force of either 8 or 20 lb, depending upon the spring size. Electrode force is applied by actuation of a foot pedal mounted beneath the welding head. A typical power source is rated at 40 watt-seconds, has 600 microfarads of capacitance, and can be adjusted for welding outputs as indicated by the curves in Fig. 3.24.

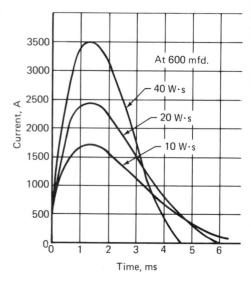

Fig. 3.24—Typical time-current wave forms of a 40 watt-second stored energy spot welding machine

ELECTRODES AND HOLDERS

The perishable tools used in resistance welding are the electrodes, which may be in the form of a wheel, roll, bar, plate, clamp, chuck, or some modification thereof. Most spot welding applications utilize electrode holders or adaptors for mounting the electrodes in the machine.

A welding electrode may perform one or more of the following functions:

(1) Conduct welding current to the parts
(2) Transmit a force to the joint
(3) Fixture or locate the parts in proper alignment
(4) Remove heat from the weld or adjacent part

The electrode design should always have sufficient mass to transmit the required welding force and current and provisions for adequate cooling when needed. High production applications sometimes involve thick sections that re-

quire special electrode designs. If it should be necessary to compromise the design, it may affect electrode life, weld quality, production rate, or all three. Consequently, selection of the electrode material is very important for good performance.

ELECTRODE MATERIALS

Standard resistance electrode materials are classified by the RWMA.[2] They are divided into two groups: copper-base alloys and refractory metal compositions. In addition to these standard materials, there are a number of proprietary alloys available from the various electrode manufacturers. For some applications, commercially

2. Standard electrode materials are described in ANSI/RWMA Bulletin No. 16, *Resistance Welding Equipment Standards*, Resistance Welder Manufacturers Association, Philadelphia, PA.

Table 3.3
Minimum properties for RWMA electrode materials

Group A copper-base alloys	Proportional limit tension, psi			Hardness, Rockwell B			Conductivity, %ᵃ			Ultimate tensile strength, psi			Elongation, % in 2 in. or 4 diameters		
	Class 1	Class 2	Class 3	Class 1	Class 2	Class 3	Class 1	Class 2	Class 3	Class 1	Class 2	Class 3	Class 1	Class 2	Class 3
Rod diam., in.							Round rod stock (cold worked)								
Up to 1	17,500	35,000	50,000	65	75	90	80	75	45	60,000	65,000	100,000	13	13	9
Over 1 to 2	15,000	30,000	50,000	60	70	90	80	75	45	55,000	59,000	100,000	14	13	9
Over 2 to 3	15,000	25,000	50,000	55	65	90	80	75	45	50,000	55,000	95,000	15	13	9
Thickness, in.							Square, rectangular, and hexagonal bar stock (cold worked)								
Up to 1	20,000	35,000	50,000	55	70	90	80	75	45	60,000	65,000	100,000	13	13	9
Over 1	15,000	25,000	50,000	50	65	90	80	75	45	50,000	55,000	100,000	14	13	9
Thickness, in.							Forgings								
Up to 1	20,000	22,000ᵇ	50,000	55	65	90	80	75	45	60,000	55,000	94,000	12	13	9
Over 1 to 2	15,000	21,000ᵇ	50,000	50	65	90	80	75	45	50,000	55,000	94,000	13	13	9
Over 2	15,000	20,000ᵇ	50,000	50	65	90	80	75	45	50,000	55,000	94,000	13	13	9
							Castings								
All	—	20,000	45,000	—	55	90	—	70	45	—	45,000	85,000	—	12	5

Table 3.3 (continued)
Minimum Properties for RWMA electrode materials

Group A copper-base alloys	Proportional limit tension, psi	Hardness, Rockwell	Conductivity, %[a]	Ultimate tensile strength, psi	Elongation, % in 2 in. or 4 diameters
Class 4 Alloys					
Cast	60,000	33C	18 (Average)	90,000	0.5
Wrought	85,000	33C	20 (Average)	140,000	1.0
Class 5 Alloys, cast					
Type H	16,000	88B	12	70,000	2
Type S	12,000	65B	15	65,000	12
Group B refractory metals				Ultimate compression strength, psi	
Class 10 – Rods, bars, and inserts		72B	45	135,000	
Class 11 – Rods, bars, and inserts		94B	40	160,000	
Class 12 – Rods, bars, and inserts		98B	35	170,000	
Class 13 – Rods, bars, and inserts		69B	30	200,000	
Class 14 – Rods, bars, and inserts		85B	30	—	

a. International Annealed Copper Standard.
b. Hot worked and heat treated but not cold worked.

available copper alloys (brasses or bronzes) or steels may be used for welding electrodes. Table 3.3 gives the minimum properties for alloys to meet the various RWMA classification requirements. The specific alloy compositions are not specified, and they will vary among the manufacturers.

Group A—Copper-Base Alloys

The copper-base alloys are divided into five classes. Class 1 alloy is a general purpose material for resistance welding applications. It may be used for spot and seam welding electrodes where electrical and thermal conductivities are of greater importance than mechanical properties. Other applications are seam welding machine shafts and welding fixtures. This alloy class is recommended for spot and seam welding electrodes for aluminum, brass, bronze, magnesium, and metallic coated steels because of its high electrical and thermal conductivity.

Class 1 alloy is not heat treatable. Its strength and hardness are increased by cold working. Therefore, it has no advantage over unalloyed copper for castings and is rarely used or fabricated in this form.

Class 2 alloy has higher mechanical properties but somewhat lower electrical and thermal conductivities than Class 1 alloy. It has good resistance to deformation under moderately high pressures, and is the best general purpose alloy. This alloy is suitable for high production spot and seam welding of clean mild and low alloy steels, stainless steels, low-conductivity copper-base alloys, and nickel alloys. These materials comprise the bulk of resistance welding applications.

Class 2 alloy is also suitable for shafts, clamps, fixtures, platens, gun arms, and various other current-carrying structural parts of resistance welding equipment. This alloy is heat treatable and may be used in both wrought and cast forms. Maximum mechanical properties are developed in wrought form by cold working after heat treatment.

Class 3 alloy is also heat treatable but it has higher mechanical properties and lower electrical conductivity than Class 2 alloy. The chief application for spot or seam welding electrodes of this alloy is the welding of heat resistant alloys that retain high strength properties at elevated temperatures. Welding of these alloys requires high electrode forces which, in turn, require a strong electrode alloy. Typical heat resistant alloys are some low alloy steels, stainless steels, and nickel-chromium-iron alloys.

Class 3 alloy is especially suitable for many types of electrode clamps and current-carrying structural members of resistance welding machines. Its properties are similar in both the cast and wrought conditions because it develops most of its mechanical attributes from heat treatment.

Class 4 alloy is an age-hardenable type that develops the highest hardness and strength of the Group A alloys. Its low conductivity and tendency to be hot-short makes it unsuitable for spot or seam welding electrodes. It is generally recommended for components that have relatively large contact area with the part. These include flash and projection welding electrodes and inserts. Other applications are part backup devices, heavy-duty seam welding machine bearings, and other machine components where resistance to wear and high pressure are important.

Class 4 alloy is available in both cast and wrought forms. Because of the high hardness after heat treatment, it is frequently machined in the solution-annealed condition.

Class 5 alloy is available principally in the form of castings with high mechanical strength and moderate electrical conductivity. It is recommended for large flash welding electrodes, backing material for other electrode alloys, and many types of current-carrying structural members of resistance welding machines and fixtures.

Group B—Refractory Metal Compositions

These materials contain a refractory metal in powder form, usually tungsten or molybdenum. They are made by the powder metallurgy process. Their chief attribute is resistance to deformation in service. They function well for achieving heat balance when two different electrode materials are needed to compensate for a difference in thicknesses or alloys being welded.

Class 10, 11, and 12 compositions are mixtures of copper and tungsten. The hardness, strength, and density increase and the electrical conductivity decreases with increasing tungsten content. They are used as facings or inserts where exceptional wear resistance is required in various projection, flash, and upset welding electrodes. It is difficult to establish guidelines for the application of each grade. The electrode design, welding equipment, opposing electrode material, and workpiece composition and condition are some of the variables that should be considered in each case.

Class 13 and Class 14 are commercially pure tungsten and molybdenum, respectively. They are generally considered to be the only electrode materials that will give good performance when welding nonferrous metals with high electrical conductivity. The welding of braided copper wire or copper and brass wires to themselves or to various types of terminals are typical uses for Class 13 and 14 materials.

Other Materials

A number of unclassified copper alloys and other materials may be suitable for resistance welding electrodes. Suitability of a particular material for electrodes will depend upon the application. Although most requirements are met by materials meeting RWMA standards, there are cases where other materials will function as well or better. For example, steel may be used for flash welding electrodes for certain aluminum applications.

Dispersion-strengthened copper is an unclassified material that may be used for electrodes. It is high purity copper that contains small amounts of submicroscopic aluminum oxide uniformly distributed in the matrix. The aluminum oxide significantly strengthens the copper matrix and raises the recrystallization temperature of cold worked material. The high recrystallization temperature of wrought material

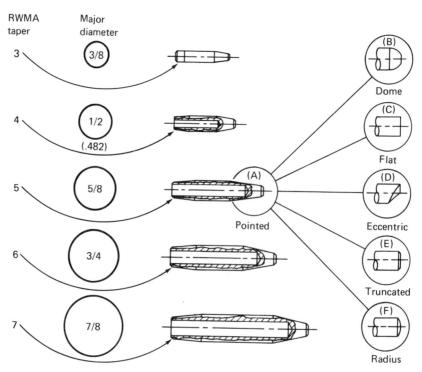

Fig. 3.25 — Standard RWMA spot welding electrode face and taper designs

sure densities in the weld zone. Figure 3.25 shows the standard RWMA electrode face and taper designs. The radius and dome contours are the most commonly used for all metals. The flat-faced electrode is used to minimize surface marking or to maintain heat balance.

The face may be concentric to the axis of the electrode as in Figs. 3.25(A), (B), (C), (E), and (F); eccentric or offset as in Fig. 3.25(D); or at some angle to the axis as in Fig. 3.26. So-called offset electrodes with eccentric faces are used to make a weld near a corner or in other less accessible areas. This is illustrated in Fig. 3.27. A facing of Group B material may be brazed to a shank of a Group A alloy to produce composite electrodes for special applications as shown in Fig. 3.28.

Shank

The shank of an electrode must have sufficient cross-sectional area to support the electrode force and carry the welding current. The shank may be straight, as in Fig. 3.25, or bent as in Fig. 3.29. The standard shank diameters are shown in Fig. 3.25.

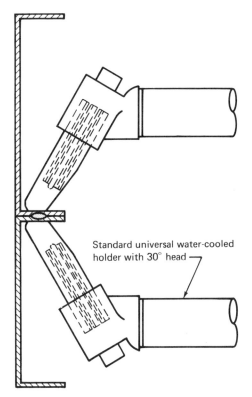

Standard universal water-cooled holder with 30° head

Fig. 3.26—Special spot welding electrodes with the faces angled at 30 degrees

provides excellent resistance to softening and mushrooming of electrodes. This significantly contributes to long electrode life. The mechanical properties and electrical conductivity of dispersion-strengthened copper bars meet the requirements for RWMA Group A, Class 1 and 2 alloys.

SPOT WELDING ELECTRODES

A spot welding electrode has four features: (1) the face, (2) the shank, (3) the end or attachment, and (4) provision for cooling.

Face

The face of the electrode is that portion which contacts the work. Its design is influenced by the composition, thickness, and geometry of the parts to be welded. In turn, the electrode face geometry determines the current and pres-

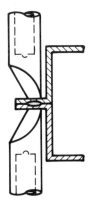

Fig. 3.27—An application of Type D offset spot welding electrodes

Tapered electrodes

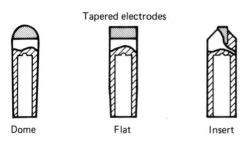

Dome   Flat   Insert

Threaded electrodes

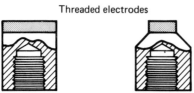

Flat   Flat

Fig. 3.28—Typical Group B electrode faces brazed to Group A alloy shanks

Attachment

The method of attaching the shank end to the holder is usually one of three general types: tapered, threaded, or straight-shank.

RWMA tapered attachments use the Jarno taper as the standard. This taper offers the following advantages:

(1) The taper number multiplied by 1/8 in. gives the nominal major diameter. For example, RWMA No. 5 taper has 5/8 in. diameter.

(2) The taper numbers progress in sequence from 3 to 7.

(3) The RWMA taper is a uniform 0.600 in./ft for all sizes.

The electrode diameter and taper length increase as the taper number increases. Longer tapers can support higher electrode forces, but there is a maximum force that should be used with each electrode size. Recommended maximum electrode forces for the various sizes are given in Table 3.4.

Threaded attachments are used where high welding forces would make removal of tapered electrodes difficult, or where electrode position is critical. Typical threaded electrodes are shown in Fig. 3.30.

Straight-shanked electrodes are used with high welding forces, especially the 5/8 and 3/4-in. diameters. The base of the electrode bears against the holder socket. The water seal is an "O" ring in a recessed groove in the holder. The electrode is mechanically held in place by a coupling or collar.

Cooling

Wherever practical, spot welding electrodes should have an internal cooling passage extending close to the welding face. This passage should be designed to accommodate a water inlet tube and provide for water out around the tube. The tube should be positioned to direct the cooling water against the tip of the electrode. In most cases, the tube is a component of the electrode holder. An exception is bent electrodes. Where internal cooling is not practical, external cooling of the electrodes by immersion, flooding, or attached cooling coils should be considered.

Two-Piece Electrodes

Two-piece or cap-and-adaptor electrodes are available with both male and female caps, as shown in Fig. 3.31. They are available with straight and bent shanks. Use of this electrode design is a matter of economics. Tip maintenance costs may be lower because only the cap needs to be replaced with wear. On the other hand, the resistance of the cap-to-adaptor interface may contribute to electrode heating and wear. Their use should be evaluated for each application compared to the one-piece design.

Table 3.4
Recommended maximum electrode force for standard spot welding electrodes

Taper no.	Shank diam., in.	Face diam., in.	Maximum electrode force, lb
4	0.482	0.19	800
5	0.625	0.25	1500
6	0.750	0.28	2000
7	0.875	0.31	2400

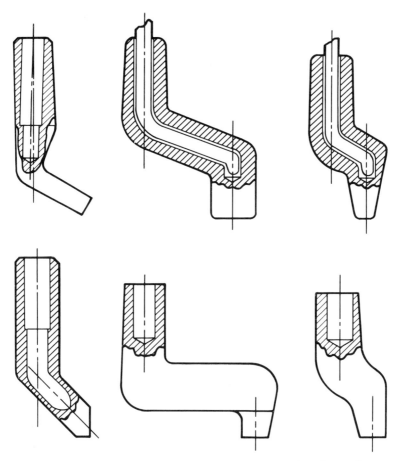

Fig. 3.29—Typical single and double bent spot welding electrodes

Method of Manufacture

Straight electrodes are machined from cold worked rods. Bent electrodes may be produced by cold forming of straight electrodes, by forging, or by casting. Forging or casting is normally used where the required shape cannot be produced by cold forming. Most bent electrodes are cold formed because they have distinct advantages over the others, including the following:

(1) The physical and mechanical properties of cold-drawn rod

(2) Placement of a water tube in the cooling hole prior to forming

(3) Lower manufacturing costs

Maintenance

A spot welding electrode has a specific face area in contact with the work. In use, this area will grow by mushrooming and the current and pressure densities will decrease at the same time. As a result, the weld will become smaller. In addition, the electrodes tend to pick up metal from the parts being welded. A small amount of pick up may not be harmful, but a considerable amount will cause the electrodes to overheat and mushroom faster.

It is not possible to predict how many welds can be made with a given setup before redressing of the electrodes is necessary. A periodic check of the weld quality as well as the electrode

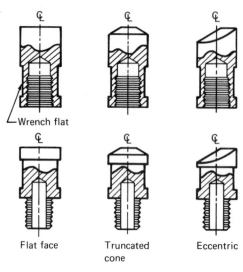

Fig. 3.30—Typical threaded spot welding electrodes

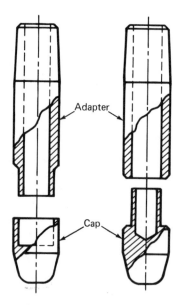

Fig. 3.31—Male and female designs of two-piece spot welding electrodes

shape will help in determining the number of welds or assemblies that can be made before redressing. Then, a schedule of electrode redressing should be set up as preventive maintenance to maintain weld quality.

A minor amount of redressing of electrodes in the machine is permissible using a plastic or metal paddle contoured on both sides to match the electrode face contour. The paddle is wrapped with fine abrasive cloth. The electrodes are brought against the abrasive cloth under a light load. The paddle is then rotated to redress the electrode faces.

Where a major amount of redressing is necessary, the electrode should be removed from the machine and refaced on a lathe. Alternatively, major redressing of the electrode may be done in the machine with a manual or power-operated dressing tool.

A file should never be used for redressing electrodes in the machine because the resulting electrode faces may be irregular in size and contour. Poorly dressed electrodes will reduce the quality of weld.

The following suggestions may be helpful in correctly applying spot welding electrodes.

(1) Use standard electrodes and holders wherever possible.

(2) Use the proper electrode material recommended for the application and the metal to be welded.

(3) Use adequate water cooling and circulate it in the correct direction in the electrodes.

(4) Align the electrodes properly. Electrodes should not skid against the parts or be out of alignment when they are in contact with the parts.

(5) Use only rawhide or rubber mallets for tapping electrodes into position and only ejector type holders or the proper tools for removing electrodes from the machine.

(6) See that the machine is set up properly. The electrodes must contact the parts with minimum impact before current flows and must remain in contact until termination of the current.

Specifications and Identification

Spot welding electrodes are covered by two standards:

(1) ANSI/RWMA Bulletin No. 16, *Resistance Welding Standards,* published by the Resistance Welder Manufacturers Association.

(2) AWS D8.6/SAE HS-J1156, *Standard for Automotive Resistance Spot Welding Electrodes,* published by the American Welding Society and the Society of Automotive Engineers.

These standards provide a code system for the various standard electrode designs. The code identifies the nose style, alloy class, shank size, and length. Methods are also given to identify bent electrode shapes, special-faced electrodes, and cap electrodes.

Straight electrodes are identified by a letter followed by four numbers with the following meanings:

(1) The letter indicates the nose style as shown in Fig. 3.25.

(2) The first digit indicates the Group A alloy class as shown in Table 3.3.

(3) The second digit indicates the taper.

(4) The third and fourth digits indicate the overall length in 0.25-in. units.

For single bent electrodes, two digits are placed ahead of the letter to indicate the bend angle in degrees. For single and double bent electrodes, two additional digits are added to indicate the offset distance in 0.062-in. units.

ELECTRODE HOLDERS

Electrodes are mounted on a spot welding machine by means of electrode holders. Various holder designs permit positioning the electrodes properly with respect to the work. The holders are clamped to the arms of the welding machine. Most of them have provisions for conducting cooling water to the electrodes, and some have an ejector mechanism for easy removal of the electrode.

There are three fundamental holder designs: straight, offset, and universal or adjustable offset. These three basic types are available in standard sizes and designs for use with standard spot welding electrodes. Similar design principles are generally employed for special holders, with or without adaptors, for use with a great variety of special or standard electrodes.

The three types of standard holders are available as nonejector and ejector types. Straight electrode holders of both types are shown in Fig. 3.32. With the ejector type, the electrode is removed by striking the ejector head or button with a hammer. With the nonejector type, the electrode taper is released by rotating the electrode with a wrench. Holders are available in different lengths and several diameters.

Offset and universal holders are produced with 90° and 30° heads as shown in Fig. 3.33. Low inertia holders which incorporate a spring for rapid follow-up are also available.

From the many available holder and electrode designs, it usually is possible to find a combination to fit most requirements. Examples of various electrode and holder combinations are also shown in Fig. 3.33.

Multiple electrode holders are available for producing two or more spot welds simultaneously in parallel. These holders have spring, mechanical, or hydraulic force equalizing systems. The lower electrode may be a flat block that opposes all upper electrodes or individual electrodes mounted in a block. Since the welds are made by parallel circuits, the proper division of current to each weld will depend upon the relative resistances of the paths. The path of lowest resistance will conduct more current than the others, and weld size may vary with the current magnitude.

PROJECTION WELDING ELECTRODES

Projection welding electrodes must have flat surfaces that are larger than the projection diameter. It is common practice to use large, flat electrodes or rectangular bar stock.

Projection welding electrodes are frequently composed of an internally water-cooled holder with replaceable inserts at the projection loca-

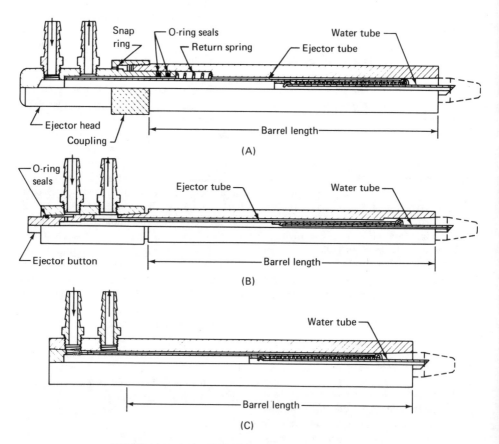

Fig. 3.32—Typical straight spot welding electrode holders: (A) and (B) ejector types, (C) nonejector type

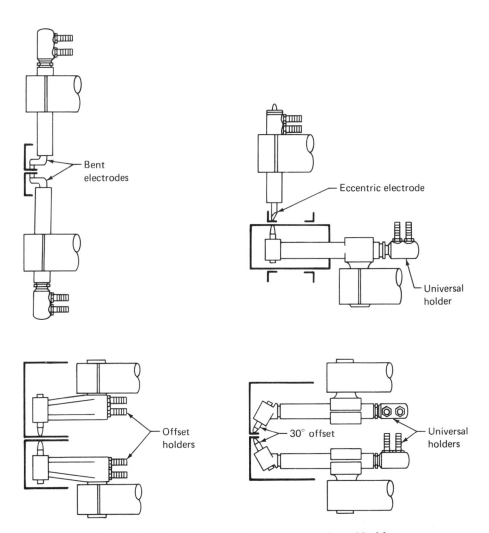

Fig. 3.33—Various combinations of electrodes and holders

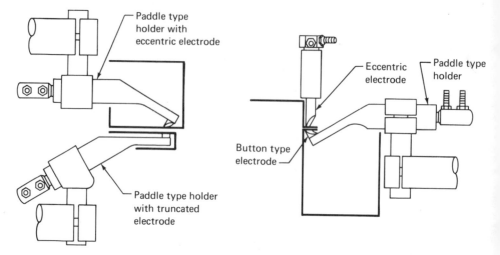

Fig. 3.33 (cont.)—Various combinations of electrodes and holders

tions. These inserts may be threaded electrodes or pieces of Group A or B electrode materials pressed or otherwise secured in the holder. An example of this design is shown in Fig. 3.34.

Since the area of contact between each electrode and the adjacent part is larger than in spot welding, current and pressure densities are lower. Therefore, electrode deterioration from wear, deformation, or pickup is not nearly so rapid as with spot welding. The electrodes do, however, eventually become pitted or deformed at the projection weld locations. When this interferes with the proper electrode contact or weld quality, the electrodes or inserts must be redressed or replaced.

Selecting the best combination of opposing electrode materials for good heat balance will minimize deterioration. Regular cleaning of the electrodes to remove grease, dirt, flash, or other contamination will prolong electrode life.

Multiple projection welding electrodes can be designed to automatically compensate for height variations or wear. These equalizing electrodes generally employ some hydraulic or mechanical method to provide automatic floating or equalizing features.

SEAM WELDING ELECTRODES

Seam welding electrodes are wheels or disks. The five basic considerations are face contour, width, diameter, cooling, and method of mounting. The diameter and width of the wheel are usually dictated by the thickness, size, and shape of the parts. The face contour depends upon the requirements for current and pressure distribution in the weld nugget and the type of drive mechanism. The four basic face contours in common use are flat, single-bevel, double-bevel, and radius as shown in Fig. 3.35.

The electrodes are usually cooled by either flooding or directing jets of water on both of the electrodes and the work from top and bottom. Where that method of cooling is unsatisfactory the electrodes and shafts can be designed for internal cooling.

Cooling by simple flooding alone is not always adequate. A steam pocket may develop at the point where the electrode meets the work and keep cooling water from the immediate area. When flood cooling is unsuitable, water mist or vapor cooling may be effective.

A seam welding electrode is generally attached to the shaft with a sufficient number of bolts or studs to withstand the driving torque. The contact area with the shaft must be great enough to transmit the welding current with minimum heat generation.

Peripheral drive mechanisms, such as knurl or friction drives running against the electrode,

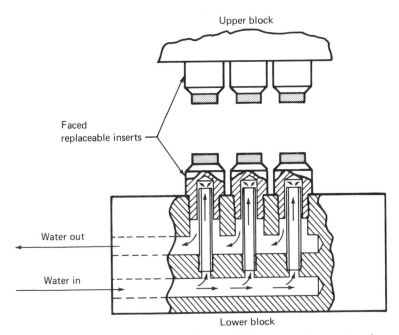

Fig. 3.34—*Typical projection welding multiple electrode construction*

require adequate work clearance. A knurl drive will mark the electrode face, which in turn will mar the surface of the weld. However, a knurl drive wheel tends to clean surface pickup from the electrode face.

Although the work and drive may require flat-faced electrodes with or without beveled edges, they are more difficult to set up, control, and maintain than radius-faced electrodes. In addition, radius faces give the best weld appearance.

Seam welding electrodes, like spot welding electrodes, have a predetermined area of contact with the parts that must be held within limits if consistent weld quality is to be maintained. Only minor dressing or touch up with light abrasives should be attempted with the electrode in the machine. Wheel dressers may be used for continuous electrode maintenance. Machining in a lathe is the preferred method of redressing an electrode to its original shape.

Precautions must be taken to prevent foreign

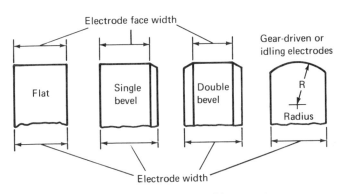

Fig. 3.35—*Seam welding wheel face contours*

materials from becoming embedded in the electrode wheel or work. Rough faces do not improve traction. Welding should be stopped while the electrodes are still on the work.

FLASH AND UPSET WELDING ELECTRODES

Flash and upset welding electrodes usually are not in direct contact with the weld area as are spot and seam welding electrodes. They function as work-holding and current-carrying clamps, and are often referred to as such. They are normally designed to contact a large area of the workpiece, and the current density in the contact area is relatively low. Accordingly, relatively hard electrode materials with low conductivity can give satisfactory performance.

Since the electrodes must conform to the parts to be welded, there are no standard designs. Two important requirements are that the materials have sufficient conductivity to carry the current without overheating, and that the electrodes be rigid enough to maintain work alignment and minimize deflection.

The electrodes are mechanically fastened to the welding machine platen. They can be solid, one-piece construction of one the RWMA Group A electrode materials of Classes 1 through 5. Service life can sometimes be increased by using Class 2, 3, and 5 material with replaceable inserts of Class 3, 4, or one of the Group B materials at the wear points.

A varying amount of wear inevitably occurs, and this may result in decreased contact area and localized burning of the work. The electrodes should be kept cool, clean, and free of dirt, grease, flash, and other foreign particles for good service. An antispatter compound may help prevent flash adherence. All fasteners and holding devices should be tight and properly adjusted, and their gripping surfaces should be properly maintained to avoid work slipping during welding.

POWER SUPPLY

Power demand from the line depends upon the welding method and the design of the welding machine. An adequate power supply is one of the prerequisites for high-production resistance welding. A major part of the power supply system for an industrial plant is within the plant itself. That part consists of the power supply transformers and conductors.

POWER SUPPLY TRANSFORMERS

In considering the installation of a resistance welding machine, it is necessary to determine if the plant supply is adequate. This includes the kVA rating of the power supply transformer and the size of the power supply conductors. The power supply transformer is usually connected to a 2300 or 4800 V primary feeder and produces 230 or 460 V power. It should not be confused with the welding transformer mounted in the welding machine. The power supply conductors are the leads between the power supply transformer and the welding machine.

The adequacy of the power supply transformer and supply conductors is governed by two factors: the permissible voltage drop and the permissible heating. The permissible voltage drop is the determining factor in the majority of installations, but consideration must be also given to heating.

It is relatively simple to determine the size of a power transformer for a single welding machine on the basis of heating alone. The welding machine transformer is rated on a heating basis and a definite ratio exists between the two ratings. Power transformers are usually rated for

Table 3.5
Equivalent continuous loading of
resistance welding machines

Type of welding	Equivalent continuous load, percent of sum of name-plate ratings
Spot, projection (single-impulse)	20
Spot, projection (multiple-impulse)	40
Flash, multipoint spot, or projection	20
Seam	70

continuous or 100 percent duty cycle operation compared to 50 percent for a resistance welding transformer. The equivalent rating of a power transformer required to supply a given welding machine on a heating basis only will be equal to 70.7 percent of the welding transformer rating at 50 percent duty cycle. For example, the size of the power transformer required for the proper operation of a 200 kVA seam welding machine is 141.4 kVA. For a single machine, the next largest size power transformer would be used.

If more than a single machine is to be served from a common power supply transformer, a study must be made of actual operating duty cycles for all machines as well as of the operating diversity factor[3] between machines. Welding machines normally are operated below their maximum thermal capacity. The approximate continuous loading for a number of similar welding machines, expressed as the percentage of the total kVA of the welding machine ratings, is given in Table 3.5. For example, if the machines are predominately spot or projection welding types producing single-impulse welds, a power supply transformer with a rating of about 20 percent of the sum of the machine nameplate ratings would, in most cases, be adequate with respect to permissible heating.

To determine the size of the power supply transformer required to serve a welding machine on the basis of voltage drop, it is first necessary to determine the maximum permissible voltage drop specified by the machine manufacturer. Normally, it should not be greater than 5 percent. When the same power transformer is used with two or more machines, the voltage drop caused by one machine will be reflected in the operation of the second. Then, it is advisable to confine the total voltage drop to not more than 5 percent for consistent weld quality. Voltage drop should be measured at the machine location. The percentage voltage drop is calculated by the following formula:

$$\text{Voltage drop, \%} = \frac{(\text{No-load voltage}) - (\text{Full-load voltage})}{\text{No-load voltage}} \times 100$$

BUS OR FEEDER SYSTEM

In general, the bus or feeder from the transformer to the machines should always be as short as possible and of low reactance design to minimize the voltage drop in the line. The simplest and most economical power line consists of insulated wires taped together in a conduit. When only two or three machines are to be served at a common location, this construction is economical and effective. Bus duct construction that permits easy tap connections at frequent intervals along its length is desirable in production plants where manufacturing layouts are continually changing.

Systems are available to interlock two or more machines to prevent simultaneous firing and the accompanying excessive voltage drop. Any scheme of interlocking will cause some curtailment in production. However, this can be minimized with a voltage monitoring type which operates only when the voltage drops below a preset value.

INSTALLATION

Resistance welding machines should be connected to the power line in accordance with the applicable electrical codes and the recommendations of the machine manufacturer. The primary cable size should be adequate on both a thermal and a voltage drop basis.

Because many control units contain phase-

3. Diversity factor is the ratio of the sum of the individual maximum demands of the various machines to the maximum demand of the whole installation.

shift heat control, the control power must be in phase with the welding power. The control power should be fused separately from the welding power.

Enclosed fusible isolation switches are frequently used for the power or welding circuit. These switches seldom have adequate interrupting capacity for safe disconnection under load. For emergency disconnecting purposes, it is advisable to use a circuit breaker. The rating of the breaker in carrying capacity should be sufficient to carry the maximum demand of the machine when its welding circuit is shorted. It may be

from two to four times the machine nameplate rating. One of the advantages of a circuit breaker is that a pushbutton can be located on the welding machine. The operator can quickly open the circuit in an emergency by hitting this button.

When fuses are used, their size should be that recommended by the machine manufacturer. The manufacturers normally provide wiring diagrams in which recommended fuse ratings are shown. The fuses should function for any normal demand or operation of the machine. The purpose of fuses is almost solely to interrupt a short circuit in the electrical system.

SAFETY

Resistance welding processes are widely used in high production operations, including the automobile and appliance industries. These processes include projection, spot, seam, flash, upset, and percussion welding in a wide range of machine types. The main hazards which may arise with the processes and equipment are as follows:

(1) Electric shock due to contact with high voltage terminals or components

(2) Ejection of small particles of molten metal from the weld

(3) Crushing of some part of the body between the electrodes or other moving components of the machine

MECHANICAL

Guarding

Initiating devices on welding equipment, such as push buttons and switches, should be arranged or guarded to prevent the operator from inadvertently activating them.

In some multiple-gun welding machine installations, the operator's hands can be expected to pass under the point of operation. These machines should be effectively guarded by a suitable device such as proximity-sensing devices,

latches, blocks, barriers, or dual hand controls.

All non-portable, single-ram welding machines should be equipped with one or a combination of the following:

(1) Machine guards or fixtures which prevent the operator's hands from passing under the point of operation.

(2) Dual-hand controls, latches, proximity-sensing devices, or any similar mechanism which prevents operation of the ram while the operator's hands are under the point of operation

All chains, gears, operating linkages, and belts associated with the welding equipment should be protected in accordance with ANSI Standard B15.1, *Safety Standard for Mechanical Power Transmission Apparatus* (latest edition).

Static Safety Devices

On press type, flash, and upset welding machines, static safety devices such as pins, blocks, or latches should be provided to prevent movement of the platen or head during maintenance or setup for welding. More than one device may be required, but each device should be capable of sustaining the load.

Portable Welding Machines

Support Systems. All suspended portable welding gun equipment, with the exception of the gun assembly, should have a support system that is capable of withstanding the total shock load in the event of failure of any component of the system. The system should be fail safe. The use of adequate devices such as cables, chains, or clamps is considered satisfactory.

Movable Arm. Guarding should be provided around the mounting and actuating mechanism of the movable arm of a welding gun if it can cause injury to the operator's hands. If suitable guarding cannot be achieved, two handles should be used. Each handle should have an operating switch that must be actuated to energize the machine. These handles must be located at safe distances from any shear or pinch points on the gun.

Stop Buttons

One or more emergency stop buttons should be provided on all welding machines, with a minimum of one at each operator position.

Guards

Eye protection against expelled metal particles must be provided by a guard of suitable fire-resistant material or by the use of approved personal protective eye wear. The variations in resistance welding operations are such that each installation must be evaluated individually. For flash welding equipment, flash guards of suitable fire-resistant material must be provided to control flying sparks and molten metal.

ELECTRICAL

Voltage

All external weld initiating control circuits should operate on low voltage. It should not be more than 120 V for stationary equipment and 36 V for portable equipment.

Capacitors

Resistance welding equipment and control panels containing capacitors involving high voltages must have adequate electrical insulation and be completely enclosed. All enclosure doors must be provided with suitable interlocks, and the contacts must be wired into the control circuit.

The interlocks must effectively interrupt power and discharge all high voltage capacitors into a suitable resistive load when the door or panel is open. In addition, a manually operated switch or suitable positive device should be provided to assure complete discharge of all high voltage capacitors.

Locks and Interlocks

All doors, access panels, and control panels of resistance welding machines must be kept locked or interlocked. This is necessary to prevent access by unauthorized persons.

Grounding

The welding transformer secondary should be grounded by one of the following methods:

(1) Permanent grounding of the welding secondary circuit

(2) Connection of a grounding reactor across the secondary winding with a reactor tap to ground

As an alternative on stationary machines, an isolation contactor may be used to open all of the primary lines.

The grounding of one side of the secondary windings on multiple spot welding machines can cause undesirable transient currents to flow between transformers when either multiphase primary supplies or different secondary voltages, or both, are used for the several guns. A similar condition can also exist with portable spot welding guns when several units are used on the same fixture or assembly or on another that is nearby. Such situations require use of a grounding reactor or isolation contactor.

INSTALLATION

All equipment should be installed in conformance with the ANSI/NFPA No. 70, *National Electric Code* (latest edition). The equipment should be installed by qualified personnel under the direction of a competent technical supervisor. Prior to its production use, the equipment should be inspected by competent safety personnel to ensure that it is safe to operate.

Additional information on safe practices for resistance welding equipment may be found in ANSI Z49.1, *Safety in Welding and Cutting* (latest edition).

Metric Conversion Factors

$$1 \text{ in} = 25.4 \text{ mm}$$
$$1 \text{ lb force} = 4.45 \text{ N}$$
$$1 \text{ psi} = 6.89 \text{ kPa}$$
$$1 \text{ W·s} = 1 \text{ J}$$

SUPPLEMENTARY READING LIST

Beemer, R. D. and Talbot, T. W., Analyzer for nondestructive process control of resistance welding. *Welding Journal,* 49(1): 9s-13s; 1970 Jan.

Blair, R. H. and Blakeslee, R. C., Half-wave and full-wave resistance welding power supplies. *Welding Journal,* 50(3): 174-6; 1971 Mar.

Dilay, W. and Zulinski, E., Evolution of the silicon-controlled rectifier for resistance welding. *Welding Journal,* 51(8): 554-9; 1972 Aug.

Johnson, K. I., ed., *Resistance Welding Control and Monitoring,* Cambridge, England: The Welding Institute, 1977.

Mollica, R. J., Adaptive controls automate resistance welding. *Welding Design and Fabrication,* 51(8): 70-72; 1978 Aug.

Parker, F., The logic of dc resistance welding. *Welding Design and Fabrication,* 49(12): 55-58; 1976 Dec.

Sherbondy, G. M. and Motto, J. W. Jr., Current ratings of power semiconductors. *Welding Journal,* 51(6): 393-400; 1972 June.

4

High Frequency Welding

Chapter Committee

W. C. RUDD, *Chairman*
Thermatool Corporation

E. D. OPPENHEIMER
Consulting Engineer

G. A. SMITH, JR.
Allegheny Ludlum Steel Corporation

R. E. SOMERS
Welding Consultant

Welding Handbook Committee Members

A. F. MANZ
Union Carbide Corporation

D. D. RAGER
Reynolds Metals Company

4

High Frequency Welding

FUNDAMENTALS OF THE PROCESS

DEFINITIONS AND GENERAL DESCRIPTION

High frequency welding includes those processes in which coalescence of metals is produced by the heat generated from the electrical resistance of the work to the flow of high frequency current with or without the application of an upsetting force. There are two processes that utilize high frequency current to produce the heat for welding: namely, (1) high frequency resistance welding (HFRW), and (2) high frequency induction welding (HFIW), sometimes called induction resistance welding.

The heating of the work in the weld area and the resulting weld are essentially identical with both processes. With HFRW, the current is conducted into the work through electrical contacts that physically touch the work. With HFIW, the current is induced in the work by coupling with an external induction coil. There is no physical/electrical contact with the work.

With conventional resistance welding processes, the current is normally 60 Hz alternating or direct current. High amperages are required to resistance heat the metal, and large electrical contacts must be in close proximity to the desired weld area. The voltage drop across the weld is low, and the current flows along the path of least resistance from one electrode to the other.

With high frequency welding, the current flow is concentrated near the surface of the part. The depth of current penetration is influenced by the current frequency and also the resistivity

and magnetic properties of the metal. As the frequency is increased, the depth of penetration decreases and the current is more concentrated. Heating to welding temperature can be accomplished at much lower amperages than with conventional resistance welding. The location of this concentrated current path in the part can be controlled by the location of the electrical contacts with HFRW or the design and position of the induction coil with HFIW.

While the welding process depends upon the heat generated from the resistance of the metal to the flow of high frequency current, other factors must also be considered for successful high frequency welding. The speed of welding depends upon the type of metal and its thickness at the joint. Flux is not normally used except when welding some brasses. Inert gas shielding of the welding area is needed only for joining metals, such as titanium, that react very rapidly with oxygen and nitrogen. Welding of carbon steels and many other alloys can usually be accomplished while water or soluble oil coolant flows over the actual weld area. High welding speeds make the process attractive for many high production applications, such as the manufacture of pipe and tubing.

The basic high frequency welding applications are shown in Fig. 4.1.

PRINCIPLES OF OPERATION

High frequency current in metal conductors tends to flow at the surface of the material at relatively shallow depths. This is commonly

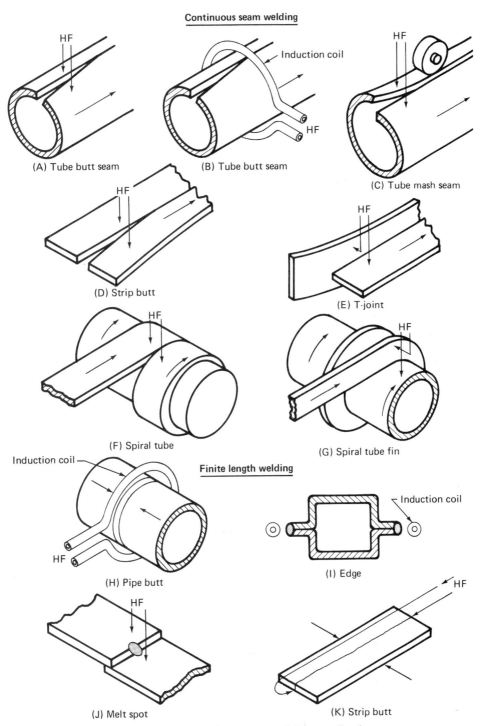

Fig. 4.1—Basic high frequency welding applications

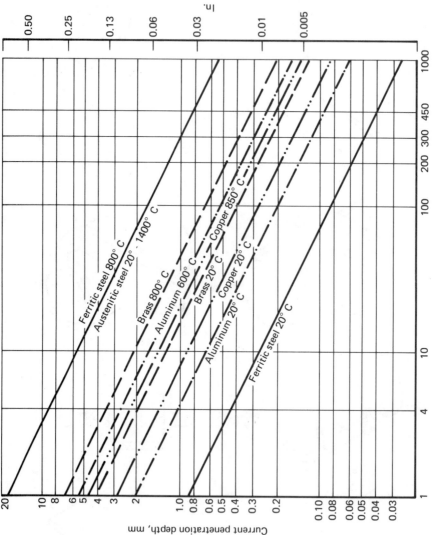

Fig. 4.2—The effect of frequency on depth of current penetration into various metals at selected temperatures

called *skin effect*. Figure 4.2 shows the decrease in depth of penetration of high frequency current into various metals with increasing frequency at room and elevated temperatures. For example, the depth of current penetration in steel at 800°C is about 0.03 in. with 500 kHz and nearly 0.25 in. with 10 kHz.

High frequency current may have its path within the workpiece controlled by either variations in the inductance of the circuit or the nearness of its own return flow path, or both. This phenomenon, called the *proximity effect,* is illustrated in Fig. 4.3.

Both the skin effect and proximity effect become more pronounced with increasing frequency. Therefore, the effective resistance of the current path in the work increases as a function of increasing frequency. As frequency is in-

creased, the current is confined to a smaller area near the surface. This concentration of current is advantageous because extremely high heating rates can be achieved in a very localized path.

Figure 4.4 compares current patterns in the work at frequencies of 60 and 10 000 Hz. It also shows the effect of the shape and location of the proximity conductor. In Fig. 4.4(A), a 1/4-in. thick steel plate is acting as one conductor for 60 Hz current. The return conductor is a copper tube positioned very close to and parallel with the plate. The 60 Hz current flowing in the steel plate travels opposite to the current in the adjacent proximity conductor. In this 60 Hz case, the size and the shape of the proximity conductor have a negligible effect on the distribution of the current in the steel plate. As a

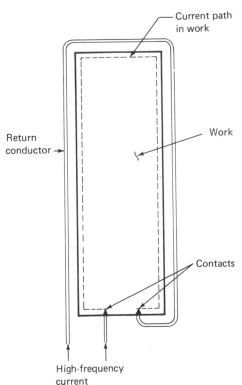

Fig. 4.3—Restriction of the flow path of high frequency current by the proximity effect of the return conductor

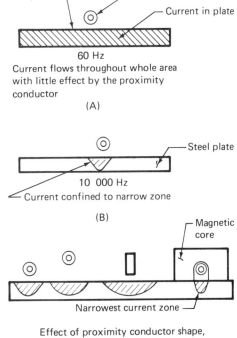

Fig. 4.4—Current depth and distribution adjacent to various proximity conductors

result, the 60 Hz current flows almost completely through the plate cross-sectional area, as shown.

When 10 000 Hz current is applied to the same system, as shown in Fig. 4.4(B), the current in the work is confined to a relatively narrow band immediately beneath the proximity conductor. This narrow band is the region of lowest inductive reactance in the plate and the closest path to the current in the proximity conductor.

The shape and magnetic surroundings of the proximity conductor have considerable effect on the distribution of the current in the work but have no effect on the depth of current penetration. Figure 4.4(C) shows two round proximity conductors at different distances from the work. The closer the proximity conductor, the more confined the current. Also shown is a rectangular proximity conductor with the narrow edge at the same distance from the work as the closest circular proximity conductor. In this case, the distribution of current in the work will be quite broad, as shown, because of the height of the conductor.

If both conductors are sheets placed edge-to-edge in a plane with a small gap between them, the proximity effect will cause the two adjacent edges to heat. The skin effect will confine the current to the metal at those edges.

Proper placement of a laminated magnetic core on three sides of the conductor confines the current in the work to a very narrow line and produces heating directly beneath the proximity conductor.

Almost all high frequency welding techniques employ some force to bring the heated metals into close contact during coalescence. During the application of force, an upset (or bulging of metal) occurs in the weld area.

PROCESS VARIATIONS

HIGH FREQUENCY RESISTANCE WELDING

High frequency resistance welding has three specific variations. These are:
(1) Continuous seam welding
(2) Finite length butt welding
(3) Melt welding

Continuous Seam Welding

Continuous HFR seam welding is generally used to weld long length products. Typical examples are shown in Fig. 4.1. High frequency current of about 400 kHz is introduced into the work with a pair of small sliding contacts. They are on either side of the seam to be welded as shown in Fig. 4.5. The faying surfaces are brought together at an angle of about 4 to 7 degrees. The apex of the vee-shaped opening is downstream from the contact, in the direction of travel.

The high frequency current follows a localized path down one side of the vee and back along the other due to the skin and proximity effects. The resistance of the metal to the flow of current heats the edges to a shallow depth. Welding speed and power level are adjusted so that the two edges are at welding temperature when brought together. At that point, pressure rolls forge the hot edges together and upset them to produce a weld. Hot metal, containing impurities from the faying surfaces of the joint, is squeezed out in both directions. The upset metal is normally trimmed off flush with the base metal.

Finite Length Welding

Techniques are available for welding the abutting ends of two strips together. This is done by passing a high frequency current through the joint area. The high frequency current is confined to the joint area by proper positioning of a return or proximity conductor over the weld joint. By selection of the proper frequency, the depth of penetration of the current can be adjusted to heat the joint uniformly through its thickness. When the joint reaches welding temperature, forging force is applied and the hot metal upsets. Joints with small upsets are made in this fashion at very high rates of speed. Figure 4.6 shows the arrangement of the welding head for this type of welding.

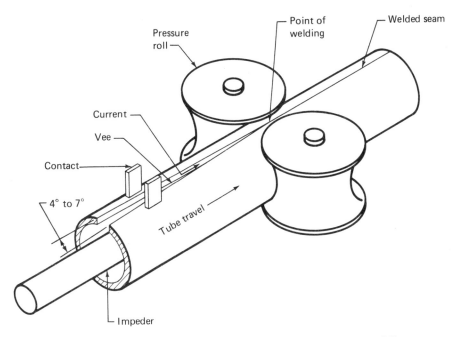

Fig. 4.5—*Joining a tube seam by high frequency resistance welding*

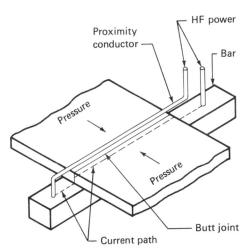

Fig. 4.6—*Joining strips together using high frequency resistance welding*

Melt Welding

Another version of the HFR welding process is called spot or line melt welding. In melt welding, the high frequency current melts a small volume of metal between electrodes that contact the parts, usually in less than 1 second. The molten metal flows together producing a welded joint between the parts. At this time, the process is in the early stages of commercialization. One of the intended uses is the welding of stacks of electrical motor or transformer laminations together. A proximity conductor placed perpendicular to the steel laminations confines current flow to a very narrow, shallow path across the stack of laminations, as shown in Fig. 4.7. Figure 4.8 shows a welded lamination stack and Fig. 4.9 is a section through a typical weld joining a stack together.

HIGH FREQUENCY INDUCTION WELDING

Tube Seam Welding

This application of HFIW is similar to continuous HFR seam welding except that the current is magnetically induced into the work. There are no electrical contacts. It can only be used where there is a complete current path or closed loop wholly within the work. The induced

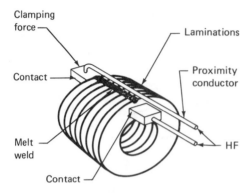

Fig. 4.7—HFR melt welding of steel motor laminations

Fig. 4.8—Melt welds on a motor lamination stack

current circulates through the weld area as well as, by necessity, through other portions of the work. Its primary application is for welding tubing. The tube edges are brought together in the same manner as in HFRW. An inductor or induction coil of copper tubing or bar circles the tube at the open end of the vee, as shown in Fig. 4.10. High frequency current in the inductor induces a circulating current in the tube. The current flows around the outside surface of the tube and along the edges of the vee. This current heats the edges to welding temperature. The weld is consummated in the same manner as with HFR tube welding. The induction method of tube welding is advantageous for coated tubing, small or thin-wall tubing, and avoids surface marking by electrical contacts. With HFIW, there is current flowing around the outside of the tube that is electrically in series with the welding current flowing in the vee. The resistive heat losses over the circumference of the tube are higher than those with HFRW. In general, the HFIW process is less efficient than the HFRW process, particularly when welding large sizes of pipe and tube.

Welding of Hollow Pieces

HFIW of individual pieces can be done only when the induced current can circulate in a closed-circuit path. A typical application is the welding of butt joints between sections of pipe

or tubing. The two ends are pressed together with a force sufficient to upset the metal when it reaches welding temperature. An induction coil is placed around the joint. High frequency current in the coil induces a circulating current in the butted pipe joint. Resistance heating due to the induced current heats the ends of both pieces very rapidly. When the metal reaches welding temperature, the metal is upset to produce a solid-phase forge weld. The upset metal is usually left in place. Figure 4.1(H) shows the placement of the coil. This process is used for welding tubes with diameters of 1 to 3 in. and wall thicknesses up to 0.375 in. for the fabrication of high-pressure boilers. Welding times may range from 10 to 60 seconds.

Welding With a Magnetic Pulse

Another form of HFIW utilizes a magnetic pulse to forge the two parts together at welding temperature. It is applicable to lap or sleeve

Fig. 4.9—Photomicrograph of section through a melt welding lamination stack (×50)

joints between two pipes. After the lap joint reaches welding temperature, a pulse of current from a capacitor bank is discharged through the inductor surrounding the joint. This current pulse of 10 kA to 150 kA lasts for approximately 50 microseconds. The rapidly rising and falling current induces a current in the outer workpiece that flows in the opposite direction to that in the coil. The two currents produce very large magnetic forces that repel each other. This motor effect drives the hot outer piece against the hot inner piece and forges them together, producing a welded sleeve joint. Steel-to-steel and copper-to-aluminum joints may be made by this process.

ADVANTAGES AND LIMITATIONS

The high frequency welding processes offer several advantages over the conventional 60 Hz resistance welding processes. One characteristic of the processes is that they can produce welds with very narrow heat-affected zones. The high frequency welding current tends to flow only near the surface of the metal because of the skin effect. Also, it flows along a narrow path because of the proximity effect. The heat for weld- ing, therefore, is developed in a small volume of metal along the edges to be joined. A narrow heat-affected zone is desirable because it tends to give a stronger welded joint than with a wider zone produced by many other welding processes. Any molten metal is squeezed out of the joint during the upsetting or forging portion of the cycle. Consequently, any low-melting phases that normally contribute to weld cracking are

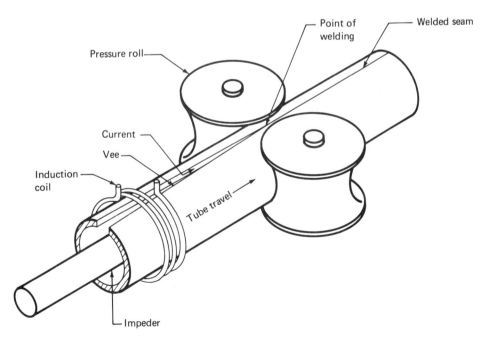

Fig. 4.10—Joining a tube seam by high frequency induction welding

eliminated. With some alloys, the narrow heat-affected zones and absence of cast structure may eliminate the need for postheat treatments. These treatments normally improve the metallurgical characteristics of welded joints.

One major advantage of HFW is the ability to weld a given tube shape with less power than required with 60 Hz power. Also, it can be used to weld very thin tube forms. Wall thicknesses of 0.030 in. and thinner can be butt welded by either the induction or the contact process.

Another advantage of a high frequency process is that it reduces electrical contact problems. Introducing currents of 1000 to 2000 A through rubbing contacts is much easier at 450 kHz than at 60 or even 10 000 Hz. Because nearly all metals have an oxide on their surfaces, commercial metal products may have an oxide film or scale of appreciable thickness. At low frequencies, the contact has to break through this skin and make contact with the metal; otherwise, considerable heating will result from the contact resistance. With high frequency power, the voltage is high and there is continuous arcing or puncturing of surface films under the contact. Most of the current is carried by the arc. In any event, the contact does not need to break through the oxide film. Only enough pressure is needed on the contacts to assure continuous contact with the surface. This feature makes it possible to weld thin materials without bending or collapsing them. Contact wear is not a major problem.

As much as 60 percent of the energy drawn from the power lines can appear as useful heat in the work. Most high frequency welding machines derive their energy from a balanced three-phase input power system.

Because the time at welding temperature is very short and the heat is localized, oxidation and discoloration of the metal as well as distortion of the parts are minimized. Metal sheaths used to shield materials that would normally be damaged from prolonged exposure to heat can be fabricated with one of the high frequency welding processes. An example of this is the continuous welding of electrical cable sheathing with the cable inside of it.

As with any process, there are also limitations. Because the process utilizes localized heating in the joint area, proper fitup is important. Equipment is usually incorporated into mill or line operation and must be fully automated. The process is limited to the use of coil, flat, or tubular stock with a constant joint symmetry throughout the length of the part. Any disruption in the current path or change in the shape of the vee can cause significant problems. Also, special precautions must be taken to protect the operators and plant personnel from the hazards of high frequency current.

Since the equipment operates in the broadcast radio frequency range, special care must be taken in its installation, operation, and maintenance to avoid radiation interference in the plant vicinity. The manufacturer's recommendations should be followed.

APPLICATIONS

GENERAL

High frequency resistance welding, where the current is introduced by means of direct contact, may be utilized for a variety of configurations and types of welding. It may be used for continuous seam welding and for finite length welding where the entire seam to be welded is heated uniformly.

Metals that may be welded with continuous high frequency resistance welding include carbon steel, alloy steel, copper, aluminum, zirconium, titanium, and nickel. Finite length welding is usually limited to the welding of carbon steels and stainless steels. The various products in Fig. 4.11 illustrate the types of joints that may be welded.

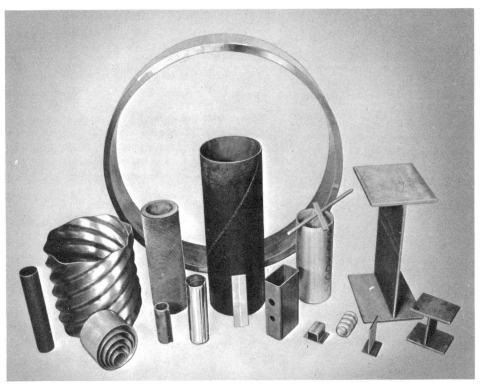

Fig. 4.11—Some products produced by high frequency welding

Numerous factors must be considered in determining whether a metal is weldable in a given configuration by any high frequency welding process. Of prime importance is the requirement that a properly shaped vee be obtainable for continuous seam welding. In general, it might appear that high-strength alloys are more difficult to weld because of problems in producing a properly shaped vee. This is not the case. Other properties of the metal being joined appear to have greater influence on weldability than its strength. In general, it appears that the carbon steels are quite weldable. On the other hand, some of the aluminum alloys appear to have rather poor weldability. Welding conditions have to be held in a narrow range to ensure good welds.

High frequency resistance welding can be used to weld dissimilar metals together, some examples of which follow:

(1) Double-walled tubing made of two types of steel to obtain unusual sonic properties

(2) Carbon steel to OFHC copper to produce a strong electrical bus bar

(3) OFHC copper to an aluminum alloy to produce a high-strength, lightweight conductor

(4) Tool steel to carbon steel for band saw blades

(5) Beryllium copper to brass for electrical contacts

(6) Type 430 stainless steel to galvanized carbon steel for automotive trim

(7) Structural shapes of carbon and alloy steels for architectural use or strength improvement

(8) Clad steel to carbon steel sheet

High frequency induction welding can be used only where there is a closed-loop path or a complete circuit for the flow of current entirely within the work. Typical applications are shown in Figs. 4.1(B), (H), and (I).

PIPE AND TUBE WELDING

One of the major applications of the continuous high frequency welding process is high-speed welding of the longitudinal seam in pipe and tubing. It is applicable to both ferrous and nonferrous pipe and tubing. HFRW and HFIW are both applicable, as shown in Figs. 4.5 and 4.10. Heating of the edges can be done so rapidly that very high welding speeds are possible, much higher than with arc welding. Welding speeds for tubing or pipe of several wall thicknesses of steel and aluminum using HFRW and a 160 kW, 400 kHz power supply are given in Table 4.1. Figure 4.12 shows a large mill for the production of 24-in. diameter pipe having a wall thickness of 0.5 inch. Pipe sections are formed to diameter and fed into the welding station.

High frequency resistance welding is utilized to produce pipe and tubing of diameters ranging from 0.5 to over 50 in. and with wall thicknesses of from 0.010 through 1 inch. Any metal can be welded with this technique. Welding speeds range from 25 to over 1000 ft/min., depending upon the metal and product size.

High frequency induction welding is suitable for tubing made of any metal. Tubes may range in diameters from approximately 0.5 to 6 in. with wall thicknesses of from 0.006 to 0.375 inch. Welding speeds range from 25 to over 1000 ft/min., depending upon the application.

STRUCTURAL BEAMS

HFRW is used for the production of structural shapes, such as I and H beams, in sizes up

Fig. 4.12—Mill for high frequency resistance welding 24-inch diameter pipe

Table 4.1
Welding speeds for pipe or tubing of various wall thicknesses using high frequency resistance welding

Wall thickness, in.	Welding speed, ft/min. Steel	Aluminum
0.03	900	1,000
0.06	500	600
0.10	300	360
0.16	175	225
0.25	100	125

to 20 in. with webs up to 3/8-in. thick. In this type of operation, strips from three coils are fed into a welding mill. Simultaneously, two high frequency resistance welding machines make the two T-joints between the web and flanges. Welding speeds range from 25 to 200 ft/min. This process can be used to manufacture structural shapes that cannot be hot rolled. It permits the joining of strips with different mechanical properties, such as high-strength steel flanges to a low carbon steel web, and also the fabrication of sections of high-strength, low alloy steels. Figure 4.13 shows the arrangement of an I or H beam mill with the location of the contact and the straightening rolls enlarged for clarity.

SPIRAL PIPE AND TUBE

Spiral pipe and tube may be manufactured by HFR welding. In this case, strip is fed continuously into a forming mill that bends it helically (like a barber pole) into a cylindrical section. The seam is continuously butt or lap welded together by HFRW process. Figure 4.11

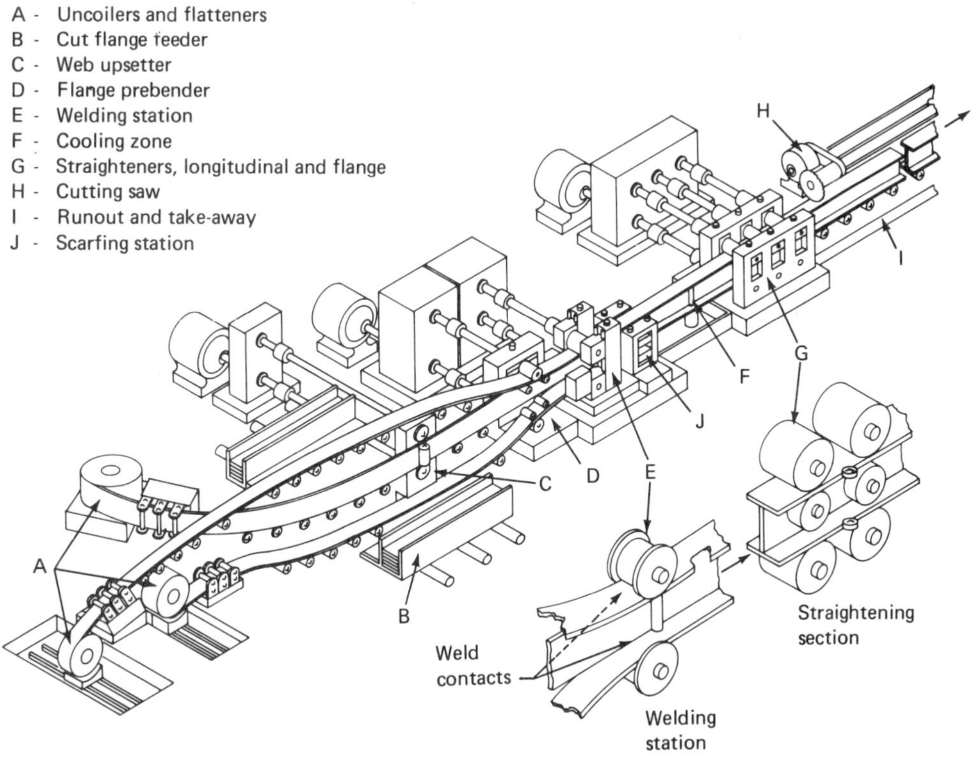
A - Uncoilers and flatteners
B - Cut flange feeder
C - Web upsetter
D - Flange prebender
E - Welding station
F - Cooling zone
G - Straighteners, longitudinal and flange
H - Cutting saw
I - Runout and take-away
J - Scarfing station
A
B
C
D
E
F
G
H
I
J
Weld contacts
Welding station
Straightening section

Courtesy of the Welding Research Council

Fig. 4.13—Mill arrangement for fabricating I or H beams by high frequency resistance welding

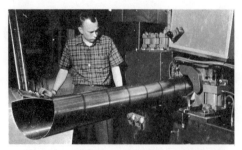

Fig. 4.14—High frequency lap welding spiral pipe

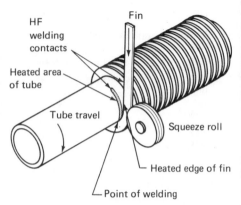

Courtesy of the Welding Research Council

Fig. 4.15—Arrangement for welding a spiral fin into tubing

shows an example of spiral welded pipe, and Fig. 4.14 shows the pipe emerging from the welding station of the mill.

FINNED TUBE

Spiral and longitudinal fins can be welded to tubing with the HFRW process. Figure 4.15 shows the arrangement for welding a spiral fin on tubing. Finned tube can be made in many combinations of metals, such as stainless steel tube with a mild steel fin, cupronickel tube with an aluminum fin, or mild steel tube with a mild steel fin. Tube diameters range from 0.625 to 10 inches. Fin pitch can vary from less than one to six turns per inch. Typical fin heights equal the radius of the tube. Fin thickness may be as large as 0.25 inch, except that it is limited by the thickness of the tube wall since the tube must resist the forging force on the fin. Various types of serrated or folded fins may also be used.

EQUIPMENT

POWER SOURCES

Units for providing high frequency power include motor-generators and solid-state inverters for frequencies up to 10 kHz, and vacuum tube oscillators for frequencies from 100 to 500 kHz. Vacuum tube oscillators are available in sizes ranging from 1 to 600 kW of power output with frequencies in the 200 to 500 kHz range. Figure 4.16 shows a basic circuit for oscillator type power sources in the 200 to 400 kHz range. The transformer and rectifier are used to convert plant line voltage to high voltage, direct current power for the oscillator circuit. The oscillator circuit converts direct current to high frequency current for the output transformer. The output transformer converts the high voltage, low am-

perage power to low voltage, high amperage power for welding.

For frequencies in the range of 3 to 10 kHz, welding power is usually produced by solid-

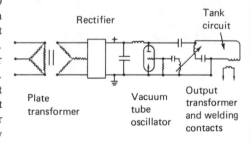

Fig. 4.16—Basic circuit of an oscillator type power source of the 200 to 400 kHz range

state inverters that operate on dc from a rectifier. These inverters have replaced motor-alternator or motor-generator power sources. The inverter portion of the equipment is a solid-state device in which silicon-controlled rectifiers are fired by a timing circuit and thus create high frequency current.

IMPEDANCE MATCHING TRANSFORMERS

Vacuum tube oscillators inherently have high output impedance and must be fed into high impedance loads. The inductors and the contact-workpiece circuits in high frequency welding are low impedance loads. An impedance matching transformer is required to transfer energy efficiently from the oscillator to the work. High frequency current can only transfer power efficiently from the high frequency generator to the work when the impedance of the work circuit matches the impedance of the power source.

The impedance matching transformer is placed as close as possible to the welding operation. It is connected by a transmission line directly to the oscillator, which is often at a convenient distance from the welding station. In some installations, capacitors are connected across the impedance matching transformer primary to improve both the power transfer and the electrical power factor.

The impedance matching transformer construction is quite simple. It usually consists of a primary winding of water-cooled copper tubing inside a cylindrical secondary winding of one or more turns of water-cooled copper sheet. To improve the magnetic coupling between the primary and secondary turns, a water-cooled, ferrite magnetic core may be used. Because the output voltage of the oscillator often is in the range of 10 to 25 kV, the insulation between the windings must be designed with care. The transformer may be insulated with air. In these instances, the magnetic coupling between the primary and secondary will be relatively poor because of the large air gap required between the transformer coils to prevent arcing. A small gap is needed to achieve good magnetic coupling between the primary and secondary coils. Good

high frequency insulation, such as teflon, transformer oil, or solid material potting compounds should be used.

The secondary winding of the impedance matching transformer, in series with the induction coil or contact system and workpiece, forms the low voltage, high current welding circuit. The connecting leads should have the lowest possible impedance to obtain high efficiency and minimize the voltage drop in the leads. This may be achieved by using short leads made of flat plates spaced 1/16 inch apart. The power losses in poorly designed leads or incorrectly matched transformers can prevent successful performance of a high frequency welding operation.

Both rotating generator and solid-state inverter power sources have output voltage typically in the 200 to 1000 V range. Their current is fed by cables to the primary coil of a laminated iron core transformer. Capacitors are connected in parallel with the primary of the transformer to improve the system power factor. The rotating generator or solid-state inverter is operated as close to unity power factor as possible for maximum efficiency and performance. These transformers lower the voltage and increase the current of the generators and inverters. The transformer provides power to an induction coil or contact system through relatively short, closely-spaced leads. The transformer winding ratio must be selected to give optimum power transfer between the generator or inverter and the welding load. These ratios commonly range from 4:1 to 13:1.

CONTACTS

The high frequency current transfer contacts, either sliding or fixed, are usually made of copper alloys or hard metallic particles in a copper or silver matrix. The contacts are silver brazed to heavy water-cooled copper mounts. Replacements can be made by exchanging the mount and contact tip assembly.

Contact tip sizes range from 0.25 to 1 in.^{2} depending upon the amperage to be carried. Welding curents are usually in the range of 500 to 5000 A. Consequently, both internal (water) and external (water or soluble oil) cooling is provided for the contact tips and mounts.

The force of the contact tips against the work is usually in the order of 5 to 50 pounds for continuous welding systems and 5 to 100 pounds for static welding operations. The required forces are dependent upon the thickness and surface condition of the parts being joined, as well as the amperage.

Contacts used for continuous seam welding of nonferrous metals may have triple the life of those used with ferrous materials. Welding of 300 000 ft of nonferrous tubing with one set of contacts is common. On stationary welding operations, contacts may last thousands of operations before requiring dressing or replacement.

INDUCTION COILS

An induction coil, also called an inductor, is generally fabricated of copper tubing, copper bar, or copper sheet. It is water cooled. The best efficiency is obtained when the induction coil completely surrounds the workpiece. The coil may have one or more turns, as required by the application. It is designed to fit in close proximity to the workpiece at the area to be heated. The strength of the magnetic field, which induces the heating current in the workpiece, diminishes rapidly as the distance between the coil and work is increased. The sharpness with which the heating pattern in the workpiece mirrors the shape of the coil improves as frequency increases and distance decreases between the coil and the workpiece. Typical spacing between the coil and workpiece is 0.08 to 0.25 inch.

IMPEDERS

When tube and pipe welding with both the HFRW and HFIW processes, current can flow on the inside surface of the tube as well as on the outside surface. This additional current, which flows in parallel with the welding current, results in power loss. To minimize this loss, a magnetic core or impeder is placed inside the tube weld area. The impeder increases the inductive reactance of the current path around the inside wall of the tube. This, in turn, reduces the unde-

sired inside current and increases the outside current in the weld zone. This results in higher welding speeds for a given power input. Impeders are usually made of one or more ferrite bodies. They are water cooled to keep their operating temperature below that at which they would lose their magnetic properties.

CONTROL DEVICES

Input Voltage Regulators

High-speed, high frequency seam welding requires accurate control of the welding power level. Relatively brief power fluctuations can result in weld imperfections. Thus, it is important that the power be automatically and continuously regulated. Several regulation systems are available. The choice depends upon the input power conditions, the welding requirements, and the power rating of the high frequency power source. The following are the most commonly used regulators.

Motor-Generators. These units provide constant voltage power by utilizing the flywheel effect of high mechanical inertia. The power is supplied to one or more high frequency power sources of the vacuum tube or solid-state type.

Electromechanical Induction Regulators. Regulators of this type may be installed on individual pieces of equipment or on a group of machines. They are used when there are no large or rapid variations in the input power. Nearby equipment that demands heavy current intermittently may cause power line voltage fluctuations that are beyond the response capability of such regulators.

Saturable Core Reactors. These magnetic devices have better response than mechanical induction regulators.

Silicon-Controlled Rectifiers. These power control devices can be used in conjunction with other solid-state devices. They normally are arranged in the circuit to regulate the input wave form by blocking a portion of the wave. Their control circuits can be designed to respond rapidly to changes in input voltage.

All of the above input voltage regulators can include weld power control. In addition, electronic and saturable core types can include overload protection, system fault detection devices, and monitoring systems for associated handling equipment employed in the welding operation.

High Frequency Generator Power Controls

A control must be provided to raise and lower the high frequency power to meet the demands of the welding operation. There are many types of high frequency power controls, all of which vary the current flowing in the contacts or the inductor coil. Controls normally used for high frequency welding are saturable reactors, silicon-controlled rectifiers, variable impedance devices, vacuum tubes, and field current regulators on generators.

A saturable reactor located in the circuit ahead of the main plate transformer provides a variable power control. With the addition of a magnetic amplifier and appropriate instrumentation, automatic control of welding power may be achieved.

Silicon-controlled rectifiers (SCR) may be placed in the primary circuit of the plate transformer for vacuum tube oscillators or in the SCR rectifier system on solid-state inverters to control welding current.

A variable impedance can be used in series with the secondary winding of the welding transformer and the inductor or contacts. It is the most direct method of varying the high frequency power level and does not alter the basic characteristics of the high frequency generator. Generally, this control is manually adjusted for the desired power level.

A controllable vacuum tube device can control power output. These operate in a manner similar to a variable resistor connected in series with the dc output of the oscillator power supply.

In generator units, power control is performed manually or automatically by varying the field strength of the generator.

For continuous operations, it is essential that the welding power be relatively ripple-free, particularly when welding thin-walled nonferrous metals at high speeds. Excessive power ripple can cause intermittent lack of fusion along the seam. These intermittent discontinuities are called "stitches." Inductive-capacitive filters are normally used on the rectifier output to reduce the ripple to less than 1 percent.

Speed-Power Control

An automatic speed-power control may be used for continuous HFRW or HFIW with a vacuum tube oscillator power source. This automatic device controls the SCR system in the primary of the main plate transformer and adjusts the power output of the oscillator with changes in welding speed. For example, assume that a weld is being produced at 200 ft/min when a problem with the mill requires that the travel speed be decreased. Without automatic speed-power control, the welding power would need to be turned off as soon as the slowdown started. However, with speed-power control, the welding power remains on during the slowdown. It is automatically adjusted to produce a good weld as the speed decreases. The control can adjust the welding power level to near zero. A good weld is achieved almost to the stop condition. Such control minimizes the loss of material in the mill.

The speed-power control automatically raises the power level as the mill speed increases. The power required for welding is not directly proportional to speed. The control is designed to automatically compensate for this.

Timing Controls

For timed high frequency resistance or induction welding, weld duration may be controlled by motor-driven or electronic timers. They actuate the main contactor in the power source. In very high-speed production using equipment with vacuum tube oscillators, it is possible to turn the grids of the oscillator tubes on and off with a small contactor. Very short weld times can be produced and controlled by this means. In some cases, movement of the workpieces during weld upset is used to actuate a switch. The switch turns off the welding current when the desired upset has occurred.

WELDING PROCEDURES

SURFACE PREPARATION

In general, a number of surface conditioning operations that aid in producing a sound weld take place during the welding cycle. First, faying surfaces are heated to high temperatures, usually above the melting point. Then relatively large amounts of plastic deformation are used to remove contaminated metal from the faying surfaces and bring clean surfaces into intimate contact. Special surface treatment operations, such as brushing or solvent cleaning, are generally unnecessary.

ALIGNMENT AND FIT-UP

In continuous seam welding, the edges of the parts to be joined are brought together in the form of a vee. The shape of the vee is important in producing satisfactory welds. Normally, the mill is set to produce a vee with an included angle of 4 to 7 degrees. If the angle is less than 4 degrees, the apex (weld point) will not stay at a fixed point, and inconsistent welding will occur along the joint. Also, arcing may occur across the vee, upstream from the weld point.

Maximum usage of the proximity effect is limited in high frequency resistance welding because the strip separation must be sufficient to prevent arcing across the vee. Arcing is undesirable because it causes nonuniform heating of the strip and may pit the edges to be joined. The tendency for arcing and the strip separation required to prevent arcing vary with the welding conditions and metals being welded.

If the angle is greater than 7 degrees, mechanical and thermal control of the edges may be lost and they may wrinkle or stretch. Alignment of the weld joint is important, and the joint must be held tightly together until it cools.

Variations in the separation and length of the vee during welding will cause corresponding changes in the amount of power drawn from the power source and in the heat developed at the edges. Variation in the width of the vee changes the current distribution at the edges, the depth of heating, and the final surface temperature of the edges. To produce consistent welds, the width of the vee and the contact-to-vee apex distance must be closely controlled.

There is no question about the need for accurate control of relative positions when welding two edges together. The ideal relationship is, of course, to have the edges parallel and matched in the vertical direction. Mismatch will cause uneven heating of the faying surfaces. Serious angular mismatch may cause excessive melting of the edges and arcing near the apex of the vee. In general, it is probably best to be sure that the faying surfaces are parallel and at the same level. In welding thick skelp into pipe, deliberate angular mismatch may be used. In this case, the inside edges of the faying surfaces are melted, and, as the edges approach the apex of the vee, they approach parallelism.

FIXTURING

The proper design of fixturing and tooling is critical to successful application of high frequency resistance welding. Precise alignment of

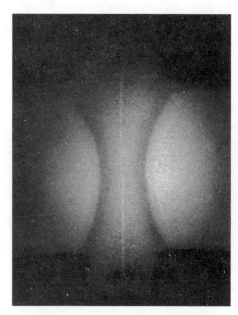

Fig. 4.17—Section through a high frequency resistance weld showing the upset

the edges of the material must be controlled throughout the welding process. Some applications of the HFW processes include the forming of flat stock into tubing that requires precision tooling to produce an acceptable welded seam. Mismatch of the edges can cause uneven heating that results in poor quality welds.

Almost all high frequency welding variations employ pressure to produce a sound weld.

In both static and continuous welding, upsetting of metal occurs in the weld area when pressure is used. A section through such a weld is shown in Fig. 4.17. The upset metal may be left in place or removed, depending upon the use of the product. In continuously welded products such as tubing, the upset is often continuously removed from both sides of the weld by scarfing or plannishing equipment.

WELD QUALITY

A number of discontinuities often found in other types of welded joints are rarely found in high frequency welded joints containing upset metal. Since metal is expelled from a properly made joint, porosity and other discontinuities produced by freezing segregation are eliminated. The weld zones are very narrow and have narrow heat-affected zones. Joint strength equal to that of the base metal can be obtained in a variety of metals. For example, welds can be made in carbon steels that have tensile strengths equal to the base metal strength.

High frequency welded joints have the same chemical composition as the metal being welded. Consequently, the weld zone responds to heat treatment in the same way as does the base metal.

High frequency welds with upset can contain discontinuities resulting from improper welding conditions (current, speed, upset, or joint preparation). For example, unwelded areas may occur in the joint as shown in Fig. 4.18.

Material-induced discontinuities may also be encountered. The most common type is the outbent fiber crack found in carbon and low alloy steel welds. Figure 4.19 shows this type of discontinuity. It is produced by a stringer inclusion in the steel being bent outward when hot metal is upset to complete the weld.

A third type of discontinuity is a small, unbonded flat spot or area along the weld line. These areas sometimes extend completely through the weld and can leave leaks in the wall of a pipe. Figure 4.20 shows an example of a flat spot. They are apparently caused by a combination of inclusions in the metal and improper welding conditions. Flat spots are formed when nonmetallic material is trapped at the bond line of the weld.

The short time at temperature and the deformation of upsetting tend to restrict grain growth in the weld zone. Rapid cooling rates may cause quench hardening of the weld zone in some carbon and alloy steels. Such hardening may decrease ductility below acceptable limits, and a postweld heat treatment may be required to obtain the desired properties. Welding of work-hardened metals that do not quench harden (stainless steels, for example) always produces a weld zone that is softer than the wrought base metal.

INSPECTION AND TESTING

Inspection of longitudinal butt joints in tubular products made by this process is directed towards in-plant quality control and is often required by some specification. Prefer-

Fig. 4.18 — Cross section of a butt joint in carbon steel containing an unwelded portion

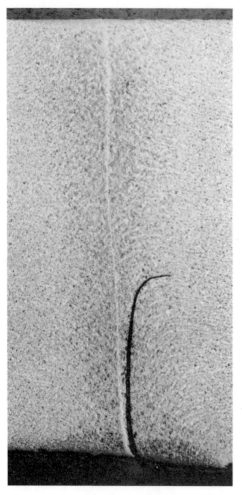

Fig. 4.19—Outbent fiber crack in a butt joint in carbon steel. Picral etch, ×30 (reduced 25%)

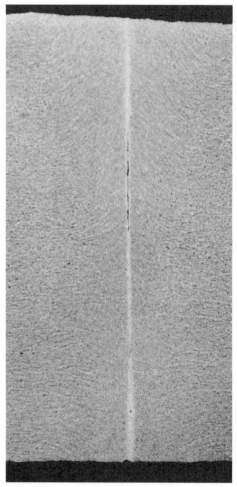

Fig. 4.20—Cross section of a butt joint in carbon steel showing a flat spot

ably, the inspection is performed continuously, immediately following the welding station. Detection of out-of-tolerance discontinuities as the tube is welded will reduce scrap and eliminate subsequent rejection, particularly if the tube is processed in fabricating a product.

Eddy current inspection is employed almost exclusively. Basically, the welded tube is passed through an encircling electromagnetic coil which induces eddy currents within the tube. Changes in the coupling between the coil and the tube and also in the electrical characteristics of the tube due to certain types of discontinuities will cause variations in the loading and tuning of the generator. The changes are detected by the differentially-wound detector coil and are reflective in meter readings or by recording devices for permanent record.

Instrumentation is also available for limit warning signals. The frequencies employed are usually in the range of 1 to 25 kHz. The method is applicable to testing of welds in both nonferrous and ferrous metals in diameters from 0.125 to 16 inches. Testing speeds range from 25 to 500 ft/min.

Typical discontinuities that are readily detected by the eddy current technique are laps, cracks, slivers, pits, unbonded areas, misaligned welds, and pin holes. A basic requirement is a reliable and consistent means for setting the degree of sensitivity of the test equipment to the required level. A standard reference specimen is required for this purpose. The reference specimen has typical discontinuities deliberately fabricated into it that represent the requirements of a particular application or specification. This testing method can also detect surface flaws in the base metal.

Ultrasonic testing is another method used for inspection of tubular products. This testing technique is normally restricted to finite tube lengths, using the immersion method automatically. This technique is more sensitive than the eddy current method to minor surface scratches and to elongated discontinuities such as the incomplete removal of the flash. As with the eddy current method, a reference specimen is required that contains precision-machined discontinuities matching the degree of acceptability required by the specification.

Magnetic particle and liquid penetrant inspection methods are slow. These methods can be justified only when the end use justifies the cost to ensure the highest possible quality. In some cases, these methods usually supplement other methods.

In-process inspection for tubular products may also include destructive testing of short sections of tube removed from production. A section is tested by placing an open end over a conical die mounted in an arbor press. Force is applied to the other end of the tube to expand it over the die. Examination of the sample may reveal unbonded areas and offset joint edges. In conjunction with this test, a finite length of tube is leak tested by pneumatically pressurizing the tube internally and immersing it in water. This test will reveal discontinuities extending through the welded seam, indicated by bubbles emerging from the flaw.

Inspection of other joint types, such as web-to-flange and end-to-end butt joints, may be accomplished by metallographic techniques and mechanical testing methods to determine joint strength and ductility.

SAFETY

Consideration must be given to the health and safety of the welding operators, maintenance personnel, and other personnel in the area of the welding operations. Good engineering practice must be followed in the design, construction, installation, operation, and maintenance of the equipment, controls, power supplies, and tooling to assure conformance to Federal (OSHA), State, and local safety regulations, as well as those of the using company.

High frequency generators are electrical devices and require all the usual precautions in handling and repairing such equipment. Voltages are in the range from 400 to 20 000 volts and are lethal. These voltages may be low frequency or high frequency. Proper care and safety precautions should be taken while working on high frequency generators and their control systems to prevent injury. Modern units are equipped with safety interlocks on access doors and automatic safety grounding devices that prevent operation of the equipment when the access doors are open. The equipment should not be operated with panels or high voltage covers removed or with interlocks and grounding devices blocked.

The output high frequency primary leads should be encased in metal ducting and should not be operated in the open. The induction coils and contact systems should always be properly grounded for operator protection. High frequency currents are more difficult to ground than low frequency currents, and grounding lines should be kept very short and direct to minimize inductive impedance. Care should be taken that the magnetic field from the output system does not heat adjacent metallic sections by induction and cause fires or burns.

Injuries from high frequency power, especially at the upper range of welding frequencies, tend to produce severe local surface tissue damage but are not likely to be fatal, since current flow in the victim's body is shallow.

Metric Conversion Factors

$t_C = 0.556(t_F - 32)$
1 in. = 25.4 mm = .025 m
1 ft = 0.305 m
1 ft/min = 5.08 mm/s
1 lbf = 4.45 N
1 psi = 6.89 kPa
1 degree (angle) = (1.75×10^{-2}) rad

SUPPLEMENTARY READING LIST

Brown, G.H., Hoyler, C.N., and Bierwith, R.A., *Theory and Applications of Radio Frequency Heating*, New York: D. Van Nostrand Co., Inc., 1957.

Dailey, R.F., Induction welding of pipe using 10 000 cycles, *Welding Journal*, 44 (6): 475-479; 1965 Jun.

Harris, S.G., Butt welding of steel pipe using induction heating, *Welding Journal*, 40 (2): 57s-65s; 1961 Feb.

Johnstone, A.A., Trotter, F.J., and a'Brassard, H.F., Performance record of the Thermatool high frequency resistance welding process. *British Welding Journal*, 7 (4): 238-249; 1960 Apr.

Koppenhofer, R. L. et al., Induction-pressure welding of girth joints in steel pipe. *Welding Journal*, 39 (7): 685-691; 1960 July.

Oppenheimer, E.D., Helical and Longitudinally Finned Tubing by High Frequency Resistance Welding, Dearborn, MI.: Soc. of Manufacturing Engineers, 1967; ASTME Tech. Paper AD 67-197.

Osborn, H.B., Jr., High frequency continuous seam welding of ferrous and non-ferrous tubing. *Welding Journal*, 35 (12): 1199-1206; 1956 Dec.

Osborn, H.B., Jr., High frequency welding of pipe and tubing. *Welding Journal*, 42 (7): 571-577; 1963 July.

Rudd, W.C., High frequency resistance welding. *Welding Journal*, 36 (7): 703-707; 1957 July.

Rudd, W.C., High frequency resistance welding of cans. *Welding Journal*, 42 (4): 279-284; 1963 Apr.

Rudd, W.C., High frequency resistance welding. *Metal Progress*: 239-40, 244; 1965 Oct.

Rudd, W.C., Current penetration seam welding —a new high speed process. *Welding Journal*, 46 (9): 762-766; 1967 Sept.

Rudd, W.C. and Udall, H.N., High frequency melt welding. *Welding Journal*, 56 (4): 28-32; 1977 Apr.

Wolcott, C.G., High frequency welded structural shapes. *Welding Journal*, 44 (11): 921-926; 1965 Nov.

5

Electron Beam Welding

Chapter Committee

D. E. POWERS, *Chairman*
  Leybold-Heraeus Vacuum Systems Inc.

S. C. BROWN
  Westinghouse Electric Corporation

H. CASEY
  Los Alamos Scientific Laboratories

J. F. HINRICHS
  A. O. Smith Corporation

J. LEE
  Chrysler Corporation

G. H. LOAN
  Nuclide Corporation

B. W. SCHUMACHER
  Ford Motor Company

C. T. SZYMKO
  Argonne National Laboratories

J. VAN EIJNSBERGEN
  Union Carbide Corporation

C. M. WEBER
  Babcock and Wilcox Company

Welding Handbook Committee Member

W. L. WILCOX
  Scott Paper Company

5

Electron Beam Welding

FUNDAMENTALS OF THE PROCESS

DEFINITION

Electron beam welding (EBW) is a process that produces coalescence of metals by the heat obtained from a concentrated beam of high velocity electrons impinging upon the surfaces to be joined. The electron, a basic particle of matter, is characterized by a negative charge and a very small mass.

The beam of electrons is produced and accelerated by an electron beam gun. Typical components of the gun are an emitter, a bias electrode, and an anode. The emitter is sometimes called the filament or cathode; the bias electrode is also called the grid or grid cup.

Auxiliary mechanical and electrical components, which may include beam alignment, focus, and deflection coils, are used in conjunction with the electron beam gun. The whole assembly is called the electron beam gun column.

GENERAL DESCRIPTION

The heart of the electron beam welding process is the electron beam gun, a simplified representation of which is shown in Fig. 5.1. Electrons are generated by heating a negatively charged emitting material to its thermionic emission temperature range. The electrons "boil off" this emitter and are then given speed and direction by their attraction to a positively charged anode. A precisely configured electrode surrounding the emitter electrostatically shapes the ejected electrons into a beam. In a diode (cathode-anode) gun, this beam-shaping elec-

trode and the emitter are both at the same electrical potential, and together are referred to as the cathode. In a triode (cathode-grid-anode) gun, the emitter is at one potential and the beam-shaping electrode can be biased to a slightly more negative potential to control beam current flow. In this case, the emitter alone is referred to as the cathode, and the shaping electrode is called the bias electrode or "grid cup." The accelerating anode is incorporated into the electron gun, making beam generation completely independent of the workpiece.

The electrons, accelerated to speeds in the range of 0.1 to 0.7 times the speed of light, pass through a small hole in the center of the anode and continue toward the workpiece. After the electrons leave the anode, they generally diverge because of random thermal radial velocity spreading and geometric mutual repulsion, as shown in Fig. 5.1. To counteract the inherent divergence effects, an electromagnetic lens system is used to reconverge the beam and focus it on the work. This beam divergence and convergence is of such a gradual nature that the useable focal range or depth of focus may extend over a distance of an inch or so.

The heat intensity imparted to the weld joint is controlled by four basic variables:

(1) The number of electrons per second impinging on the workpiece (beam current)

(2) The kinetic energy of these electrons (accelerating voltage)

(3) The diameter of the electron beam at the workpiece (beam spot size)

(4) The travel speed with which the work-

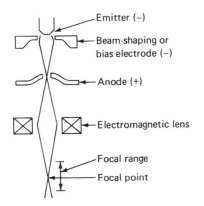

Emitter (−)

Beam-shaping or bias electrode (−)

Anode (+)

Electromagnetic lens

Focal range

Focal point

Fig. 5.1—A simplified representation of an electron beam gun column

piece or electron beam gun is being translated (welding speed)

Accelerating voltages and beam currents employed for typical electron beam gun systems vary over the ranges of 30 to 175 kV and 50 to 1000 mA, respectively. The electron beam produced by these guns can be focused to diameters in the range of 0.01 to 0.03 inch. The resulting power density attainable can reach values up to 10^6 W/in.^{2}, a level significantly higher than that possible with arc welding processes.

A comparative measure of electron beam systems, generally used to indicate their probable welding capability, is the maximum power density that a system is capable of delivering to the workpiece. This comparison factor is totally dependent upon the maximum beam power (current × voltage) and minimum beam spot size attainable with each system. Electron beam welding systems capable of producing beam power levels up to 100 kW and power densities in excess of 10^7 W/in.^{2} have been built, but they are not yet used commercially.

At very high power densities, the electron beam is capable of instantly penetrating into the workpiece and forming a "vapor hole." The walls of the vapor hole are molten. The molten metal from the forward portion flows around the periphery of the hole and solidifies at the rear to form weld metal as the beam advances along the joint. Hence, the weld metal is much deeper

than it is wide, and the heat-affected zone is very narrow. For example, the width of a butt weld in 0.5-in. thick steel plate may be as small as 0.060 in. when made in a vacuum. This stands in remarkable contrast to the weld zone in arc and gas welded joints.

Because of the deep penetrating capability of an electron beam, the angle of incidence with which the beam impinges on the surface of a workpiece will affect the final angle at which the resulting weld zone is formed with respect to that surface.

PROCESS VARIATIONS

There are three variations or modes of the electron beam welding process: high vacuum (EBW-HV), medium vacuum (EBW-MV), and nonvacuum (EBW-NV). The principal difference between these process modes is the ambient pressure at which welding is done. With the high vacuum mode, welding is done in the pressure range of 10^{-3} to 10^{-6} torr.[1] For medium vacuum, the pressure range is 10^{-3} to 25 torr. Within this range, the pressure span from about 10^{-3} to 1 torr is commonly called a "soft" or "partial" vacuum, and from about 1 to 25 torr, a "quick" vacuum. Nonvacuum electron beam welding is done at atmospheric pressure. In all cases, the electron beam gun must be held at a pressure of 10^{-4} torr or less for stable and efficient operation.

High vacuum and medium vacuum welding are done inside a vacuum chamber. This imposes an evacuation time penalty to take advantage of the "high purity" atmosphere. The medium vacuum welding machine retains most of the advantages of high vacuum welding but with improved production capability. Chamber evacuation times are much shorter, resulting in higher production rates. Nonvacuum welding, although it incurs no pumpdown time penalty, is unsuitable for some applications because the welds are generally wider and shallower than welds made with equal power in vacuum.

1. A torr is the accepted industry term for a pressure of one millimeter of mercury. One standard atmosphere can be expressed as 760 torr or 760 mm of mercury.

With medium vacuum operation, the beam is generated in high vacuum and then projected into a welding chamber operating at higher pressure. This is accomplished through an orifice of special design that is large enough to pass the beam but too small to allow significant back diffusion of gases into the gun chamber.

In nonvacuum electron beam welding equipment, the beam is generated in high vacuum and then projected through a series of orifices and differentially-pumped chambers. It finally emerges into a work environment that is at atmospheric pressure. Figure 5.2 shows the three basic modes of electron beam systems.

High Vacuum Welding

High vacuum (10^{-3} torr or lower) is the natural environment for all electron guns. Although special devices enable the electron beam to be brought into an environment of higher pressure, the gun itself cannot operate effectively at pressures greater than 10^{-3} torr.

The principal advantages which accrue as a result of welding in high vacuum are as follows:

(1) Maximum weld penetration and minimum weld width can be achieved, thereby producing a minimum of weld shrinkage and distortion.

(2) Maximum weld metal purity is possible due to the absence of contaminating gases.

(3) The relatively long gun-to-work distances possible with this method enhance the ability to observe the welding process and provide the ability to weld relatively inaccessible joints.

An electron beam is scattered by collision of the electrons with any gas molecules that may be present in its path. The frequency at which these collisions occur will vary directly with both the concentration of gas molecules present and the total distance traveled.

The high vacuum minimizes exposure of the weld zone to oxygen and nitrogen contamination while hot. It also causes evolved gases to rapidly move away from the weld metal, thereby improving weld metal purity. Thus, high vacuum welding is better suited for welding highly reactive metals than the other two process variations.

Production of high vacuum requires pumping time which significantly limits production rates. This can be offset somewhat by welding a number of assemblies in a single load and by keeping the chamber volume to a minimum. However, chamber size limits the number of parts per batch that can be accommodated for welding. Consequently, high vacuum welding is

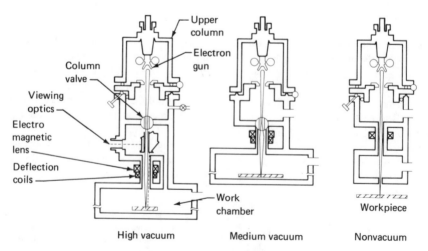

Fig. 5.2—The basic modes of electron beam welding

Fig. 5.3—Electron beam welding a gear in medium vacuum

generally considered to be a relatively low production type of operation.

Medium Vacuum Welding

A principal asset of medium vacuum welding is the fact that its requirements can be met without pumping the welding chamber to very low pressure (high vacuum). With small welding chambers, pumping times may be a matter of seconds. This is of major importance in terms of commercial and economical processing. This variation is ideally suited for the mass production of parts where repetitive production tasks are involved. A welding chamber of minimum volume is used. For example, gears can be successfully welded to shafts in their final machined or stamped condition with no subsequent finishing needed. Such an operation is shown in Fig. 5.3. Close tolerances can be maintained.

Because welding is done at pressures where the concentration of air is significant (100 ppm), medium vacuum welding is less desirable than high vacuum welding for reactive metals. However, medium vacuum welding may be acceptable for refractory metal welding where absorbtion of very small amounts of oxygen and nitrogen can be tolerated.

The higher concentration of air will also tend to scatter the electron beam, producing an increase in beam diameter and a decrease in beam power density. This results in welds that are slightly wider, more tapered, and less penetrating than similar type welds produced under high vacuum conditions.

As with high vacuum welding, a limitation is imposed on the work size by the vacuum chamber, and the chamber must be evacuated prior to welding. Therefore, production rates are generally lower than those attainable with nonvacuum welding.

Nonvacuum Welding

The major advantage of nonvacuum welding is that the work is not enclosed in a vacuum chamber. Therefore, production rates can be higher and costs lower than with the other variations because time is not needed for chamber evacuation. Also, the size of the weldment is not limited by a chamber.

These advantages are gained at the expense of low weld depth-to-width ratios, reduced weld penetration, and small gun-to-work distances. The welding atmosphere is not as "pure" as with high and medium vacuum welding, even when inert gas shielding is used.

Operating conditions for nonvacuum welding differ from the other two variations. Beam dispersion increases rapidly with ambient pressure, as shown in Fig. 5.4. Therefore, at atmospheric pressure the gun-to-work distance must be less than 1.5 inches. This restriction limits the shape of the workpiece near the weld joint.

The depth of penetration in nonvacuum electron beam welds is affected by beam power level, travel speed, gun-to-work distance, and the ambient atmosphere through which the beam passes. Figure 5.5 shows penetration as a function of travel speed for three different power levels, indicating the significant increase in travel speed that can be used by increasing power for a given penetration.

Figure 5.6 shows the effect of the ambient atmosphere and the gun-to-work distance on the penetration-travel speed relationship. Penetration is greater with helium, which is lighter than air, and lower with argon, which is heavier than air. For a given penetration and gun-to-work distance, helium shielding will permit welding at a significantly higher travel speed.

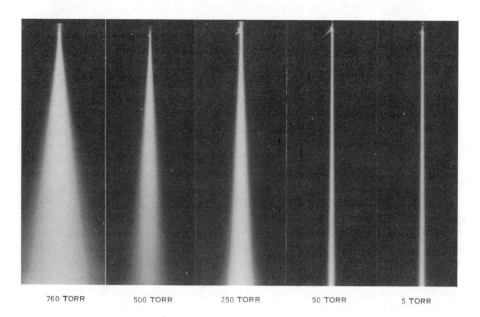

760 TORR   500 TORR   250 TORR   50 TORR   5 TORR

Fig. 5.4—Electron beam dispersion characteristics at various ambient pressures

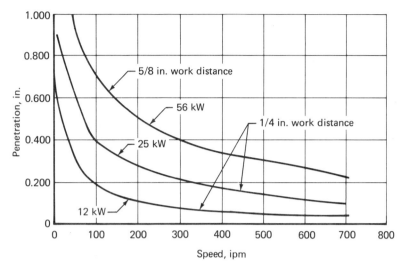

Fig. 5.5—Effect of travel speed on penetration of nonvacuum electron beam welds in steel (175 kV in air)

Nonvacuum electron beam welding appears to demonstrate more efficient penetrating capability at power levels above 50kW. This apparent capability probably results from the decrease in ambient air (or gas) density produced by local heating of the air (or gas) as the beam passes through it.

Many materials have been welded successfully using the nonvacuum technique. They include carbon, low alloy, and stainless steels; high temperature alloys; refractory alloys; and copper and aluminum alloys. Some of these metals can be welded in air while others require inert gas protective atmospheres to avoid excessive oxygen and nitrogen contamination.

With 60 kW nonvacuum equipment, it is possible to produce single-pass welds in many metals of 1-in. thickness at relatively high speeds. Figure 5.7 is a cross section of a nonvacuum weld in 3/4-in. Type 302 stainless steel plate.

ADVANTAGES AND LIMITATIONS

Electron beam welding has some unique performance capabilities, such as its environment, high power densities, and outstanding control capabilities, that permit the solution of a wide range of joining problems.

The following are some of the definite advantages of electron beam welding:

(1) The ability to produce weld metal with a cross section that is deeper and narrower than those of arc welds in most metals. This ability to attain an extremely high weld depth-to-width ratio permits single-pass welding of joints that would normally require multipass arc welds.

(2) The total heat input per unit length for a given depth of penetration can be much lower than with arc welding. This gives a much narrower head-affected zone, noticeably less distortion, and fewer thermal effects compared to arc welding.

(3) A high purity environment (vacuum) for welding minimizes contamination of the metal by oxygen and nitrogen.

(4) The ability to project the beam over a distance of several feet in vacuum often allows welds to be made in otherwise inaccessible locations.

(5) Rapid travel speeds are possible because of the high melting rates associated with this concentrated heat source. This reduces the time required to accomplish welding and increases the productivity and energy efficiency of the process.

(6) Reasonably square butt joints in both

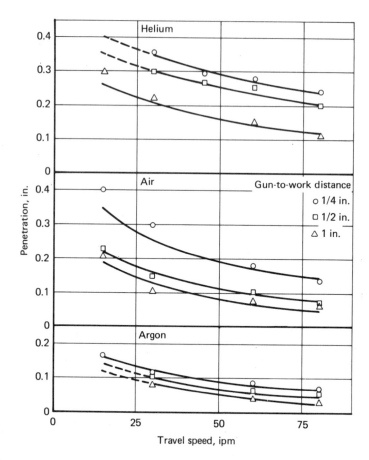

Fig. 5.6—Penetration vs travel speed for nonvacuum electron beam welds in AISI 4340 steel in helium, air, and argon with three gun-to-work distances (175 kV, 6.4 kW)

thick and relatively thin plates can be welded in one pass without filler metal addition.

(7) Hermetic closures can be welded with the high or medium vacuum modes of operation while retaining a vacuum inside the component.

(8) The beam of electrons can be magnetically deflected to produce various shaped welds and magnetically oscillated to improve weld quality or increase penetration.

Some of the probable limitations of elec-

tron beam welding are as follows:

(1) The equipment costs are generally higher than those of arc welding equipment. However, final system cost depends strongly upon the particular welding requirements; final part costs can be highly competitive with arc welding.

(2) Proper utilization of the high weld depth-to-width ratio can require precision machining of the joint edges, exacting joint alignment, and good fit-up. In addition, joint gap

must be minimized to take advantage of the small size of the electron beam. However, these precise part requirements are generally not mandatory where high depth-to-width ratio welds are not required.

(3) For high and medium vacuum welding, work chamber size must be large enough to accommodate the assembly operation. The time needed to evacuate the chamber will influence production costs.

(4) With the nonvacuum mode of electron beam welding, the restriction on work distance from the bottom of the electron beam gun column to the work will impose limitations on the product design in areas directly adjacent to the weld joint.

(5) Because the electron beam is deflected by magnetic fields, nonmagnetic or properly degaussed metals should be employed for tooling and fixturing that will come into close proximity to the beam path.

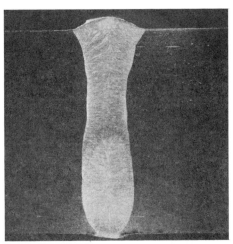

Fig. 5.7—A cross section of a nonvacuum electron beam weld in 3/4-in. stainless steel plate made in air with 12 kW of power

CHARACTERISTICS OF WELDS

The electron beam welding process produces weld metal geometries that differ significantly from those made by conventional arc welding processes. Typical transverse cross sections through electron beam and gas tungsten arc welds are shown in Fig. 5.8. The geometry of a typical electron beam weld exhibits a weld depth-to-width ratio that is very large in comparison to that of arc welds. This feature is due to the high power density of the electron beam. The beam is concentrated in a small area. Electron beam power density can exceed the power densities obtained with arc welding by several orders of magnitude.

The high depth-to-width ratios of electron beam welds account for two important advantages of the process. First, relatively thick joints can be welded in a single pass. Figure 5.9 is an example of a single-pass electron beam weld in a 4-in. thick carbon steel section. It was made with a beam power of 33 kW and a travel speed of

about 5 in./min. Thick weld joints, which would require multiple-pass arc welding procedures, can be made with a single-pass electron beam welding procedure in considerably less time. Second, the rate (travel speed) at which welding can be accomplished is much greater than with arc welding. Thus, the electron beam welding process is a very rapid one.

High vacuum welds with depth-to-width ratios of 50:1 are possible in a number of alloys. The welding of heavy sections in a single pass is practical using a square-groove butt joint. For example, aluminum plates up to 18 in. thick have been welded. While some problems remain with joining thick sections, production applications in steel extend up to 6 inches. The medium vacuum mode sacrifices some penetration in comparison to the high vacuum mode. With the nonvacuum mode, the maximum penetration attainable is generally about 2 inches.

The ability to produce welds with these

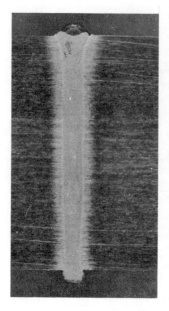 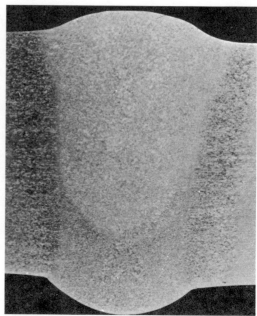

Welding conditions

| | Electron beam weld | Gas tungsten arc weld | | |
		1st Pass [a]	2nd Pass [b]	Total
Voltage, V	30×10^3	11.7	13.0	...
Current, A	200×10^{-3}	270.	270.	...
Welding speed, in./min	95	6.5	7.5	...
Power, kW	6	3.2	3.5	6.7
Energy, kJ/in.	3.8	29.6	28.0	57.6

a. Without wire
b. With 0.062 in. dia. Type 2319 Al filler metal

Fig. 5.8—Comparison of electron beam (left) and gas tungsten arc (right) welds in 1/2-in. thick Type 2219 aluminum alloy plate

characteristics depends upon the process mode: high, medium, or nonvacuum. In all cases, it is highly dependent upon the beam spot size and the total beam power. Figure 5.10 gives a general indication of the effect of ambient pressure on weld metal profile and penetration when all other operational variables are essentially constant, but not necessarily optimum. The depth-to-width ratio is also critically dependent upon base metal physical properties, such as melting point, heat capacity, thermal diffusivity, and vapor pressure. This process, particularly in its high vacuum mode, is an excellent tool for the welding of parts of dissimilar metals or different masses, and the repairing of welded components impossible to salvage with other processes. The low total heat input to the workpiece noticeably minimizes weld joint distortion. Here again, the high and medium vacuum modes are the most advantageous, although the nonvacuum

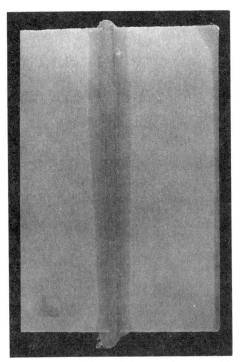

Fig. 5.9—Single-pass electron beam weld in a 4-in. thick carbon steel section

mode offers some notable advantages over conventional arc welding processes.

Because the high power density produces welds that are not controlled by thermal conduction, metals of significantly different thermal conductivities can be welded together. In welding a joint between two pieces of different thickness, the unequal heat loss due to the difference in mass is not a significant problem. Also, fusion can be accomplished without significant need for heat sinks that would be required for arc welding. The beam energy is usually concentrated on the thick section at the joint, and the power is adjusted for penetration through the thin section.

Arc welding procedures often require the use of preheat for thick sections of a metal having a high thermal conductivity, such as aluminum or copper. Little or no preheat is required for electron beam welding high conductivity metals with the available power density.

Reactive and refractory metals are detrimentally affected by very small amounts of gaseous contaminants, such as oxygen, nitrogen, and hydrogen. However, these metals may be electron beam welded without introducing these contaminants. Such metals include tungsten, molybdenum, columbium, tantalum, zirconium,

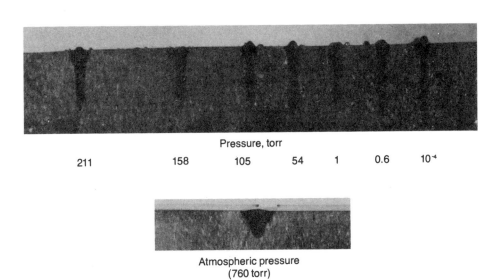

		Pressure, torr				
211	158	105	54	1	0.6	10^{-4}

Atmospheric pressure
(760 torr)

Fig. 5.10—Electron beam weld profile change with increasing ambient pressure when all other operating variables are essentially constant

and titanium. The high vacuum mode of welding is the most suitable one for joining these metals. The other two modes offer decreasing weld performance capabilities, although they still permit satisfactory application in select cases.

Although electron beam welding is a high power density process, it is also a low energy process. That is, the energy required to weld a joint of a given thickness is considerably less than that required by more conventional arc welding processes. Two advantages can result from the low energy input: First, it minimizes distortion and reduces the size of the weld heat-affected zones. Second, the high cooling rates associated with narrow electron beam welds can affect metallurgical reactions, such as phase changes. However, the fundamental rules of metallurgy regarding cooling rates and the resulting microstructure still apply. The weld metal will have mechanical properties normally associated with the bulk properties of the microstructure.

Another aspect of electron beam welding involves the control of the process. Accurate control of the electron beam has always permitted a high degree of reliability and the result is reproducible welding. At present, incorporation of minicomputers and microprocessors offers additional control of the welding conditions.

Control of the environment where welding is taking place can affect the composition of the weld metal. Electron beam welding in a high vacuum will permit gases to escape and high vapor pressure metals to evaporate. At the other pressure extreme, nonvacuum electron beam welding in air may increase the nitrogen and oxygen in the weld metal.

Since the electron beam welding process is capable of producing weld metal with a width as small as 1/10 or 1/20 of the thickness of the part, the process is ideally suited to weld components of a critical nature or components that must utilize maximum material and design allowances.

The weld metal geometry produced by this process can be used to control distortion in weldments. In this type of weld (Fig. 5.8 [left]), the weld metal is essentially parallel-sided except where the electron beam first impinges on the top surface of the abutted members. Contraction of the metal during cooling is fairly uniform through the joint. When the weld metal has a characteristic V-shape, as in arc welding, there is significant warpage from unequal thermal contraction across the joint.

Since the electron beam can penetrate through extremely thick sections, beveling or chamfering one or both edges of abutting members is not necessary. However, closer machining tolerances may be required for electron beam welding than for arc welding.

In medium vacuum welding, weld metal cross section configurations are similar to those of high vacuum welding, but the depth-to-width ratios are somewhat smaller. With nonvacuum electron beam welding, the weld bead may be as wide as a typical gas tungsten arc weld. The weld metal may possess all the characteristics of that produced by the more conventional welding processes if insufficient power or large gun-to-work distances are employed. However, at high power and speed, depth-to-width ratios on the order of 5 to 1 are feasible with the nonvacuum welding mode.

EQUIPMENT

Electron beam welding equipment is classified as high vacuum, medium vacuum, and nonvacuum, depending upon the environment in which welding takes place. All three types employ an electron beam gun column, one or more vacuum pumping systems, and a power supply. High and medium vacuum equipment operate with the work in an evacuated welding chamber. Although the work does not need to be placed in an evacuated welding chamber for nonvacuum electron beam welding, a vacuum environment must be provided for the electron beam gun column portion of the system.

All three modes of welding can be performed with high voltage equipment. The gun column may be fixed in position or have limited movement. Low voltage equipment is only suitable for high and medium vacuum operations. With high vacuum, the gun may be fixed in position on the chamber or be mobile inside the chamber. With medium vacuum, the gun must be fixed in position on the chamber.

A mobile, low voltage gun column can usually be moved in the y and z directions during operation while the work carriage moves in the x direction. Gun movement is achieved by action of the welding operator or by automatic programming.

ELECTRON BEAM GUNS

In general, electron beam welding guns are operated in the space-charge-limited condition. If the gun is operated in this condition, the beam current produced at any accelerating voltage is proportional to the 3/2 power of the accelerating voltage $(I = KV^{3/2})$, where the constant of proportionality, K, is a function of gun geometry.

Besides acceleration voltage, a broad range of conditions must be satisfied if an electron gun is to deliver the required power and power density. Design, configuration, emitter characteristics, total power capabilities, and focusing provisions contribute to optimum gun characteristics. For a given metal and joint thickness, characteristically narrow welds can be made if (1) sufficient beam power is available to permit rapid travel speed, and (2) the beam power density is great enough to develop and continuously maintain a vapor hole to the depth of penetration required.

An electron beam gun generates, accelerates, and focuses a beam of electrons. The components of a gun can logically be divided into two categories: (1) the elements necessary for the generation of free electrons (the emitter portion) and (2) the field-shaping elements for the production of a useful beam (the beam-shaping electrode and the anode). Although

major design effort must be given to the proper electrode configuration, and, hence, beam-shaping, careful attention also must be paid to the emitter portion. The emitter may be either (1) a directly (resistance) heated wire or ribbon type filament or (2) a rod or disc type filament indirectly heated by an auxilliary source, such as electron bombardment or induction heating. The choice of emitter design will affect the characteristics of the final beam spot produced on the work.

Only self-accelerated guns, similar to the Pierce and Steigerwald telefocus gun configurations, are used for electron beam welding. They have superior focusing and power capabilities, and also permit placing the gun anode and workpiece at earth ground potential.

The Pierce gun originally was designed as a diode configuration capable of producing a rapidly converging beam with the primary focal point close to the anode. Beam divergence was uniform thereafter. The Steigerwald telefocus gun was designed as a triode configuration which produced a gradually converging beam with the primary focal point some distance from the anode. Current designs of Pierce and telefocus type guns are modifications of the original design configurations. Present day reference to use of a Pierce or telefocus type gun in any particular type of electron beam equipment generally indicates only whether a low voltage diode gun or a high voltage triode gun is being employed.

When a change in beam current is desired at a given accelerating voltage in a diode gun, mechanical change of the cathode-to-anode spacing of the gun is necessary. This changes the proportionality constant, K, of the gun. Several spacers are supplied by the manufacturer to provide a wide range of operating conditions. Operation with the proper spacer permits adequate beam power, beam spot size, and control sensitivity. Diode guns can also be operated in the temperature-limited condition: that is, the beam current at a given voltage is determined by controlling the power to, and thus the temperature of, the electron emitter. Electron emission is related to both emitter temperature and accelerating voltage.

The triode type gun is similar to the diode

gun except that the beam-shaping electrode ("grid cup") is biased with a variable negative voltage relative to the emitter. This makes it possible to easily vary the beam current at any constant accelerating voltage; thus, both the accelerating voltage and beam current can be varied independently within limits.

Full control of beam current with a bias voltage permits rapid change in beam current. Electronic switching circuits in some equipment provide rapid pulsing of the beam. Beam current pulsing can minimize heat input and yet produce deep weld penetration, but it is not used extensively for welding. Accurate control of current slope rates to and from welding power level may be extremely useful in many applications.

All guns focus the electron beam to a small spot on the workpiece with an electromagnetic lens. Furthermore, the beam can be oscillated in a repetitive or nonrepetitive manner by means of a suitable set of magnetic deflection coils and controls. The coils are positioned immediately below the electromagnetic focusing lens where they can deflect the electron beam from its normal axial position. Simultaneous use of two sets of deflection coils at 90 degrees allows the formation of the classical lissajous figures on the work surface. Sinusoidal beam deflection perpendicular to the direction of welding permits broadening of the weld bead to simplify manual tracking of weld seams. Circular and elliptical deflection tends to reduce weld porosity. More complex deflection patterns may be employed to enhance penetration and improve weld quality.

Electron beam welding guns are designed for either low voltage or high voltage operation. Low voltage guns operate below 60 kV and are generally modified Pierce types. High voltage guns operate above 60 kV and are generally the modified telefocus types. Similar power levels are available with either low or high voltage guns. Beam power is the product of the accelerating voltage and the beam current. Therefore, operation at high voltage requires less beam current than operation at low voltage for equivalent power. In general, the use of either low voltage or high voltage equipment in high or medium vacuum applications will produce

similar quality welds in most metals. However, differences in weld geometry can exist because one system operates with low voltage and high amperage and the other with high voltage and low amperage. High voltage equipment is the only suitable type for nonvacuum electron beam welding applications.

High voltage equipment requires special insulating techniques that limit the mobility of the electron beam gun. Although for some applications a sliding vacuum seal permits limited movement, the gun is usually fixed on the welding chamber. Low voltage guns may be mounted on the chamber in fixed position or on a travel carriage inside the chamber for mobile operation.

POWER SUPPLIES

The power supply of an electron beam welding machine is an assembly of at least one main power supply plus one or more auxiliary power supplies. It produces high voltage power for the gun and auxiliary power for the emitter and beam control. Depending upon whether a diode type or a triode type of gun is used, the high voltage power supply consists of two or more of the following components:

(1) A main high voltage dc power source that delivers the electron beam accelerating voltage and the total beam current

(2) An emitter power supply with either ac or dc output

(3) An electrode bias supply that impresses a voltage between the emitter and the bias electrode (grid cup) to control beam current

(4) Special electronic supplies, such as a beam pulser or a beam current regulator

The main high voltage power source and the auxiliary power supplies are frequently placed together in a common oil-filled tank. A high purity, electrical grade transformer oil serves both as an electrical insulating medium and as a heat transfer agent to carry heat from the electrical components to the tank walls. The components are typically supported from the cover plate of the tank so that they can be removed from the tank with the cover plate. The oil. need not be removed except for periodic filtering or replacement.

Another high voltage insulating material, used infrequently in a power supply, is sulfur hexafluoride gas at a pressure of 45 psi. A power supply with this gas insulation is considerably more compact and lighter than an oil-insulated unit of the same rating.

The components in both the main high voltage power supply and the auxiliary supplies are primarily transformers, diodes (rectifiers), capacitors, and resistors. Electron tube diodes were initially used in the main power supply by some manufacturers, but solid-state diodes, either selenium or silicon, are usually employed now. Cost, regulation, physical size, and ability to absorb transients of voltage and current are some of the considerations affecting the choice of components.

Main High Voltage Power Source

This unit converts three-phase line power to high voltage dc power for the electron beam gun. Common power ratings are in the range of 3 to 35 kW; the maximum is 60 kW. Units are designed to provide the output voltage and beam current for a particular electron beam gun type (high or low voltage). Typical machine ratings are given in Table 5.1.

The requirements for maximum voltage ripple in the dc output vary depending upon the desired quality of beam focus. Excessive voltage ripple will produce undesirable ripple in the beam current that may be reflected in weld quality. Therefore, maximum ripple voltage is usually one percent or less.

The inherent decrease in output voltage with increasing load is typically in the range of 15 to 20 percent. Various controls and regulators are used to compensate for this voltage decrease and to minimize the effects of line voltage variations as well as the effects of temperature and other factors influencing the output voltage. Sophisticated controls are designed to eliminate all of the effects mentioned and to maintain a stable accelerating voltage to within 1 percent of the selected value. Other less costly controls eliminate only some of the effects and are adequate for less critical applications. Some of the controls used, in approximate order of increasing sophistication and performance, include:

(1) Line voltage regulator (constant voltage transformer)

(2) Servo-operated variable transformer with feed-back from the high voltage output

(3) Motor-generator with electronic exciter and feedback from the high voltage output

(4) Electronic regulation with both current and voltage feedback available

Emitter Power Supply

Directly heated wire or ribbon emitters are typically formed in a hairpin-like shape. The heating current can be either ac or dc. Direct current is preferred because the magnetic field created by the heating current can influence the direction of the beam. The cyclic nature of ac heating will cause the beam spot to oscillate with a small, but significant, amplitude about a fixed point.

Since the magnitude of any such magnetic effect will increase with the amount of heating current employed, it is necessary to use filtering even with dc heating currents to reduce any ripple to 3 percent or lower.

Current and voltage ratings of a filament power supply depend upon the type and size of directly heated filament used. For 0.020-in. diameter tungsten wire filaments, the supply would be rated for 30A at 20V. Ribbon type

Table 5.1
Typical electron beam welding machine ratings

Rating, kW	Output	
	kV, max	mA, max
1.25	25	50
3	30	100
3	60	50
6	30	200
7.5	150	50
9	30	300
15	30	500
15	60	250
15	150	100
17.5	175	100
25	175	144
30	60	500
35	150	230
35	200	175
60	175	345

filaments are employed to provide a much larger emitting area than that of a wire type filament. Ribbons require power supplies rated for higher currents and lower voltages (30 to 70A at 5 to 10V).

Indirectly heated emitters are also in common use. Here, an auxiliary source of either the bombardment or inductive type is used to indirectly heat the gun emitter to electron emission temperatures. The power supplies, therefore, differ from those employed with directly heated filaments. A typical rating of a power supply for an auxiliary electron gun that heats a disc emitter by bombardment would be 100 to 200 mA at several kilovolts.

Bias Voltage Supply

The bias voltage supply for a triode type gun is usually designed to give complete control of beam current from zero to maximum. To do this, the dc power supply applies a variable voltage on the beam-shaping electrode (grid cup), negative with respect to the emitter. A voltage in the range of 1500 to 2000V is needed to cut off the beam current. For maximum beam current, the voltage requirement decreases to the 200 to 300V range. As with the high voltage power supply, the bias supply must also produce well-filtered voltage that does not introduce more than one percent ripple in the beam current.

Some electron beam equipment uses a self-biasing system. The bias voltage is derived partially from the main accelerating voltage through a voltage divider, and partially from the voltage across a series resistor in the main power circuit. This system does not have a separate bias supply per se.

Electromagnetic Lens and Deflection Coil Power Supplies

The electromagnetic lens (focusing coil) is generally powered by a constant current power source of solid-state design. The strength of the magnetic field is directly related to the current flowing through the coil. The current provided to the coil must remain constant for consistent beam spot size, even when the voltage drop

across the coil changes with temperature variations.

Beam deflection coils are also powered by solid-state devices. Two sets of coils at 90 degrees are usually placed at the base of the gun column for x and y deflection of the beam. Programming of the power sources for the two sets of coils can provide beam movement along either axis or both axes simultaneously. Geometric beam patterns (circle, ellipse, square, and rectangle) can be produced by electronic control. The ripple on the dc input to both the deflection coils and the electromagnetic lens must be low to minimize any adverse effects of beam instability on weld quality.

VACUUM PUMPING SYSTEMS

Vacuum pumping systems are required for evacuation of the electron beam gun chamber, the work chamber for high and medium vacuum modes, and the orifice assembly at the exit of the gun column for medium vacuum and nonvacuum welding. Two basic types of vacuum pumps are used in these systems. One is a mechanical piston or vane type used to obtain a pressure reduction from 1 atmosphere to about 0.1 torr. For medium vacuum welding, these mechanical pumps are generally operated in conjunction with a Roots type blower, another kind of mechanical pump. The other type is an oil diffusion pump used to reduce the pressure to 10^{-4} torr or lower. The sequencing of these pumps to produce the needed vacuum can be accomplished by either manual or automatic operation of valves in the system. In commercial electron beam welding equipment, automatic valve sequencing is virtually standard.

The vacuum system for an electron beam gun chamber consists of a mechanical roughing pump and a diffusion pump. A similar system is used to evacuate the work chamber in the high vacuum mode of operation. In both these cases, the system may be ducted to the chamber through a water- or liquid nitrogen-cooled (optically dense) baffle, if necessary, to minimize backstreaming of diffusion pump oil into the chambers. A combination diffusion and mechanical pumping system is shown in Fig. 5.11.

The vacuum pumping system can be mounted on the same base as the vacuum chamber and connected with rigid ducting, except for the mechanical roughing pump. The roughing pump is connected to the system with a flexible tube to minimize vibration of the welding chamber. A large diameter vacuum valve isolates the diffusion pump from the chamber during the roughing portion of the pumping cycle. A small mechanical pump holds the isolated diffusion pump at low pressure.

The roughing and diffusion pumping periods are normally controlled by automatic sequencing of pneumatic or electric vacuum valves. Automatic evacuation cycles are accomplished with pressure-sensing relays that activate the appropriate valves in the preprogrammed sequence. The control units are designed to protect the vacuum system in case of accidental pressure rise in the chamber.

Cryogenic traps, such as liquid nitrogen (LN), are not usually used on electron beam welding systems unless pressures of 10^{-6} torr or lower are required.

For medium vacuum welding, the work chamber is evacuated with a mechanical vacuum pumping system of high capacity. The types and sizes of mechanical pumps used in the system will depend upon the work chamber size, the workload, and the desired production rate. Automatic evacuation cycles are used to enhance high speed production capability.

With nonvacuum welding equipment, the electron beam gun chamber is evacuated with a combination mechanical-diffusion pumping system. The various pressure stages in the orifice assembly through which the electron beam exits from the gun are pumped with a series of mechanical vacuum pumps.

The evacuation process and its rate depend

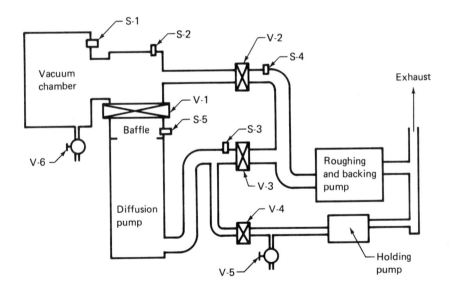

Valves
V-1 — high vacuum
V-2 — roughing
V-3 — backing
V-4 — holding
V-5 — vacuum release
V-6 — vacuum release

Vacuum sensors
S-1 — ion type
S-2 — thermocouple type
S-3 — thermocouple type
S-4 — thermocouple type
S-5 — ion type

Fig. 5.11—Vacuum pumping system for high vacuum operation

upon the capacities of the pumps, the work and fixturing load, the size of the chamber, and the total leakage rate of the system. The total leakage rate is the increase in chamber pressure per unit of time attributed to both real leaks and virtual leaks in the system.

Real leaks are actual holes or voids in the chamber capable of passing air or gas. A virtual leak is the semblance of a real leak being present somewhere in the vacuum system resulting from the outgassing of adsorbed or occluded gases on the interior surfaces of the system when under vacuum. For satisfactory system operation, no in-leakage (real leaks) should be detectable when a helium mass spectrometer leak detector, having a sensitivity of 1×10^{-7} standard cm^3/s of helium, is employed to leak-check the system.[2] A customary limiting value for an acceptable rise rate test is in the range of 1 to 2×10^{-2} torr per hour, averaged over a 10-hour test period. This test is conducted by isolating the chamber to be tested from the pumping system (without exposing it to atmosphere) immediately after a four-hour preparatory pumpdown of the chamber. The 10-hour period starts when the vacuum level reaches 1×10^{-3} torr.

LOW VOLTAGE SYSTEMS

Work chambers of low voltage machines are usually made of carbon steel plate. The thickness of the plate is designed to provide both adequate x-ray protection and the structural strength necessary to withstand atmospheric pressure. Lead shielding may be required in certain areas of the chamber to ensure total radiation tightness of the system.

The weldment inside the chamber may be observed by direct viewing through leaded glass windows. However, the effectiveness of this technique depends upon the distance between the operator and the weld joint and the shape of the workpieces. When direct viewing is difficult, an optical viewing system may be provided to give the operator a magnified view

of the weld seam. It is used for setup operations, inspection of the weld, alignment of the weld joint with respect to the electron beam, and positioning of the gun to center the sharply focused electron beam on the weld seam.

Closed-circuit television is used to provide a better viewing system. It may have both a light source and a television camera mounted in a serviceable location outside the chamber. An optical protection system is mounted inside the chamber to shield the viewing equipment from metal splatter and metal vapor deposition. The closed-circuit television system permits continuous monitoring of welds and minimum exposure to the intense light from the weld.

Computer control and numerical control are available for automatic programming of beam power and travel speed as a function of gun position.

HIGH VOLTAGE SYSTEMS

High voltage electron beam systems operate above 60 kV. The equipment is designed to operate at voltages between 100′ kV and 200 kV, with beam powers of up to 100 kW. The electron beam gun of a high vacuum welding machine is housed in an external vacuum chamber mounted on top of the welding chamber with either stationary or sliding vacuum seal, as shown in Fig. 5.12. In the latter case, motion of the gun is limited to the y axis, with x axis motion provided by the weldment carriage. Other motions, such as z axis movement, are provided by special jigging of the work. Computer- and numerically-controlled electron beam power and travel speed are also available.

The external location of the gun reduces its maneuverability, but provides ready access to the gun components for service. This arrangement also provides the operator with a view of the beam spot and the weld through optics that are coaxial with the electron beam (Fig. 5.12). A view through a coaxial optical system is shown in Fig. 5.13. In addition, direct viewing is provided through lead glass windows in the chamber walls.

Work chambers for this equipment are usually carbon steel boxes of ribbed design with

___
2. See ASTM E498 for a description of this method of leak testing.

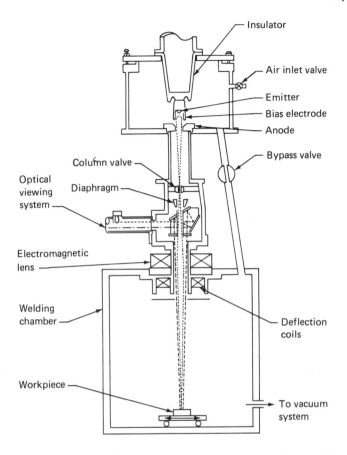

Fig. 5.12—A typical high voltage electron beam welding machine arrangement

an external cladding of lead for adequate x-ray protection.

Nonvacuum electron beam welding machines do not require a vacuum chamber around the workpiece. The gun column may be fixed atop an x-radiation shielding box containing the workpiece and travel carriage. Another arrangement is to place the gun column and the workpiece in an x-ray shielded room where both the gun and workpiece may be traversed. The equipment is operated remotely from outside the room.

SEAM TRACKING METHODS

With electron beam welding, as with other automatic welding processes, it is important that the relative positions of the beam spot and the weld joint be established accurately prior to initiation of the welding cycle. This relationship must then be accurately maintained throughout the entire welding cycle. This total requirement is somewhat complicated in electron beam welding because (1) the beam spot is very small and produces a relatively narrow weld bead; (2)

Fig. 5.13—A sighting on an electronic component through a coaxial optical system of a high voltage electron beam gun

welding is performed at relatively high travel speeds; and (3) the workpiece is contained in a vacuum chamber or radiation enclosure making continuous observation of welding somewhat difficult.

As previously mentioned, most high vacuum electron beam machines are equipped with some type of viewing system that permits the operator to observe the weld joint and the beam spot. The initial correct position of the electron beam in relation to the joint can easily be established with a viewing system. On medium vacuum and nonvacuum machines, where operator viewing of some sort is not normally provided, this initial beam-to-joint alignment is accomplished through precise handling (tooling and fixturing) of the part.

For welding long or slightly irregular joints, a means for automatically maintaining proper beam-to-joint alignment is desirable. Direct optical viewing of welding and manual correction for deviations in the joint path is, at best, a difficult task, although some equipment is utilized in this fashion.

There are two methods for maintaining beam position on a nonlinear joint. The first involves programming by analog means or continuous path numerical controls. This method is applicable if the parts are machined precisely to the required contour and are accurately positioned for welding.

A second method for tracking the joint is an adaptive electromechanical control. This control uses a tracking device that follows the weld joint and signals the control to adjust the work or gun position to keep the beam on the joint. Two seam tracking devices that employ the same electrical circuitry and are quickly interchanged to accomplish various tracking requirements are the stylus and contour types.

The stylus type seam tracking system is based on a probe, or stylus, that rides in the joint. Lateral (cross seam) movements of the probe, resulting from a change in joint position, are converted to electrical signals by a transducer. The electrical signals drive a positioning servomotor that maintains preset alignment of the joint-to-gun focus. The electrical signals from this system define a right-error, a left-error, and the null or correct gun position. Alternatively, the electron beam can be deflected electronically to accommodate changes in the location of the joint.

A simple modification of the stylus type seam tracking system permits the edge welding of certain types of assemblies by using the weldment as a cam. A preloaded ball type stylus rides against the edge of the weldment as it is rotated or driven linearly. In some cases, the device may trace a replica of the joint configuration. As a proximity control to accommodate changes in vertical position of the weld joint, the system is modified to maintain a constant gun-to-work distance by applying the tracking signal to a servomotor drive on the z axis.

An electronic seam-locating system is also available to provide seam tracking capability that can be used in a two-step operation. The first step is to visually verify that the weld joint and weld travel direction are parallel with the workpiece clamped into position. This is a mechanical adjustment of the fixturing in the welding chamber.

The vacuum chamber is then closed and evacuated to the welding pressure. A finely focused electron beam of less than 1 mA is aimed at the weld joint and oscillated across the joint at 60Hz. The reflected electrons from this scanning beam are deflected by a pickup plate, and the resulting signal is displayed as a visible trace on an oscilloscope. A discontinuity in the oscilloscope line trace indicates the seam location, which is used for the initial electron beam column centerline-to-joint alignment. Any shift in the discontinuity position relative to a null location, when traversing the weld joint, indicates a change in joint position with respect to the electron beam column centerline. From the oscilloscope display, the operator can tell quickly and precisely the position of the electron beam column centerline with respect to the joint. From this information, necessary adjustments can be made to assure proper positioning of the electron beam with respect to the joint. Excessive weld joint gap and runout are also easily detected with this system.

All of the above seam tracking devices can be used in conjunction with a record-and-playback device that allows the joint to be traced and its location recorded. Then the joint can be welded using playback programmed control of the beam or work position.

WORK HANDLING EQUIPMENT

Once the relative positions of the electron beam and the weld joint are established, it is important that this relationship be maintained during the entire welding operation. Therefore, the work motion mechanisms must be accurate and well defined. Their design and manufacture should follow good machine tooling practices. Ruggedness, repeatability, smoothness, and accuracy are prime requirements. Since travel speed affects weld geometry, it is important that this variable be controlled accurately and be repeatable. In general, electric motor drives having an accuracy of about $\pm$ 2 percent of set speed are adequate.

Most electron beam welding machines provide standard mechanisms for linear and rotary motion of the workpiece relative to the electron beam. Linear motion in the horizontal plane is usually provided either by movement of a work table or by movement of the electron gun. Rotary motion about a vertical axis is achieved with a motor-driven horizontal rotary table. Machines can be equipped with an external platform so that the work table and any work-handling mechanisms can be withdrawn from the vacuum chamber for ease of loading and fixturing.

Figure 5.14 shows an x-y work table on its external platform. An adjustable (0 to 90 degree) rotary work positioner and tailstock are mounted on the table. Rotary motion about the horizontal axis can be accomplished with the rotary positioner and tailstock. The positioner is power-

Fig. 5.14—An X-Y work table with a rotary positioner on the run-out platform of the electron beam welding machine

driven. Simple linear, rotary, and circular welds can be made with these mechanisms.

It is often desirable to produce circular welds in several parts as a single load in the chamber. In this case, the components are arranged on an eccentric table that travels in a circular manner. The circle diameter can be changed by a simple adjustment of the table. The eccentric table is positioned on the work table. Each piece in turn is moved into position, and the circular weld motion is made with the eccentric table. Programmed indexing from piece to piece can be added. This eccentric table can be used for single circular welds in large or odd-shaped pieces where it is inconvenient or diffi-

cult to rotate the entire piece about the center of the weld.

Multiple-spindle rotary fixtures are frequently used to make circumferential welds in a group of similar parts. The parts are successively indexed into welding position by a motor drive. The joint on each part can be positioned for welding by linear movement of the work table on which the rotary fixture is mounted. It is possible to automate the entire operation. An example of a special purpose rotary fixture is shown in Fig. 5.15.

Since all operating variables of electron beam welding are machine controlled, this process is readily adaptable to computerized numerical control. Movement of the workpieces or the gun, as well as electron beam deflection, can be preprogrammed in any combination. The beam current itself is also programmable. It can be changed from one discrete level to another, or changed at a specified rate. Other variables, such as accelerating voltage, emitter power, and chamber pressure as well as auxiliary functions, can be part of the program for either control or monitoring. Electron beam welding systems use this computerization capability for contour welding of intricately-shaped parts, or for monitoring the welding variables of parts requiring precise process control.

MEDIUM VACUUM EQUIPMENT

Equipment for medium vacuum electron beam welding is basically a modification of standard high vacuum equipment. A diffusion pump and an "aperture" tube (an orifice that allows beam passage, but impedes gas flow) are added to the beam column. A column valve is used to isolate this separately pumped gun region from the work chamber region, so that the gun remains under vacuum when the chamber is vented to atmosphere. Once a part is in place, the work chamber is rapidly pumped down to some medium vacuum level and then the column valve is opened. In this manner, a high vacuum of 10^{-4} torr or better is always maintained in the beam-generating section of the column, but the beam can be impinged on the workpiece in its medium vacuum environment. Low and high voltage systems are produced for this mode of welding.

General purpose medium vacuum machines,

Fig. 5.15—A loaded rotary welding fixture in location beneath the electron beam gun

Fig. 5.16—A general purpose medium vacuum electron beam welding machine

such as the one in Fig. 5.16, are employed advantageously in short production runs. However, most medium vacuum machines are special purpose ones tooled for particular assemblies. Figure 5.17 illustrates some typical medium vacuum tooling concepts. In each case, the work chamber and tooling are an integral assembly, specifically designed for a single part design.

Medium vacuum welding machines of a great variety are in operation for high production applications. For example, a machine with a single welding station and multiple loading stations can have production capability in the region of 200 parts per hour. A dual welding station machine, on the other hand, could increase the production capability up to 500 parts per hour. The production rates in the final analysis are dependent upon the design of the parts.

Another method for achieving high production with medium vacuum equipment is shown in Fig. 5.18. Here, a sliding seal is used to provide intermediate vacuum zones before and after the separately pumped medium vacuum welding chamber. This method of maintaining a series of continuously pumped vacuum zones eliminates the need for evacuation time and enables the high production capability of a dial feed table to be fully utilized.

NONVACUUM EQUIPMENT

A beam of electrons passing through a gas is primarily scattered by the shell electrons of the gas atoms or molecules. As the gas pressure increases, scattering becomes more severe (Fig. 5.4). This effect produces a noticeable broadening of the beam profile and a decrease in beam power density, but not necessarily a loss in total beam power.

An electron beam must be generated in high vacuum. To weld with the beam at atmospheric pressure, the beam must pass through a series of chambers or stages operating at progressively higher pressures. In addition, the electron vel-ocity (accelerating voltage) must be high to minimize the scattering effect of the gas atoms or molecules.

A series of chambers operating at successively higher pressures is obtained by staging (differentially pumping) the vacuum over some finite length with an orifice system through which the electron beam passes. The orifice sys-

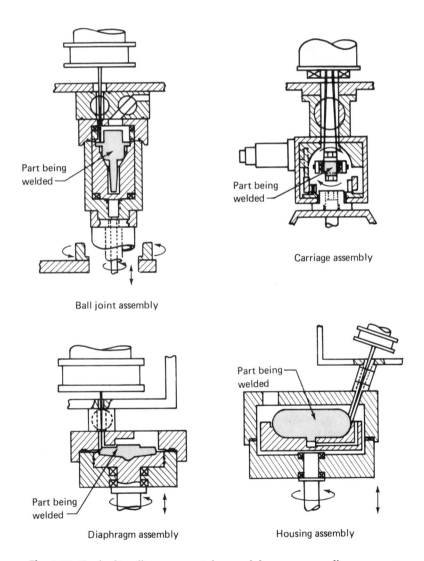

Ball joint assembly

Carriage assembly

Diaphragm assembly

Housing assembly

Fig. 5.17—Typical tooling concepts in special purpose medium vacuum electron beam welding machines

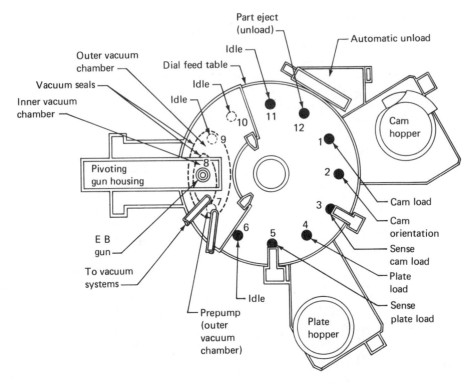

Fig. 5.18—A medium vacuum electron beam welding system with a prepumping zone for continuous part feed capability

tem design must be capable of producing not only the atmospheric-to-high vacuum gradient required, but also the type of gas dynamic characteristics needed in this type of staging system. A beam of relatively high speed electrons is produced with high accelerating voltage; about 150 kV minimum is required to produce a practical working distance between the final orifice and the workpiece. In addition, the beam power level and the ambient atmosphere through which the beam passes can greatly influence the useful working distance attainable.

Figure 5.19 illustrates a nonvacuum electron beam gun column assembly with an orifice system. The electron gun is typical of those used for the other variations of electron beam welding. Accelerating voltages are in the range of 150 to 200 kV. Beam current and, thus, power are controlled by the voltage on the bias elec-

trode of the gun. The beam is focused by an electromagnetic lens to the minimum diameter of the orifice system shown in the lower part of the figure. It emerges from the vacuum environment into air at atmospheric pressure through the lower orifice. Inert gas shielding can be added, if desired. The workpiece is placed near the lower orifice. A high vacuum is maintained continuously in the gun area during operation by an oil diffusion or turbomolecular pump. Higher pressures are maintained in the interim pressure stages by mechanical pumps. In most cases, the work is moved in front of the gun column. The gun can be placed in either a vertical or a horizontal position. The welding area must be shielded to prevent external x-radiation as with high vacuum and medium vacuum modes.

Figure 5.20 shows another type of nonvacuum electron beam welding gun column and its power supply and controls. This particular type features a gas-filled, high voltage power supply that can be mounted directly on the gun column assembly during operation. The gun and power supply can be traversed along a weld joint during operation.

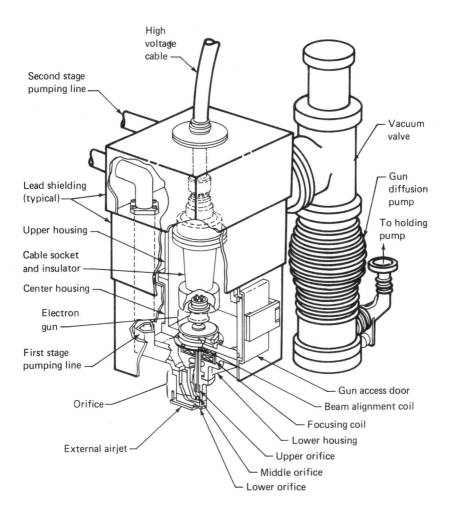

Fig. 5.19 — A nonvacuum electron beam gun column assembly

Fig. 5.20—Nonvacuum electron beam gun column and its gas-filled power source, disassembled

WELDING PROCEDURES

JOINT DESIGNS

Butt, corner, lap, edge, and T-joints can be made by electron beam welding using square-groove or seam welds. Fillet welds are difficult to make and are not generally used. Typical electron beam weld joint designs are shown in Fig. 5.21. Modifications of these designs are frequently made for a particular application.

Square-groove welds require fixturing to maintain fit-up and alignment of the joint. They can be self-aligning if a rabbet joint design is used. The weld metal area can be increased using a scarf joint, but fit-up and alignment of the joint are more difficult than with a square-groove weld. Edge, seam, and lap fillet welds are primarily used to join sheet gage thicknesses.

JOINT PREPARATION AND FIT-UP

For most applications, the fit-up of parts must be more precise than for arc welding processes. This is due to the very small diameter of the electron beam as compared to a welding arc. The beam must impinge on and melt both members simultaneously, except for seam welds where the beam penetrates through the top sheet. Also, in most cases, filler metal is not needed to fill a joint gap or provide reinforcement to the weld. To avoid underfill or incomplete fusion, joints must be carefully prepared to provide good fit-up and alignment.

A metal-to-metal fit between parts is desirable but difficult to obtain. Normally, the gap or opening between the faying surfaces should

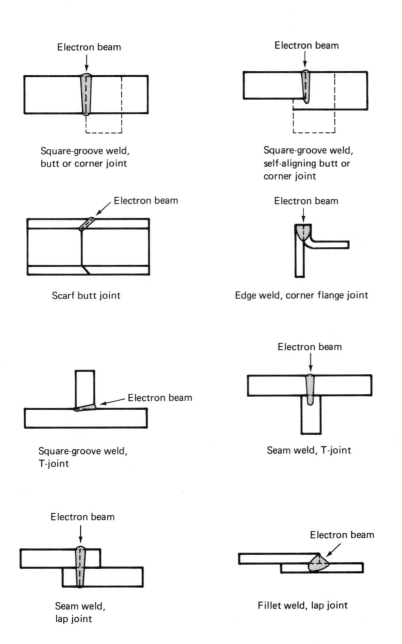

Fig. 5.21—Typical electron beam weld joint designs

not be more than 0.005 inch. The acceptable gap for a particular application will depend upon the process mode employed, the type of base metal, the thickness of the joint, and the required weld quality. In general, sheet sections require a fit-up of less than 0.005 in.; plate sections may tolerate somewhat more than this amount. Aluminum alloys can tolerate somewhat larger gaps than steel. Beam deflection or oscillation with high and medium vacuum welding, to widen the fusion zone, and nonvacuum welding may permit larger gaps. However, the acceptable joint gap and tolerance should be determined for each application to avoid unnecessary costs.

In general, roughness of the faying surfaces is not critical so long as the surfaces can be properly cleaned to remove any contamination. Burrs on the sheared edges of sheet are not detrimental unless they separate the faying surfaces of lap joints.

CLEANING

Cleanliness is a prime requisite for any welding process where high standards are to be met. The particular requirement will depend upon the end use of the welded product. Contamination of the weld metal is likely to cause porosity or cracking, or both, as well as a deterioration of mechanical properties. Improper cleaning of the components to be welded may cause chamber evacuation time to be excessive with the vacuum modes.

Acetone is a preferred solvent for cleaning electron gun components and workpiece parts, but it is highly flammable and must be handled accordingly. Chlorinated hydrocarbon solvents should not be used because of their detrimental effect on high voltage equipment operation. If a vapor degreaser containing a chlorinated hydrocarbon solvent must be used for heavy degreasing, the parts should be thoroughly washed in acetone afterward. Another alternative is to use a fluorocarbon type solvent in the degreaser. After final cleaning, the joint area should not be touched by hand or tools to avoid contamination.

Surface oxides and other forms of contamination that solvents will not dissolve should be removed by mechanical or chemical means. Flat surfaces of soft metals, such as magnesium, aluminum, and copper, can be scraped by hand. Machining without coolant is preferred for all but very hard metals where grinding must be used. Surfaces that are not prepared by machining should be chemically cleaned. Wire brushing is not generally recommended because it tends to embed contaminants in the metal surface. Grit blasting and grinding are not recommended for soft metals, including soft steels, because the grit may be embedded in the surfaces. Nonvacuum welding will generally require less stringent precleaning than vacuum welding.

FIXTURING

All electron beam welding is done by either machine or automatic operation. The parts must be fixtured to align the joint, unless the design is self-fixturing, and then either the assembly or the electron beam gun column must be moved to accomplish the weld.

Where practical, self-fixturing joint designs should be used. A pressed or shrink fit can be used to position circular parts for welding. However, these methods require close tolerance machining, which may not be economical for high production welding.

Fixturing for electron beam welding does not have to be as strong and rigid as that required for automatic arc welding. The reason is that electron beam welds are generally made with much lower power than arc welds. Therefore, stresses in the weldment caused by thermal gradients are significantly lower.

The close joint fit-up and alignment for electron beam welds generally require that the fixturing be made to close tolerances. If necessary, copper chill blocks may be used to remove heat from the joint. A backup of the same metal as the workpieces should be positioned at the back side of the weld to protect the part or fixture from an exiting electron beam.

Work tables and rotating positioners should have smooth and accurate motion at the required travel speeds. All fixturing and tooling should be made of nonmagnetic metals to prevent magnetic deflection of the beam. All magnetic metals should be demagnetized before welding them.

The entry and exit of the electron beam tends to produce underfill at both ends of the welded joint. To minimize or eliminate this defect, tabs of the same metal as the workpieces are fitted tightly against both ends of the joint. The beam is initiated on the starting tab, traversed across the weld joint, and terminated on the runoff tab. Later, the tabs are cut off flush with the ends of the workpiece by machining.

FILLER METAL ADDITIONS

When the faying surfaces of butt joints are fitted together with acceptable tolerances, filler metal is not normally needed to obtain a full thickness weld. As welding processes along the joint, weld metal flows from the leading edge to the trailing edge of the vapor hole. This action, combined with thermal contraction as welding progresses, usually produces a welded joint free of underfill when proper welding procedures are used. However, there are applications when it is desirable or necessary to add filler metal to obtain an acceptable welded joint.

Filler metal may be added to produce certain physical or metallurgical characteristics in the weld metal. Some weld metal characteristics that may be altered or improved by filler metal addition are ductility, tensile strength, hardness, and resistance to cracking. For example, addition of a small amount of aluminum wire or shim can produce a "killing" action in steel, which will reduce porosity.

Filler metal is often added to the joint for metallurgical purposes. Wire feed is not employed exclusively. The dilution obtained from a dissimilar filler metal added as wire at the surface does not occur uniformly to any great depth. In the case of heavy plate, filler metal in the form of a thin shim must almost always be employed. The presence of the filler shim requires that beam oscillation or a large diameter spot be used to melt the shim and the base metal on both sides of the joint. This is not the case with thin metal weldments where filler wire can be added and dilution will occur throughout the entire joint. Typical examples of filler metal addition for metallurgical reasons are the welding of Type 6061 aluminum alloy using Type 4043 aluminum filler metal and the welding of beryllium using aluminum or silver filler metal.

Filler metal may be added to fill the joint during a second pass after the penetration pass has been made. This is done to provide a full thickness weld.

Filler wire feeding equipment is usually either a modified version of that used for gas tungsten arc welding or specially designed units for use in a vacuum chamber. Filler wire diameters are generally small, 0.020 in. and under. The wire feeder must be capable of uniformly feeding small diameter wire into the leading edge of a small molten weld pool. The wire feeding nozzle should be made of a heat-resistant metal.

For welding in a vacuum chamber, the filler wire drive motor must be sealed in a vacuum-tight chamber. Otherwise, the motor will contribute significantly to the outgassing load and increase evacuation time. Provisions for adjustment of the wire feed nozzle to position the wire with respect to the electron beam and the weld joint must be located externally.

SELECTION OF WELDING VARIABLES

The expression most widely accepted for the rate of energy input to the workpiece is joules per inch.[3] The equation for this expression is:

$$\text{Energy input, J/in.} = \frac{EI}{S} = \frac{P}{S}$$

where: E = beam accelerating voltage, V
I = beam current, A

---

3. Energy input to the weld from a heat source is discussed in Ch. 2, Welding Handbook, Vol. 1, 7th ed., p.35-6.

P = beam power, W or J/s

S = travel speed, in./s

The above equation and data for welding various thicknesses of a metal provide for interpolation of welding variables for that metal by graphical means. A curve relating energy input per unit length with thickness for a particular family of alloys can be determined from a few tests to establish the welding conditions for typical metal thicknesses. Figure 5.22 shows several such curves. Such graphs are useful to determine initial welding power and travel speed settings for a particular alloy and thickness. This is possible because of three factors. First, the welding machine settings are usually regulated by closed-loop servocontrols that ensure stability and reproducibility. Second, the adjustment of each variable is independently controlled, thus permitting flexibility in selection. Third, there are only four basic variables to adjust: accelerating voltage, beam current, travel speed, and beam focus. These factors combine to make the process of establishing the welding schedule relatively simple.

Once the required energy input per unit length is determined for a given metal thickness, the travel speed can be selected and the required welding power defined or vice versa. The beam voltage and current can now be selected to produce the required power.

The beam size selected will be dependent upon the desired weld bead geometry. To maintain a selected beam spot diameter at the surface of the workpiece, it is necessary to correspondingly increase the focus coil current as the accelerating voltage is increased, since the beam spot size is a dependent function of the accelerating voltage. If the accelerating voltage is maintained constant but the gun-to-work distance is increased, a corresponding decrease in focus coil current is necessary to maintain a selected beam spot diameter at the surface of the workpiece.

Changes in individual welding variables will affect the penetration and bead geometry as follows:

(1) Accelerating Voltage. As the accelerating voltage is increased, the penetration of the

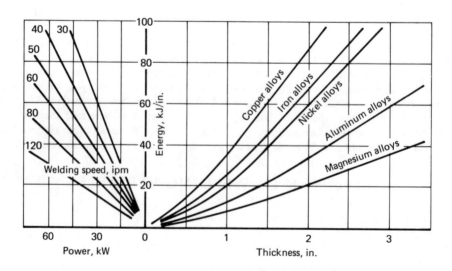

Fig. 5.22—High vacuum electron beam welding energy requirements for complete penetration welds in several metals as a function of joint thickness

weld will also increase. For long gun-to-work distances or the production of narrow, parallel-sided welds, the accelerating voltage should be increased and the beam current decreased to obtain maximum focal range (see Fig. 5.1).

(2) Beam Current. As the beam current is increased, penetration of the weld bead will also increase.

(3) Travel Speed. As the travel speed is increased, the weld bead will become narrower and the penetration will decrease.

(4) Beam Spot Size. Sharp focus of the beam will produce a narrow, parallel-sided weld geometry because the effective beam power density will be maximum. Defocusing the beam, either by overfocusing or by under-focusing, will increase the effective beam diameter and reduce beam power density. This, in turn, will tend to produce a shallow or V-shaped weld bead. These effects are shown in Fig. 5.23.

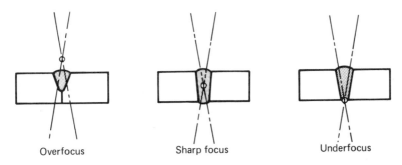

Overfocus   Sharp focus   Underfocus

Fig. 5.23—Effect of electron beam focusing on weld bead geometry and penetration

METALS WELDED

In general, metals and alloys that can be fusion welded by other welding processes can also be joined by electron beam welding.[4] This includes similar and dissimilar metal combinations that are metallurgically compatible. The narrow weld metal geometry and thin heat-affected zones, especially in the high vacuum process mode, favor welded joints with better mechanical properties and fewer discontinuities than arc welded joints. However, electron beam welds in alloys that are subject to hot cracking or porosity will likely contain such discontinuities. The weldability of a particular

alloy or of combinations of alloys will depend upon their metallurgical characteristics, part configurations, joint design, process variation, and welding procedure.

STEELS

Rimmed and Killed Steels

In rimmed steel, the chemical reaction that occurs between carbon and oxygen to form a gas (CO) will occur in the molten weld pool. As a result, violent weld pool action, spatter, and porosity in the weld metal are expected with this type of steel.

Electron beam welds in rimmed steel can be improved if deoxidizers, such as manganese,

---

4. The weldability of various metals is covered in Sec. 4, Metals and Their Weldability, *Welding Handbook*, 6th ed. This section will be revised as Vol. 4, 7th ed.

silicon, or aluminum, are incorporated through filler metal additions. Deoxidizers can also be added locally to the joint area by painting, spraying, and shim inserts.

Welding of rimmed steel can also be somewhat enhanced by the proper selection of electron beam welding conditions, such as slow travel speed, to produce a wide and shallow weld cross section. The conditions should be selected to provide time for gases to escape from the molten weld metal. Then a weld of reasonable quality can be obtained. The use of beam deflection may also be effective in reducing weld porosity.

Hardenable Steels

Thick sections of hardenable steels may crack when electron beam welded without preheat. Very rapid cooling in the fusion and heat-affected zones will result in the formation of hard martensite. The combination of a hard, brittle microstructure and residual stresses can cause crack formation. Cracking can be prevented by preheating, which can be done with a defocused electron beam in many applications. However, careful programming and monitoring are necessary to achieve the proper preheat temperature with this method of heating.

Stainless Steels

Austenitic Stainless Steels. The high cooling rates typical of electron beam welds help to inhibit carbide precipitation because of the short time that the steel is in the sensitizing temperature range.

Martensitic Stainless Steels. Although these steels can be welded in almost any heat-treated condition, a hard martensitic heat-affected zone will result. Hardness and susceptibility to cracking increase with increasing carbon content and cooling rate.

Precipitation-Hardening Stainless Steels. These types of steels can, in general, be electron beam welded to produce good mechanical properties in the joints. The semiaustenitic types, such as 17-7PH[5] and PH14-8 Mo,[5] can be welded as readily as the 18-8 types of austenitic stain-

5. Trademarks

less steel. The weld metal becomes austenitic during welding and remains austenitic during cooling. In the more martensitic types, such as 17-4 PH[5] and 15-5 PH,[5] the low carbon content precludes formation of hard martensite in the weld metal and heat-affected zone.

Some of the precipitation-hardening stainless steels have poor weldability because of their high phosphorus content. Steels 17-10 P and HNM are not usually electron beam welded.

ALUMINUM ALLOYS

In general, aluminum alloys that can be readily welded by gas tungsten arc and gas metal arc welding can be electron beam welded. Two problems that may be encountered in some alloys are hot cracking and porosity.

The non-heat treatable series of alloys (1xxx, 3xxx, and 5xxx) can be electron beam welded without difficulty. Welded joints will possess mechanical properties similar to annealed base metal.

The heat treatable alloys (2xxx, 6xxx, and 7xxx) are crack sensitive to varying degrees when electron beam welded. Some may also be prone to weld porosity. Aluminum alloy Types 6061-T6 and 6066-T6, which are considered difficult alloys to join, can be successfully welded by the electron beam process. Best results with these alloys are obtained by using small amounts of Type 4043 aluminum filler metal.

As-welded joints in 1.5-in. thick Type 7075-T651 aluminum alloy exhibit lower mechanical properties than unwelded plate. The low weld properties are caused by overaging in the heat-affected zone. Postweld solution treating and aging will produce heat-treated properties in the welded joint. At high travel speeds, weld porosity may result from vaporization of certain elements in the alloy. The high zinc content of Type 7075 aluminum alloy is responsible for vapor formation. At low travel speed, the vapor has time to escape to the surface before the weld metal solidifies. Zinc-free aluminum alloys can be welded at higher speeds without encountering a severe porosity problem. It is advantageous to weld thermally hardened aluminum alloys at high travel speed to minimize the width of the softer weld and heat-affected zones.

TITANIUM AND ZIRCONIUM

Titanium and zirconium absorb oxygen and nitrogen rapidly at welding temperatures and this reduces their ductility. Acceptable levels of oxygen and nitrogen are quite low. Therefore, these materials and their alloys must be welded in a relatively inert environment. High vacuum electron beam welding is best for both metals, but medium vacuum and non-vacuum with inert gas shielding may be acceptable for some titanium applications. Most zirconium applications require welding in a vacuum or inert gas because of corrosion resistance requirements.

REFRACTORY METALS

Electron beam welding is an excellent process for joining the refractory metals because the high power density allows the joint to be welded with a minimum of heat input. This is important with molybdenum and tungsten because fusion and recrystallization raise the ductile-to-brittle transition temperatures of these metals above room temperature. The short time at temperature associated with electron beam welding minimizes grain growth and other reactions that raise transition temperatures.

The refractory metals can be divided basically into two groups: rhenium, tantalum, and columbium (readily welded) and molybdenum and tungsten (difficult to weld).

Molybdenum and tungsten can be successfully electron beam welded provided the joints are not restrained during welding. Thin sections are easy to handle, and it may be better to fabricate a composite structure of thin welded sections rather than to weld a single thick section. Freedom from impurities such as oxygen, nitrogen, and carbon is important. Alloys of metals containing rhenium are best suited for welding because they remain ductile at lower temperatures than the pure metals do.

DISSIMILAR METALS

Whether two dissimilar metals or alloys can be welded together successfully depends upon their physical properties, such as melting points, thermal conductivities, atomic sizes, and thermal expansions. Weldability is usually predicted by empirical experience in this area. A generalization of weldability can be made by examining the alloy phase diagram of the metals to be welded. If intermetallic compounds are formed by the metals to be joined, the weld will be brittle.

Information on the relative weldability of some dissimilar metals is given elsewhere in the Handbook.[6] However, the information must be reviewed in the light of each particular application with regard to joint restraint and service environment.

The problem of metallurgical incompatibility can sometimes be solved by the use of a filler metal shim or by welding each of the materials to a compatible transition piece. Examples are given in Table 5.2. Table 5.3 presents a summary of the weldability of various metal combinations derived from phase diagram information and accumulated practical experience.

6. Sec. 4, Metals and Their Weldability, *Welding Handbook*, 6th ed. This section will be revised as Vol. 4, 7th ed.

Table 5.2
Examples of filler metal shims for electron beam welding

Metal A	Metal B	Filler shim
Tough pitch copper	Tough pitch copper	Nickel
Tough pitch copper	Mild steel	Nickel
Hastelloy X[a]	SAE 8620 steel	321 stainless steel
304 stainless steel	Monel[a]	Hastelloy B[a]
A-286 stainless steel	SAE 4140 steel	Hastelloy B[a]
Inconel 713[a]	Inconel 713[a]	Udimet 500[a]
Rimmed steel	Rimmed steel	Aluminum

a. Tradenames

Table 5.3
Weldability of dissimilar metal combinations

	Silver	Aluminum	Gold	Beryllium	Cobalt	Copper	Iron	Magnesium	Molybdenum	Columbium	Nickel	Platinum	Rhenium	Tin	Tantalum	Titanium	Tungsten
Aluminum	2																
Gold	1	5															
Beryllium	5	2	5														
Cobalt	3	5	2	5													
Copper	2	2	1	5	2												
Iron	3	5	2	5	2	2											
Magnesium	5	2	5	5	5	5	3										
Molybdenum	3	5	2	5	5	3	2	3									
Columbium	4	5	4	5	5	2	5	4	1								
Nickel	2	5	1	5	1	1	2	5	5	5							
Platinum	1	5	1	5	1	1	1	5	2	5	1						
Rhenium	3	4	4	5	1	3	5	4	5	5	3	2					
Tin	2	2	5	3	5	2	5	5	3	5	5	5	3				
Tantalum	5	5	4	5	5	3	5	4	1	1	5	5	5	5			
Titanium	2	5	5	5	5	5	5	3	1	1	5	5	5	5	1		
Tungsten	3	5	4	5	5	3	5	3	1	1	5	1	5	3	1	2	
Zirconium	5	5	5	5	5	5	5	3	5	1	5	5	5	5	2	1	5

1. Very desirable
   (solid solubility in all combinations)
2. Probably acceptable
   (Complex structures may exist)
3. Use with caution
   (Insufficient data for proper evaluation)
4. Use with extreme caution
   (No data available)
5. Undesirable combinations
   (Intermediate compounds formed)

APPLICATIONS

Electron beam welding is primarily used for two distinctly different types of applications: high precision and high production.

High precision applications require that the welding be accomplished in a high purity environment (high vacuum) to avoid contamination by oxygen or nitrogen, or both, and under minimum heat effects and maximum reproducibility conditions. These types of applications are mainly in the nuclear, aircraft, missile, and electronic industries. Typical products produced include nuclear fuel elements, special alloy jet engine components, pressure vessels for rocket propulsion systems, and hermetically-sealed vacuum devices.

High production applications can take advantage of the low heat input and the high reproducibility and reliability of electron beam welding if a high purity environment is not required. These desired characteristics permit welding of components in the semifinished or finished condition using both medium and nonvacuum equipment. Typical examples are gears, frames, steering columns, and transmission and driveline parts for automobiles; thin-wall tubing; bandsaw and hacksaw blades, and other bimetal strip applications. Figure 5.24 shows a bimetallic strip welding machine where individual strips are fed continuously into and out of the weld chamber through a series of pressure zones.

Nonvacuum electron beam welding has found its major application for high volume production. One example is the automotive industry where nonvacuum welding is employed for many large volume applications, two of which are shown in Figs. 5.25 and 5.26. The manufacture of welded tubing is another ex-

ample. Integrated welding machine/tube mill units have been built to weld copper or steel tubing continuously at speeds up to 100 ft/min.

Fig. 5.24—A continuous process strip electron beam welding machine

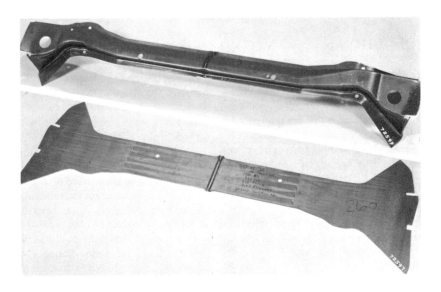

Fig. 5.25—Nonvacuum electron beam welded automotive frame (top) formed from a welded blank (bottom)

Fig. 5.26—Nonvacuum electron beam welded torque converter assembly

WELD QUALITY

INTRODUCTION

To produce welds that meet the requirements of specifications set forth in the welding industry, it is necessary to exercise good control over the three factors that are primarily responsible for electron beam weld quality: (1) joint preparation, (2) welding procedure, including provisions for keeping the beam on the seam, and (3) characteristics of the metals being welded. The first two of these are covered in other sections of this chapter.

The third factor relates to the physical and mechanical properties of the metals being welded, as well as to their metallurgical characteristics. Weld discontinuities of metallurgical origin are cracking and porosity.

Weld zones constitute regions of different microstructures within the base metal structure.

Unlike a cast ingot, weld metal grains usually grow from partially melted grains at the fusion line. The phenomenon is called epitaxial solidification. The nature of the weld metal structure is controlled by the size and orientation of the base metal grains and by the thermal gradients in and the shape of the weld pool.

The nature of the stress resulting from fusion welding is also important. Metal immediately adjacent to the moving weld pool is first heated; it expands against the restraining forces of the surrounding cold base metal; then it cools and contracts. In effect, metal is plastically deformed (upset) during the heating cycle and restrained in tension during cooling. A series of residual tensile and compressive stresses surround the weld zone, often resulting in warpage of the welded assembly.

In considering these factors, it would appear that electron beam welding offers the following unique characteristics for controlling the weld joint properties:

(1) Base metal recrystallization and grain growth can be minimized.

(2) Beam oscillation and travel speed can be used to control the shape of and the temperature gradients in the weld pool.

(3) Low heat inputs result in low thermal stresses in the base metal and, hence, low distortion. Also, residual stresses are symmetrically distributed due to the characteristic two-dimensional symmetry (parallel sides) of the electron beam weld zone.

It is, however, not always possible to realize the full potential of the process, and the weldability of the metal ultimately is controlled by the metallurgical factors. In this respect, electron beam welds exhibit most of the common discontinuities associated with fusion welding. A possible exception is hydrogen-induced cold cracking of carbon steel weldments because normally there is no source of hydrogen in an autogenous electron beam weld.

One type of discontinuity sometimes found in partial joint penetration welds is large voids at the bottom of the weld metal. In all probability, a large number of these voids will be aligned and appear as linear porosity rather than scattered porosity. When the weld just penetrates through the joint, root porosity will appear as a lack of fill accompanied by spatter on the back side of the weld.

Another occurrence peculiar to the vacuum mode of welding is the release of trapped air through the molten weld metal. This sometimes creates a defect. It only happens when one attempts to weld a gas-filled container that is not properly vented to vacuum.

Other discontinuities are generally the same as those found in other types of fusion welds. Electron beam weld discontinuities include:

(1) Porosity and spatter
(2) Shrinkage voids
(3) Cracking
(4) Undercutting
(5) Underfill
(6) Missed joints

(7) Lack of fusion

The probability of encountering these types of discontinuities is more pronounced in welding thick sections. A knowledge of the causes of the discontinuities and a means of avoiding them are essential for the production of high quality welds in thick sections. As an example, in welding thick sections in the horizontal position, holes and porosity can be avoided by tilting the beam axis a few degrees in the plane of welding. Equally important is a reliable non-destructive testing method, such as ultrasonic inspection, to determine the presence of certain types of defects that are not detectable by radiography.

POROSITY AND SPATTER

Porosity in electron beam welds is caused by the evolution of gas as the metal is melted by the beam. The gas may form as a result of (1) the volatilization of high vapor pressure elements in the alloy, (2) the removal of dissolved gases, or (3) the decomposition of compounds such as oxides or nitrides. Copper-zinc (brasses) and aluminum-magnesium alloys are difficult to electron beam weld because of metal vapor formation. Both zinc and magnesium have low boiling points. Dissolved gases and compounds are more likely to be present in alloys originally melted in air or other protective gas atmospheres.

Spatter is caused by the same factors as porosity. The rapid evolution of gas or metal vapor causes the ejection of drops of molten weld metal that scatter over the work surface and interior of the chamber. Spatter and porosity can even occur in vacuum remelted alloys when a residual phase volatizes under the intense heat of the electron beam. Examples of spatter and porosity in HP 9-4-25 steel[7] are shown in Figs. 5.27 and 5.28 respectively.

An effective means of preventing porosity and spatter is the use of vacuum melted or fully deoxidized metals for electron beam welding applications. When gas-emitting or high vapor pressure alloys must be welded, special

---

7. A high-strength alloy steel.

Fig. 5.27—Spatter on the surface of a weldment

techniques are required to minimize porosity. The addition of filler metal containing a deoxidizer may be helpful in welding metals that are not completely deoxidized. Slow welding speeds will provide time for gas bubbles to escape from the molten metal.

The use of oscillatory beam deflection may be effective in reducing porosity. In extreme cases, it may not be possible to avoid porosity completely, but remelting the joint a second or third time will reduce it. However, these techniques reduce joint strength in age-hardening alloys that are heat-treated prior to welding.

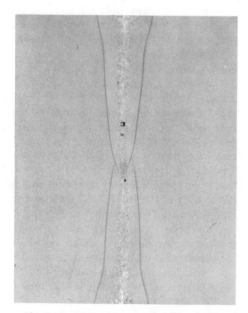

Fig. 5.28—Porosity near the bottom of electron beam weld beads

SHRINKAGE VOIDS

Shrinkage voids may occur between the dendrites near the center of the weld metal. These voids are characterized by irregular outlines in contrast to the generally round outlines of porosity. Shrinkage voids usually occur in alloys having high volumetric shrinkage on solidification. In electron beam welds where the bond lines are essentially parallel, solidification proceeds uniformly from the base metal to the center of the weld. When solidification shrinkage of the metal is great, voids will form if the face and root surfaces freeze before the center of the weld. An example of shrinkage voids in an electron beam weld in 15-7Mo PH stainless steel is shown in Fig. 5.29. Low travel speed or beam oscillation may eliminate or minimize shrinkage voids by increasing the volume of molten metal and decreasing the solidification rate. However, these conditions will generally produce a wider fusion zone than would be obtained otherwise.

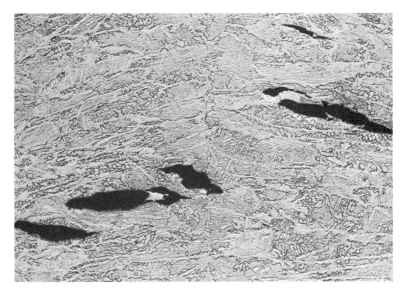

Fig. 5.29 — Shrinkage voids in an electron beam weld

CRACKING

Hot or cold cracks may occur in electron beam welds in alloys that are subject to these types of cracking. Hot cracking is generally intergranular and cold cracking is transgranular. Hot cracks form in a low-melting grain boundary phase during solidification of the weld metal. Cold cracks form after solidification as a result of high internal stresses produced by thermal contraction of the metal during cooling. A crack originates at some imperfection or point of stress concentration in the metal and propagates through the grains by cleavage.

Hot cracking may be minimized by welding at high travel speeds with minimum beam energy. Cold cracking may be overcome by redesigning the joint to eliminate points of stress concentration or by changing the welding procedures to minimize porosity. Preheating quench-hardenable steels to a suitable temperature to control the formation of martensite in the weld zone will prevent cracking.

UNDERCUTTING

Undercutting refers to grooves produced in the base metal at the edges of the weld bead, as shown in Fig. 5.30(B). The defect occurs when the weld metal does not wet the base metal. Undercutting is promoted by very high travel speeds, improper cleaning procedures, or beam asymmetry (usually occurs on one side only). Alloy additions that reduce surface ten-

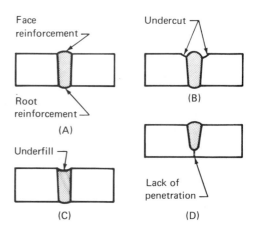

Fig. 5.30 — Correct (A) and incorrect (B, C, and D) electron beam weld geometries

sion or increase fluidity, such as aluminum additions to carbon steel welds, have a beneficial effect.

UNDERFILL

Electron beam welds with good bead geometry have essentially parallel bond lines with a uniform crown or buildup of weld metal on the top surface, as shown in Fig. 5.30(A). Full (100 percent) penetration welds will have a uniform root surface, the width of which is dependent upon the welding conditions. In thick sections of metal, such as 3-in. stainless steel, the face and root surface shapes are dependent upon the surface tension supporting the column of molten metal as it is being transported along the weld joint. At slow welding speeds, the molten weld metal will sag due to insufficient surface tension and the force of gravity. This will form an extremely heavy root reinforcement and the weld face will show severe underfill (concavity), as shown in Fig. 5.30(C). Various techniques can eliminate this condition. These include the use of a backing strip, a step joint, or welding in the horizontal or the vertical position.

Excessive sagging of the root surface usually results when the beam energy is too high or the molten weld metal is too wide. This can be reduced by proper adjustment of the welding variables. If underfilling persists at the best beam operating conditions, filler metal must be added to fill the groove. A number of techniques are effective in providing the required filler metal. One is to place a narrow strip over the face of the joint and then weld through it. The thickness of the strip must be slightly greater than the depth of any undercut, so that the undercut will be entirely in the strip. Filler metal wire may similarly be added to the face of the weld as it is being made or during a subsequent smoothing pass made with a defocused beam.

LACK OF PENETRATION

There are numerous applications of electron beam welding in which full penetration of the joint is not required. These applications generally use seal welds or welds subjected to shearing forces only. In these cases, the sharp notch at the root of the weld is acceptable. However, when a welded joint must support a transverse tensile stress at the root of the weld, full joint penetration is required.

Lack of penetration may be caused by low beam power, high travel speed, or improper focusing of the beam. This condition is shown in Fig. 5.30(D).

MISSED JOINTS

When a small diameter electron beam is used to make a long joint in a thick section, the beam axis must be in the same plane as the joint faces and aligned with the joint along its entire length of travel. Otherwise, the possibility of missing the joint at some location is great. Even when the beam is properly aligned with the joint, electrostatic or magnetic forces can cause beam deflection, resulting in portions of the joint being missed. An electrostatic field can be generated by the accumulation of an electrical charge on an insulated surface, such as the glass in the vacuum chamber windows. The electron beam will be deflected away from the charged surface if the beam passes too close to it.

Residual magnetism in a ferromagnetic base metal or in the fixturing can cause unexpected beam deflection. For example, a steel part may be magnetized during grinding if it is held by a magnetic chuck, and the residual magnetism in the part will cause the beam to deflect and miss the joint. This can be avoided by demagnetizing all ferromagnetic parts before welding and by using nonmagnetic materials for fixturing.

Unexpected beam deflection can occur when welding dissimilar metals, especially when one is ferromagnetic. An example of this is shown in Fig. 5.31, a weld between a nonmagnetic nickel-base alloy and a magnetic maraging steel. Residual or induced magnetism in the steel deflected the electron beam and caused lack of fusion at the root of the joint. If dissimilar materials are to be welded in produc-

Fig. 5.31—Beam deflection when welding dissimilar metals

tion, it is important that test welds be made and examined to determine whether beam deflection will occur.

LACK OF FUSION

In most cases, lack of fusion occurs in partial penetration welds. However, it can also occur near the root of full penetration welds

made with insufficient beam power. Figure 5.32 shows an example of this in an electron beam weld in a titanium alloy. Lack of fusion generally can be avoided by using properly adjusted welding variables. There are circumstances, however, where partial penetration welding is required. One example is the welding of circular joints where the beam power and penetration must be decreased as the end of the weld overlaps the start to avoid crater formation. A partial penetration weld is formed in the overlap and lack of fusion can occur. Another example is the welding of thick sections. Two partial penetration weld passes, one from each side of the joint, may be required when the metal thickness is too great to be penetrated in a single pass.

Welding with a slightly defocused beam and low travel speed (to compensate for the lower energy density) is effective in eliminating lack of fusion. Beam oscillation, either circular or transverse, is sometimes effective. Preheating is helpful because it reduces the thermal gradient at the root of the weld.

Lack of fusion is difficult to locate nondestructively because it is similar to fine cracks and usually can not be detected with x-ray inspection. Penetrant tests are ineffective because the unfused areas do not usually extend to the surface.

Ultrasonic testing is the only nondestructive test method that can detect lack of fusion in electron beam welds. Experienced nondestructive test personnel are required to perform the test and interpret the results. Even then, the test method is not suitable for many applications.

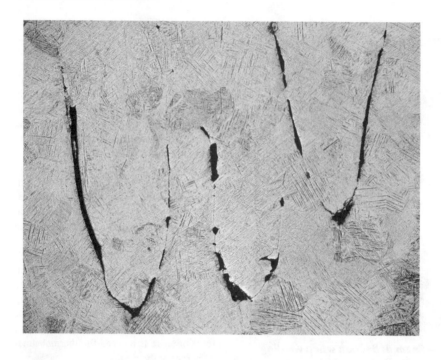

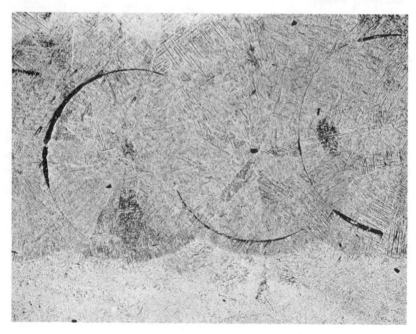

Fig. 5.32—Lack of fusion or spiking in vertical (top) and horizontal (bottom) sections through an electron beam weld in a titanium alloy

SAFE PRACTICES

Since electron beam welding machines employ a high energy beam of electrons to provide the thermal energy for welding, the process requires the user to observe various safety precautions not normally necessary with other types of welding equipment. The four primary potential dangers associated with electron beam equipment are: electric shock, x-radiation, fumes and gases, and damaging visible radiation. Precautionary measures must be taken at all times to assure that proper protective procedures are always observed. AWS F2.1, *Recommended Safe Practices for Electron Beam Welding and Cutting,* and ANSI Z49.1, *Safety in Welding and Cutting,* (latest editions) give the general safety rquirements that should be strictly adhered to at all times.

ELECTRIC SHOCK

Every electron beam welding machine operates with high voltage that can cause fatal injury, regardless of whether it is referred to as a low voltage or a high voltage machine. The manufacturers of electron beam equipment, in meeting various underwriter requirements, produce machines that are well-insulated against the danger of high voltages. However, all precautions should be exercised with all systems in which high voltages are present.

X-RADIATION

The x-rays generated by an electron beam welding machine are produced when electrons, traveling at a high velocity, collide with matter. The majority of x-rays are produced when the electron beam impinges upon the workpiece. Substantial amounts are produced when the beam strikes gas molecules or metal vapor in both the gun column and work chamber. Underwriters and Federal regulations have established firm rules for permissible x-radiation exposure levels, and producers and users of equipment must observe these rules.

Generally, the steel walls of the chamber are adequate protection in systems up to 60 kV, as-suming proper design. All high voltage machines utilize lead lining to block x-ray emission beyond the chamber walls. Lead glass windows are employed in both high and low voltage electron beam systems. Generally, the shielded vacuum chamber walls provide adequate protection for the operator.

In the case of nonvacuum systems, a radiation enclosure must be provided to assure the safety of the operator. Thick walls of high-density concrete or some other similar material may be employed instead of lead, especially for large radiation enclosures on nonvacuum installations. In addition to the normal precautions necessary for ensuring that large radiation enclosures are properly shielded, special safety precautions should be imposed to prevent personnel from accidently entering or being trapped inside these enclosures when equipment is in operation.

A complete x-ray radiation survey of the electron beam equipment should always be made at the time of installation and at regular intervals thereafter. This should be done by personnel trained in the proper procedures for doing a radiation survey to assure initial and continued compliance with all radiation regulations and standards applicable to the site where the equipment is installed.

FUMES AND GASES

It is unlikely that the very small amount of air left in a high vacuum electron beam chamber would be sufficient to produce ozone and oxides of nitrogen in harmful concentrations. However, nonvacuum and medium vacuum electron beam systems are capable of producing these by-products, as well as other types of airborne contaminants, in concentrations well above acceptable levels.

Adequate area ventilation should be employed to reduce concentrations of any airborne contaminants around the equipment below permissible exposure levels, and proper exhausting techniques should be employed to maintain residual concentrations in the chamber or enclosure below these same limits.

VISIBLE RADIATION

Direct viewing of visible radiation emitted by the molten weld metal can be harmful to eyesight. In the presence of intense light, proper eye protection is necessary. Optical viewing should be done through filters in accordance with ANSI Z87.1, *Occupational And Educational Eye and Face Protection* (latest edition).

Metric Conversion Factors

1 in. = 25.4 mm
1 torr = 133.3 Pa
1 W/in.^{2} = 1.55×10^3 W/m^2
1 in./min = 0.423 mm/s
1 kJ/in. = 39.4 kJ/m
1 STD · cm^3/s = 1.04×10^{-5} micron · ft^3/h
1 psi = 6.89 kPa

SUPPLEMENTARY READING LIST

Bench, F. K. and Ellison, G. W. EB welding of 304L stainless steel with cold wire feed. *Welding Journal*, 53 (12): 763-766; 1974 Dec.

Ben-Zvi, I., Bogart, L., and Turneaure, J. P., Simple device for controlling 100% penetration in electron beam welds. *Welding Journal*, 51 (12): 842-843; 1972 Dec.

Bibly, M. J., Burbridge, G., and Goldak, J. A., Gases evolved from electron beam welds in plain carbon steels. *Welding Journal*, 51 (12): 844-847; 1972 Dec.

Bibly, M. J., Burbridge, G., and Goldak, J. A., Cracking in restrained EB welds in carbon and low alloy steels. *Welding Journal*, 54 (8): 253s-258s; 1975 Aug.

Dixon, R. D. and Pollard, L. Jr., Effect of accurate voltage control on partial penetration EB welds. *Welding Journal*, 53 (11): 495s-497; 1974 Nov.

Duhamel, R. F., Nonvacuum electron beam welding technique development and progress. *Welding Journal*, 44 (6): 465-474; 1965 June..

Fink, J. H., Analysis of atmospheric electron beam welding. *Welding Journal*, 54 (5): 137s-143s; 1975 May.

Hinrichs, J. F., et al., Production electron beam welding of automotive frame components. *Welding Journal*, 53 (8): 488-493; 1974 Aug.

Lubin, B. T., Dimensionless parameters for the correlation of electron beam welding variables. *Welding Journal*, 47 (3): 140s-144s; 1968 Mar.

Mayer, R., Dietrich, W., and Sundermeyer, D., New high-speed beam current control and deflection systems improve electron beam welding applications. *Welding Journal*, 56 (6): 35-41; 1977 June.

Metzger, G. and Lison, R., Electron beam welding of dissimilar metals. *Welding Journal*, 55(8): 230s-240s; 1976 Aug.

Murphy, J. L. and Turner, P. W., Wire feeder and positioner for narrow groove electron beam welding. *Welding Journal*, 55 (3): 181-190; 1976 Mar.

O'Brien, T. B., el al., Suppression of spiking in partial penetration EB welding. *Welding Journal*, 53 (8): 332s-338s; 1974 Aug.

Passoja, D. E., Heat flow in electron beam welds. *Welding Journal*, 45 (8): 379s-384s; 1966 Aug.

Privoznik, L. J., Smith, R. S., and Heverly, J. S., Electron beam welding of thick sections of 12% Cr turbine grade steel. *Welding Journal*, 50 (8): 567-572; 1971 Aug.

Sandstrom, D. J., Bucken, J. F., and Hanks, G. S., On the measurement and interpretation and application of parameters important to electron beam welding. *Welding Journal*, 49(7): 293s-300s; 1970 July.

Schumacher, B. W., Atmospheric EB welding with large standoff distance. *Welding Journal*, 52 (5); 312-314; 1973 May.

Schwartz, M. M., *Electron Beam Welding*, New York: Welding Research Council, 1974 July; Bulletin 196.

Tews, P., et al., Electron beam welding spike suppression using feedback control. *Welding Journal*, 55(2): 52s-55s; 1976 Feb.

Tong, H. and Geidt, W. H., A dynamic interpretation of electron beam welding. *Welding Journal*, 49 (6): 259s-266s; 1970 June.

Weber, C. M. and Funk, E. R., Penetration mechanism of partial penetration electron beam welds. *Welding Journal*, 51 (2): 90s-94s; 1972 Feb.

Weidner, C. W. and Schuler, L. E., Effect of process variables on partial penetration electron beam welding. *Welding Journal*, 52 (3): 114s-119s; 1973 Mar.

6

Laser Beam Welding and Cutting

Chapter Committee

R. E. SOMERS, *Chairman*
Welding Consultant

C. M. BANAS
*United Technologies
Research Center*

S. R. BOLIN
Raytheon Laser Center

J. I. NURMINEN
Westinghouse Electric Corporation

Y. C. JOHN PENG
GTE Sylvania

R. J. SAUNDERS
Coherent, Inc.

Welding Handbook Committee Member

R. E. SOMERS
Welding Consultant

6

Laser Beam Welding and Cutting

FUNDAMENTALS OF THE PROCESS

DEFINITION AND GENERAL DESCRIPTION

The word laser is an acronym for "*l*ight *a*mplification by *s*timulated *e*mission of *r*adiation." From an engineering standpoint, the laser may be considered as an energy conversion device in which energy from a primary source (electrical, chemical, thermal, optical, nuclear) is transformed into a beam of coherent electromagnetic radiation at ultraviolet, visible, or infrared frequency. Transformation is facilitated by certain solids, liquids, or gases which, when excited on a molecular or atomic scale by special techniques, produce coherent "light." Such light is monochromatic (single wavelength) and coherent; i.e., all waves are in phase. Because high-energy laser beams are coherent, they can be highly concentrated with transmitting or reflective optics to provide the high-energy density required for welding, cutting, and heat treating.

The first laser beam was produced in 1960 using a ruby crystal pumped by a flash lamp. Such lasers can produce only short pulses of light energy at a repetition frequency limited by heating of the crystal. Therefore, even though individual pulses may exhibit instantaneous peak power levels in the megawatt range, pulsed ruby lasers are limited to low average powers. Currently, pulsed and continuously-operating solid-state lasers capable of welding and cutting thin sheet metal are available. The latter utilize neodymium-doped, yttrium aluminum garnet

(Nd-YAG) crystal rods to produce a continuous monochromatic output; power levels to 1 kW have been attained. Also, electrically excited gas lasers that produce continuous-wave (cw) energy have been developed. Multikilowatt laser beam equipment based on CO_2 is capable of full penetration, single-pass welding of steel to 3/4-in. thickness.

PRINCIPLES OF OPERATION

Solid-State Lasers

Solid-state lasers are based on single crystals or glasses doped with small concentrations of transition elements (such as chromium in ruby) or rare earths (such as neodymium in glass and YAG). Electrons in the atoms of the transition or rare earth element can be selectively excited to higher energy levels upon exposure to intense incoherent optical radiation (white light), as shown in Fig. 6.1. Selective excitation of a specific energy level (pumping) is the key to laser operation since it permits establishment of a "population inversion." A population inversion exists when the number of particles at a given energy level substantially exceeds that corresponding to thermal equilibrium. Without such an inversion, photons generated by stimulated emission would be absorbed within the medium and lasing could not occur. Once excited, the electrons return to their normal energy state in one or more steps, each involving loss of a discrete quantity (quantum) of energy. Lasing occurs if one of these energy losses results in the emission of a quantum of electromagnetic radiation (photon).

An example of a solid-state laser is a ruby crystal consisting of aluminum oxide with a small concentration of chromium atoms in solution. When ruby is exposed to the intense radiation from one or more xenon or krypton flash lamps, some of the electrons in the chromium are excited to a high energy level. These electrons immediately drop to an intermediate energy level, resulting in the evolution of heat. Then, they return to the ground energy state on emission of a photon of red light having a wavelength of 0.69 μm. An electron at the intermediate energy level returns to the ground state faster when it is stimulated by the red light output of its neighbors. When a large number of electrons is involved, amplification of the red light output occurs.

The amplification effect is accentuated by shaping the ends of the crystal and mirror-coating them (or by using external mirrors) so that the emitted light is reflected back and forth along the crystal. If some of the light is then allowed to escape from one end through a partially transmitting mirror, a nearly monochromatic and nondivergent output beam is obtained. Several closely-

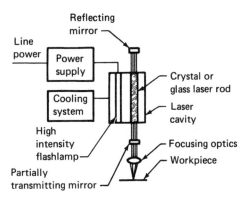

Fig. 6.2—Elements of a solid-state laser system

spaced wavelengths may actually be seen due to the fine structure of the energy levels in the lasing medium. Beam divergence occurs due to diffraction and to system imperfections (multiple-mode operation, scattering, etc.). The former is inherent in any optical system of finite size, while the latter can be reduced with improved components and techniques. At the energy levels used for welding, typical pulsed ruby laser beam divergence falls within an included angle of 0.25 to 0.75 degrees.

The wavelength of all solid-state laser light is determined by the fluorescence spectrum of the dopant element. Both Nd-glass and Nd-YAG lasers emit near infrared radiation with a wavelength of 1.06 μm. Simple optical systems can be used to concentrate the beam into a small spot. The components of a solid-state laser system are shown in Fig. 6.2, and an operating unit is shown in Fig. 6.3.

The laser cavity that couples the output from the flash lamp to the lasing medium is an interesting feature of the apparatus. In many systems, the cavity comprises a highly polished elliptical cylinder with the flash lamp and laser medium at alternate foci. In this arrangement, the flash lamp output is focused on the laser crystal. Sometimes multiple elliptical cylinders are employed so that more than one lamp can be used to pump a single crystal. Arrangements of one, two, or four lamps in conjunction with a single crystal are common. Other reflective cavities are used in which helical, hollow flash tubes sur-

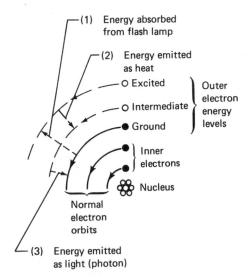

Fig. 6.1—Electron energy absorption and emission during laser action

round the crystal. Inert gas discharge or quartz-iodine ("sun") lamps are used to generate a continuous output.

Because solid-state lasers are, at best, only about 3 percent efficient and may have average powers of a few hundred watts, large quantities of heat are generated as an adjunct to the laser process. Therefore, a high capacity cooling system is needed to remove the heat developed. Typically, a closed-loop system is used with deionized water as the cooling agent. The water flows around the laser rod and flash lamps and additionally cools the cavity walls. For a solid-state laser with a 400 W average output, the cooling system must remove approximately 15 kW of waste heat.

Gas Lasers

Figure 6.4 is a diagram of a simple gas laser. It consists basically of a tube through which the lasing gas mixture is circulated. The

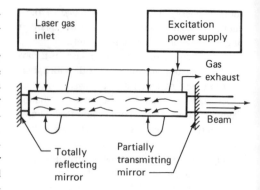

Fig. 6.4—Basic gas laser system

gas lases when it is selectively excited between a totally reflecting mirror at one end of the tube and a partially transmitting mirror at the other. Many gas lasers have been developed; the most common of these is the red output helium-neon which is used in myriad alignment, measurement, and other tasks. Most gas lasers are limited

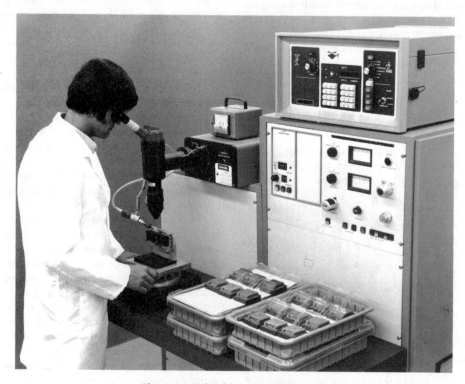

Fig. 6.3—Pulsed laser welding system

to milliwatt continuous power levels. A few, such as the argon laser, have been developed with cw outputs of tens of watts, and one, the CO_2 laser, is capable of durable multikilowatt operation.

In a CO_2 laser suitable for industrial use, CO_2 molecules are vibrationally excited in an electric discharge. Because the direct vibrational excitation of CO_2 by an electric discharge is relatively inefficient, N_2 is added to the medium. Nitrogen accepts energy effectively from the discharge and, because the vibrational energy levels of nitrogen are very close to some of those of carbon dioxide, the CO_2 is excited by resonant energy exchange with N_2; this two-step process is much more rapid and efficient than the direct CO_2 process. Transition from the upper vibrational energy state in CO_2 to an intermediate level then occurs with the emission of a photon; the energy difference between the two levels corresponds to a characteristic wavelength of 10.6 μm in the infrared. Carbon dioxide molecules at the intermediate level must then return to the ground state in order to complete the process. Helium added to the lasing mixture expedites this transition. Collisions between CO_2 molecules and helium result in the transfer of residual excitation energy to the helium; this energy is then removed as waste heat.

As with solid-state lasers, the operation of a gas laser requires the establishment of a nonequilibrium, population inversion condition. This is accomplished electrically with a high voltage glow discharge. Unfortunately, a glow discharge tends toward instability at current levels above approximately 300 mA independent of chamber size. If the glow discharge transforms to an arc, thermodynamic equilibrium conditions are established and lasing cannot occur. It has therefore been necessary in some cases to stabilize the glow discharge in high power systems by an auxiliary ionization means. Radiofrequency (r-f) electric power and high voltage electron beams have been utilized for this purpose. Currently, high power CO_2 lasers are available that operate solely with a dc electric discharge requiring no auxiliary ionization provisions.

Other means of generating high laser power

have been developed. Notable among these is the gas dynamic laser represented by Fig. 6.5. In such a system, fuel and oxidizer are chosen to yield combustion products suitable for lasing. For example, combustion of CO and CH_4 with O_2 and N_2 yields CO_2, N_2, and H_2O, the latter substituting for helium in the carbon dioxide lasing process. Vibrational excitation of N_2 and CO_2 occurs due to the elevated temperature of the products. On rapid expansion through a supersonic nozzle, the static gas temperature is quickly reduced. A set of small nozzles is used rather than a large single unit to decrease expansion time. Due to differences in the times required for various species to assume the energy distribution appropriate to the reduced temperature, selected high-energy states may be momentarily "frozen." A nonequilibrium condition occurs which promotes establishment of the population inversion necessary for laser action.

Gas dynamic laser systems developing short duration continuous power outputs to 100 kW have been reported. One such unit, operating at 90 kW, was used to demonstrate the capability for full penetration, single-pass welding of 1.5-inch thick steel at 120 in./min. Due to the high cost of operation, however, the gas dynamic laser is not suitable for industrial welding and cutting applications.

The convectively cooled, electrically excited CO_2 laser is currently the basis for high power, production rated equipment to 20 kW continuous output. Rapid flow of the laser gases

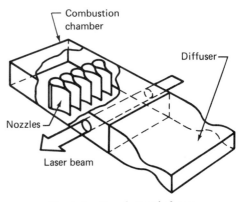

Fig. 6.5—Gas dynamic laser

is utilized to remove waste heat from the laser cavity. Since such lasers typically exhibit an electrical-to-optical conversion efficiency of 10 to 15 percent, effective waste heat removal is essential to continuous operation. To minimize operating costs, a gas-to-liquid heat exchanger is used and the laser gases are recirculated through the system. Only a small quantity of gas is consumed due to the requirement for continual removal and replenishment of a small amount of the laser gas mixture to prevent build-up of contaminants generated by dissociation of CO_2 and N_2 in the discharge.

Gas flow in the CO_2 laser may be coaxial with or transverse to the laser beam. Typical operating pressure is 80 torr or less, although pulsed CO_2 laser operation has been attained at atmospheric pressure. With coaxial flow, several tubes may be arranged electrically in parallel and optically in series to produce high output power. Physically parallel tube arrays are used to reduce system length. Such configurations are optically connected in series by means of mirrors. With transverse flow, mirrors are used to reflect the beam back and forth within a single discharge chamber. Multiple beam passes are required to effectively remove available optical energy from the entire lasing medium. In both cases, a totally reflecting mirror is utilized at one end and a partially transmitting (at the laser wavelength of 10.6 μm) mirror is used as the output window. For power levels to several kW, coated semiconductor materials, notably zinc selenide (ZnSe), serve effectively as output windows. However, such materials are not durable at higher power levels; hence, an annular metal mirror is used to facilitate partial transmission. Output is then beamed through an "aerodynamic window," an opening across which a controlled high velocity flow of compressed air maintains the pressure difference between the laser chamber and the atmsophere.

PROCESS VARIATIONS

Lasers can be arbitrarily divided into low and high power types on the basis of their average power output. From a welding viewpoint, a logical division is at 1 kW since continuous power levels of approximately 1 kW or above are required for deep penetration (keyhole) welding. Below approximately 1 kW, laser welding is currently limited to the conduction mode.

Low Power Lasers

Most solid-state lasers are operated in the pulsed mode and most gas laser systems may also be operated in this manner. Outputs may be in the magnitude of hundreds of kilowatts during a millisecond pulse at a pulse repetition frequency of 1 Hz, yielding average powers of several hundreds of watts. For either solid-state or gas lasers, peak power decreases as the pulse length and repetition frequency increase; i.e., average power remains essentially unchanged. Carbon dioxide lasers are available that will produce pulse peaks of 3 kW at frequencies of up to 2.5 kHz from a nominal 500 W average power unit.

While their usual operating mode is pulsed, neodymium-YAG lasers can also be designed for continuous power output. In this case, a crystal medium such as yttrium aluminum garnet (YAG) is a more suitable host for neodymium (Nd) than glass because a crystal can better withstand the temperature gradient and heat dissipation requirements of cw operation. A steady source of incoherent light is used instead of a flash lamp. Commercial versions of the Nd-YAG laser are available to power levels of several hundreds of watts in continuous operation. Another version of the Nd-YAG laser can emit a continuous train of extremely short pulses. An example of this performance is the generation of pulses of 25 picosecond (25×10^{-12}s) duration at intervals of 2.5 nanoseconds (25×10^{-9}s).

Solid-state lasers dominate the commercial field in numbers of units in operation in the low average power pulsed range. In low power cw applications, CO_2 laser systems are predominant as they are in the high power area.

High Power Lasers

At present, all high power laser systems are of the continuous CO_2 type. Commercial units based on either coaxial or transverse flow range from about 2 to 10 kW output. Experimental transverse flow units capable of outputs to 20 kW are in laboratory use.

CAPABILITIES AND LIMITATIONS

A laser beam and an electron beam produce similar welding behavior and exhibit characteristic similarities and differences. The laser beam consists of a stream of photons moving at the velocity of light, and the electron beam is a stream of electrons moving at a speed proportional to the square root of the accelerating voltage.[1] The energy of a beam photon corresponds to the wavelength and is of the order of 0.1 to 2.0 eV, while that of a beam electron is numerically equivalent to the beam voltage (typically 30 to 150 keV). The laser beam can be focused and directed by optical means (mirrors and lenses) and the electron beam by electrostatic and magnetic means. Both beams can be concentrated into a very small area, producing power densities greater than 1 MW per square inch. For reference, it is noted that a power density of 1

---

1. An electron beam and the devices used to generate and manipulate it are discussed in Chapter 5.

MW/in.² corresponds to a thermal source temperature of about 13 000 K.

A laser beam can be transmitted for appreciable distances through air without serious power attenuation or degradation of optical quality. On the other hand, an electron beam is rapidly dispersed by electron collisions with air molecules. Out-of-vacuum working distances for electron beams are therefore relatively short.

A laser beam is partially reflected or deflected by smooth metallic surfaces while an electron beam is not. Significant laser beam reflection may occur, thereby inhibiting energy transfer to a workpiece. Conversely, an electron beam is deflected by electrostatic or electromagnetic fields or a magnetic material, while the laser beam is not.

When a high power electron beam strikes a metal surface, x-rays are generated. Adequate x-ray shielding is therefore required to protect personnel from exposure. Laser beams do not generate x-rays during normal metal processing operations. However, they do produce high-intensity light which can damage eyesight or cause severe burns. Appropriate safety provisions and procedures must therefore be established.

LASER WELDING

For welding, the laser beam must be focused to a small diameter to produce high power density. This is accomplished with transmitting optics for low power lasers and with reflective optics for high power lasers. Minimum beam size can be adjusted by the optics design. Although variable focus optics can be designed, customary practice involves control of the incident spot diameter by varying the position of the workpiece surface relative to the plane of optimum focus of fixed optics. The depth of focus determines the permissible tolerance on the work-to-lens or mirror distance.

As with electron beam welding, laser welds can be produced by the conventional conduction limited manner and by the keyhole technique. In conduction limited welding, the beam impinges on and is absorbed by the metal surface. The inner portions of the material are heated entirely by conduction from the surface. This type of welding is frequently done with low power cw and pulsed laser beams.

The keyhole (deep penetration) welding mode, illustrated in Fig. 6.6, is used to produce high depth-to-width ratio welds with pulsed or cw laser systems. In this welding mode, the intense energy concentration at the workpiece surface induces local vaporization. A vapor cavity surrounded by molten metal is formed as the beam starts to move along the joint. The cavity is maintained against the fluid dynamic forces of the liquid metal surrounding it by the

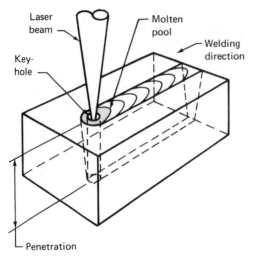

Laser beam — Molten pool — Welding direction — Key-hole — Penetration

Fig. 6.6—Characteristics of deep penetration laser welding

pressure of the vaporized metal. Metal is progressively melted at the leading edge of the moving molten pool and flows around the deep penetration cavity to the rear of the pool where it solidifies in a characteristic chevron pattern. In the keyhole mode, penetration is not limited by the thermal diffusivity of the material because beam energy penetrates directly into the cavity. The process may be visualized in terms of a taut hot wire being drawn through a cake of ice. Due to beam penetration into the material, high depth-to-width ratio welds are formed rather than the characteristic hemispherical ones formed by conduction welding processes.

Effective laser beam welding depends upon absorption of beam energy by the workpiece. Unfortunately, shiny metal surfaces at room temperature are quite reflective to light at laser wavelengths. For example, absorption of a low-intensity CO_2 beam may be only 40 percent for stainless steel and as low as 1 percent for polished aluminum or copper. Absorption levels are higher for Nd-YAG and even higher for ruby, but significant reflection still occurs.

Despite the above, acceptable absorption levels occur during welding. Absorption increases as metal temperature and incident beam power intensity increase. A sharp increase in

absorption occurs for pulsed lasers when power density exceeds a certain threshold value. Typical threshold levels are 10^8 W/m^2 for carbon steel, 10^9 W/m^2 for aluminum and copper, and 10^{10} W/m^2 for tungsten.

In high power CO_2 laser welding, absorption is enhanced by the deep penetration cavity which forms at power densities exceeding approximately 10^8 W/m^2. As the beam passes into the material, multiple reflections and absorptions occur. Cumulative absorption is therefore quite high due to this geometric blackbody effect. Direct calorimetric measurements have shown absorption levels above 90 percent in alloy steels, and levels greater than 50 percent have been noted in aluminum alloys.

PULSED POWER

With a pulsed solid-state laser system, the joint penetration is primarily determined by the pulse energy and duration. Overlapping spots are used to form a seam weld at a speed governed by the pulse repetition frequency. To generate a sound weld, the laser beam must deliver sufficient energy to the workpiece to melt adequate material for the fusion joint. Pulse duration must be long enough to enable conduction and melting to the desired depth. Since pulse energy and duration specifically fix beam power, power intensity at the workpiece surface must then be controlled by the focusing optics and the location of the surface relative to the plane of optimum focus. Conditions must be chosen to avoid vaporization. This requirement is complicated by the fact that the energy distribution across the incident spot may not be uniform. Often, the energy peaks at the center cause local vaporization. Under these conditions it is necessary to defocus the beam; a slight increase in pulse energy may then be required to compensate for the reduced average power intensity. In usual practice, such adjustments can be made quickly by trial and error procedure.

Pulsed weld penetration is governed by laser process and material variables. Increased pulse energy and duration lead to increased penetration. The limit on penetration occurs with incipient vaporization at the material surface. For

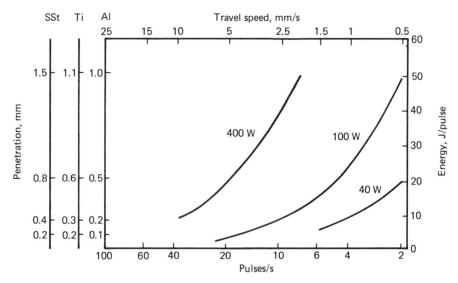

Fig. 6.7—Dependence of penetration in three metals on the operating conditions, for pulsed lasers of three power ratings

a given pulse energy and duration, shallow penetration is attained in a material with high thermal diffusitivity. High pulse energy and short pulse duration (higher power) are appropriate for such materials and the converse applies for materials of low thermal diffusivity. Maximum penetration for present pulsed solid-state laser systems is approximately 1/16 inch.

Individual spot weld diameter is determined by the energy absorbed, the rate at which it is delivered, and the material thermal diffusivity. Seam welding speed is given by the product of the weld spot diameter, the fraction of overlap required, and the pulse repetition frequency. The process factors are illustrated in Fig. 6.7 for stainless steel, titanium, and aluminum representing low, medium, and high thermal diffusivity, respectively. The interdependence of average power, seam welding speed, pulse energy, penetration, and repetition frequency is shown.

CONTINUOUS POWER

Continuously operating, low power lasers can be used to produce conventional conduction-limited welds at low speeds in sheet gages. With high power lasers, narrow welds can be formed in sheet metals at high speeds using the keyhole mode. The latter mode can also be used at lower speeds to generate high depth-to-width ratio welds in plate. Usually, keyhole welding cannot be done at welding speeds below about 15 in./min.

Within appropriate welding conditions, penetration in the keyhole mode is directly related to power when speed and other process variables are constant. This behavior is illustrated in Fig. 6.8 for rimmed steel. Below the threshold level of about 1 kW, penetration will decrease sharply at the speeds indicated.

A penetration of 3/4 inch is attainable in alloy steel at the 15 kW maximum power level of current industrially suited systems. Heavier sections may be welded in two passes, one from each side, although special precautions must be taken to avoid defects at the root of the second weld.

At a given power level, weld penetration decreases with increasing travel speed when the other variables remain unchanged. This relationship for rimmed steel is illustrated in Fig. 6.9.

In high power welding, ionization of metal vapor atoms may lead to the formation of a plasma above the workpiece. Such a plasma

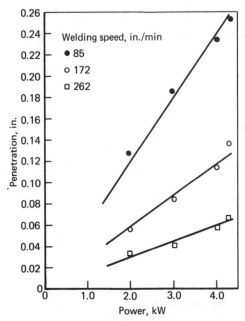

Fig. 6.8—Dependence of weld bead penetration in rimmed steel on CO_2 laser power at three travel speeds

V-groove may be utilized to provide for the desired filler addition. Conventional wire feed devices may be used.

As with other fusion welding processes, abrupt termination of power can leave a crater in the weld metal. Conversely, instantaneous exposure of the workpiece to full beam power may cause local internal porosity. Programmed upslope and downslope of welding power or the addition of end tabs can be used to overcome such problems.

APPLICATIONS

Many different metals can be satisfactorily welded with a laser beam. With pulsed systems, these include copper, nickel, iron, zirconium, tantalum, aluminum, titanium, columbium, and their alloys. Continuous systems are applicable to most of these metals except that capabilities are limited in copper and aluminum alloys. Although the laser provides precise energy control and certain unique process features, it is a fusion welding process. As such, it is constrained by the same principles of good welding practice

effectively absorbs the laser beam and drastically reduces beam energy input to the workpiece. This can be prevented by sweeping the ions from the interaction region with a flow of gas, preferably helium. This flow is usually integrated with weld zone shielding requirements. If adequate suppression is not attained, the plasma expands to the atmosphere above the workpiece and a luminous plume is formed as shown in Fig. 6.10.

Joint designs and fit-up requirements are generally similar to those for electron beam welding, as discussed in Chapter 5. A joint gap in excess of 3 percent of the material thickness will normally cause joint underfill. Similar conditions will prevail if too much energy is used for welding, resulting in drop-through. Underfill can be eliminated by the addition of filler metal during either the primary weld pass or a "cosmetic" second pass. Filler metal may also be added to modify weld metal chemistry. In this case, a square-groove with a narrow gap or a

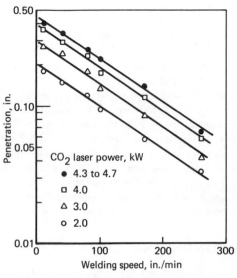

Fig. 6.9—Dependence of weld bead penetration in rimmed steel on travel speed at several CO_2 laser power levels

Fig. 6.10—Undesirable plasma formation above the workpiece during laser welding

and the requirements for metallurgical compatibility common to more conventional processes.

Since laser welding equipment is more costly than conventional systems of equivalent power, selection of applications must be based on unique laser capabilities. Some of these capabilities that may be used as a guideline for selection of laser applications are as follows:

(1) The specific energy input to the workpiece is very small. This means that the extent of the heat-affected zone and any thermal damage to material adjacent to the weld are minimized. The rapid cooling of the weld pool may have metallurgical advantages for some applications.

(2) The high power density of the laser beam can be used to make difficult welds. These may involve welds between metallurgically compatible dissimilar metals with widely different physical properties, between metals of high electrical resistivity, or between parts varying

greatly in mass and size.

(3) Electrical contact with the workpiece is not necessary since the heat source is a light beam. Joints in restricted locations can be welded provided a line of sight to the weld point is available. Further, absence of contact requirements makes the laser ideal for use in high speed automated welding systems. Spot welding can take place "on the fly" with parts moving at 60 to 120 in./min. Seam welding in thin sheet can be accomplished at speeds up to 250 ft./min. Additionally, the part may be held stationary and the laser beam moved along the weld seam, or a combination of beam and part motion can be used. This flexibility often simplifies parts fixturing.

(4) Precision welding can be done with a well-defined focused spot. Spot weld sizes on the order of a few thousandths of an inch in diameter can be achieved with accurate positioning.

(5) Fusion zone purification occurs under certain conditions during the welding of steels. Preferential absorption of the beam by nonmetallic inclusions in the metals leads to their vaporization and removal from the weld zone.

(6) Laser welding is ideally suited to automation.

Pulsed Power

Several joint configurations can be used for welding wires together with pulsed power. These include butt (end-to-end), lapped, tee, and cross wire joints.

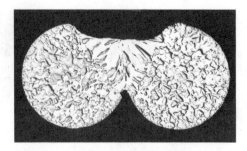

Fig. 6.11—Cross section through a weld between parallel 0.02-in. diameter nickel wires

The optimum configuration for joining electrical wires is a butt joint, but lap joints and cross wire configurations may prove easier to fixture and weld in small diameters. For lap joints, the two wires should be parallel and in contact along the overlap. The laser beam is directed at the joint to melt and fuse the two wires together as shown in Fig. 6.11. When a cross wire configuration is used, the laser beam should be directed at the point of contact between the wires.

Spot welds between overlapping sheets can be produced with a pulsed laser as illustrated in Fig. 6.12. Seam welds between sheets may be a series of overlapping spots produced with pulsed power or a continuous weld bead made with continuous power. A protective gas shield may be necessary with seam welding to minimize oxidation of the weld area. Other examples of welded assemblies produced with pulsed laser

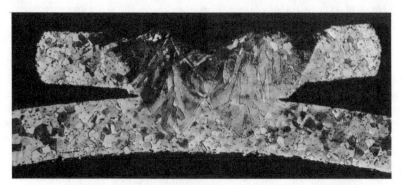

Fig. 6.12—Cross section of laser spot welds between 0.005-in. nickel ribbons

Fig. 6.13—Hermetic seal weld on an electronic module made with a pulsed laser beam

equipment are shown in Figs. 6.13, 6.14, and 6.15.

One prominent application area for pulsed laser welding is in the electronics industry. Miniaturization dictates stringent joining requirements that very often cannot be met by conventional welding techniques. The limited accessibility to the joint and the need for precise application of energy to avoid thermal damage to components can be met with laser welding. Laser encapsulation of microelectronic pack-

ages is an example.

Laser welding is being used to salvage parts. The low heat input and the resulting low distortion are ideal for making small repairs on machined parts. Accessibility into confined

Fig. 6.14—Titanium tube with 0.01-in. wall welded to a 0.025-in. thick titanium disc with a pulsed laser beam

Fig. 6.15—A titanium capsule hermetically sealed by laser welding

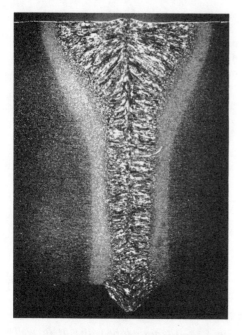 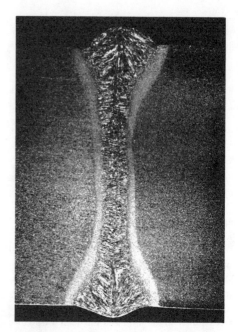

Fig. 6.16—Cross sections of butt joints in 0.52 in. X-80 steel plate welded from one and both sides with 12 kW of CO_2 laser power

space, good viewing and welding optics, and precise control of small amounts of energy are also attributes in many repair applications.

Continuous Power

Continuous power laser systems, principally of the CO_2 type, are gradually moving from the laboratory into industrial applications. Numerous potential production applications are being evaluated.

High power CO_2 lasers can generate single-pass, full penetration welds in some metals up to 0.75-in. thickness. Representative laser welds in 0.52-inch thick X-80 steel, produced with 12 kW of power, are shown in Fig. 6.16. The dual-pass weld exhibited smaller grain size than the single-pass weld due to the lower specific energy input and the resultant higher weld zone cooling rate. A high power beam can produce deep spot welds in a stationary workpiece that have characteristics similar to continuous welds (see Fig. 6.16).

Other examples of the characteristics of high power laser welds are the three welds in Ti-6A1-4V alloy shown in Fig. 6.17. These full penetration welds in 0.23-inch thick plate were made at 40, 60, and 80 in./min, respectively, at 5.5 kW of power. The corresponding specific energy inputs were 8250, 5500, and 4125 J/in., which resulted in decreasing fusion zone width as the energy decreased.

Typical power requirements for a nominal welding speed of about 50 in./min in many metals are 3.0, 6.0, and 12 kW for 1/8, 1/4, and 1/2-in. joint thicknesses, respectively. This approximate guideline for laser welding is applicable to steels, titanium alloys, and nickel-base alloys. Somewhat higher power is required for copper-nickel alloys, zirconium, and the refractory metals (Cb, Mo, Ta, and W).

Only limited performance has been demonstrated in laser welding of aluminum alloys. Traditional difficulties in aluminum welding are compounded for the laser by the high initial re-

40 in./min   60 in./min

80 in./min

Fig. 6.17—CO₂ laser welds in 0.23-in. Ti-6Al-4V alloy plate produced with 5.5 kW power at three travel speeds

flectivity of the material. High power density is required to overcome this problem, and the level required often leads to overheating of the weld metal, sporadic vaporization, and intermittent plasma formation. These conditions will produce weld discontinuities. Beam spot oscillation may alleviate these conditions somewhat. Fillet and tee joints, which tend to trap the incoming beam, are easier to weld. In general, however, laser welding of aluminum alloys and copper

requires further process development. The latter will reflect even a 15 kW continuous beam; but 1/4-inch penetration has been demonstrated with a 4 kW cw beam "pulsed" at 24 kHz.

In addition to welding, the laser beam may be rapidly scanned over a metal surface to produce a thin melted layer that solidifies rapidly. This procedure may be incorporated with filler wire or powder additions for surfacing applications.

Generally speaking, established procedures for weld joint preparation apply to laser welding. Flat (downhand) position welding is preferred, but out-of-position (horizontal, overhead, vertical-up, vertical-down) welds can be made under conditions well within the keyhole welding mode.

TOOLING AND ACCESSORIES

Tooling for laser welding includes positioning devices similar to those used for electron beam welding discussed in Chapter 5. The purpose of tooling is to provide positioning of the weldment relative to the laser beam. Gas coverage, operator protection, and joint fit-up and alignment can also be provided by the tooling. Welding motion is established by movement of the workpiece or the beam, or by a combination of the two.

One difference between laser and electron beam tooling is that transparent and magnetic materials can be used with a laser beam. Both glass and clear plastic throwaway devices are frequently employed to hold workpieces in position for pulsed laser welding. They are then discarded if the metal plasma destroys their optical properties.

Inert gas coverage of the molten pool is usually provided for continuous power welding. The gas may fulfill several functions, including atmosphere protection for reactive metals such as titanium and tantalum; better wetting, flow, and weld bead appearance; and plasma suppression above the weld. Inert gas coverage may be provided by trailing and backup shielding techniques. For extremely reactive materials, a flowing gas dry box may be utilized with the beam entering through one of the exhaust gas ports.

Closed-circuit TV systems may be used to position the beam on the workpiece and to view the welding process. This is useful for continuous welding if the workpiece location is manually controlled by the operator.

LASER BEAM CUTTING AND DRILLING

A laser beam can be used as a heat source for straight or contour cutting of sheet or plate as is done with a plasma arc or oxyfuel gas torch. It can also be used to drill holes. Continuous power is preferred for the former, while pulsed power is used almost exclusively for the latter. Metals, ceramics, and a wide variety of other materials can be cut or drilled with a laser beam.

The mechanics by which a pulsed laser beam removes metal during drilling involves a combination of melting and vaporization. Sufficient vaporization is established to eject liquid metal from the hole. With a cw beam, the extremely high power density required for such a process cannot be attained; hence, a high-velocity gas jet is used to augment the laser beam.

Some of the advantages of laser beam cutting (LBC) compared to oxyfuel gas (OFC) or plasma arc cutting (PAC) are:

(1) The laser beam generator need not be in close proximity to the workpiece surface. This permits cutting or drilling in locations of limited accessibility.

(2) The process does not require that the workpiece be a part of an electric circuit as with plasma arc cutting.

(3) A laser beam provides very high power density in a small spot so that high-speed cutting is attained with a very narrow kerf and heat-affected zone. These factors, together with the ease with which laser systems can be automated, indicate application to computer numerical control cutting of contoured parts.

(4) Hole characteristics can be controlled by appropriate selection of optics. The beam can be shaped with optics of long focal length to drill deep, small diameter holes.

Some limitations on application of the process are:

(1) The laser beam generator is costly compared to oxyfuel gas and plasma arc cutting equipment, and the application must justify the

cost differential.

(2) Although steel up to 2 inches in thickness has been cut with a laser system, the process is currently best suited for metal thicknesses of 0.5 in. and under.

GAS-ASSIST CUTTING

Since the power density attainable with cw laser systems is insufficient to promote cutting by the vaporization-liquid metal expulsion process, an assist gas jet is normally used to blow the molten metal from the kerf. With low power cw systems, oxygen is generally used as an assist gas to take advantage of the exothermic reaction with those metals which can be oxygen cut. Similar behavior occurs in high power cutting; however, a wide variety of other assist gases such as compressed air, helium, argon, carbon dioxide, and nitrogen may also be effectively used. Cuts obtained with inert gas exhibit clean, nonoxidized edges but sometimes develop a tenacious lower edge burr. With oxygen-assisted cutting, the burr is usually quite brittle and therefore easily removed.

As in deep penetration welding, excellent coupling occurs with the workpiece despite the high reflectivity of metals at room temperature to light at laser wavelengths. For example, 0.062-in. thick aluminum alloy sheet can be cut at 300 in./min with a 3 kW CO_2 laser using compressed air assist. Only a slight increase in cutting speed is attained if the surface is initially anodized to provide an absorptive surface. Kerf width with these conditions is about 0.02 in., and the heat-affected zone extends only about 0.001 in. into the base material.

In most metals, cutting speed at constant power is inversely proportional to thickness, as indicated in Fig. 6.18 for Type 302 stainless steel. Within certain limits, cutting speed is directly proportional to laser power, as shown in Fig. 6.19. Representative laser beam cutting performance with several metals is given in Table 6.1. Although cutting performance improves with laser power, the improvement is generally not directly proportional to power.

DRILLING

On interaction of a pulsed laser beam with a metal, some energy is reflected, particularly

Table 6.1
Typical laser beam cutting performance

Metal	Thickness, in.	Power, kW	Cutting speed, in./min
Steel	0.051	0.5	142
	0.063	0.5	98
	0.090	0.6	70
	0.125	4.0	160
	0.660	4.0	45
	2.125	6.0	13
Stainless steel	0.012	0.35	173
	0.039	0.5	65
	0.090	0.6	70
	0.125	3.0	100
Nickel alloys	0.059	0.85	90
	0.125	4.0	120
Titanium	0.040	0.23	193
	0.200	0.6	130
	1.25	3.0	50
	2.00	3.0	20
Aluminum	0.125	4.3	100
	0.250	3.8	40
	0.500	5.7	30

at the onset of the pulse. Of the absorbed energy, most is used in melting, a small fraction is used for vaporization, and only a small amount is conducted into the base material. The momentum of the vaporized material is sufficient to blow the molten material out of the hole. If the pulsed power density exceeds 65 MW/in.^{2} and the pulse duration is longer than 0.5 ms, the volume of metal removed is directly proportional to the total pulse energy. The proportionality constant depends upon the material; values of 160 kJ/in.^{3} and 50 kJ/in.^{3} pertain to steel and aluminum, respectively.

In drilling, as in cutting, the low power density outer fringes of a focused beam do not contribute favorably to the process. Rather, they may inhibit metal expulsion and increase overall energy input leading to a relatively wide heat-affected zone. For precision processing, it is therefore desirable to remove the outer low-intensity fringes of the laser beam. This can be accomplished by suitable aperturing within the beam-shaping optics.

Pulsed ruby, YAG, and CO_2 lasers are applicable to drilling. Figure 6.20 shows a gas turbine blade with cooling holes drilled with a ruby laser

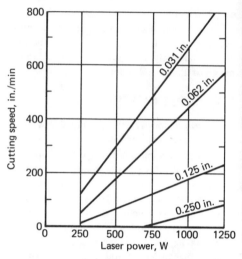

Fig. 6.19—Dependence of cutting speed on CO_2 laser beam power with oxygen assist for Type 302 stainless steel

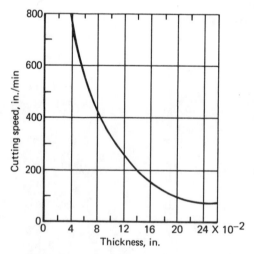

Fig. 6.18—Dependence of cutting speed on thickness for Type 302 stainless steel using an oxygen assisted 1.25 kW CO_2 laser beam

beam. Deep, narrow holes can be drilled in metals, including high temperature nickel-base alloys. Holes are drilled in production as deep as 1/4 inch. A typical machine designed for drilling will fixture the part, move it in three linear and two rotational axes of motion, and automatically adjust laser and optical variables so that many holes may be drilled in one setup. Figure 6.21 shows a 0.060-in. thick nickel alloy part that was cut and drilled with a 400 W pulsed YAG laser beam and a gas-assist nozzle.

Typical drilling conditions for a pulsed ruby laser are given in Table 6.2. The range of practical hole sizes for production drilling with a pulsed laser is shown in Fig. 6.22. Larger holes can be generated by cutting out circles (trepanning) with a rotating continuous or pulsed beam. Laser drilling is a thermal process subject to uncontrollable variations. Consequently, the wall of a drilled hole may show random recast zones, and exact dimensional control is not possible. On the other hand, drilling performance is fast, easily automated, and essentially independent of metal hardness.

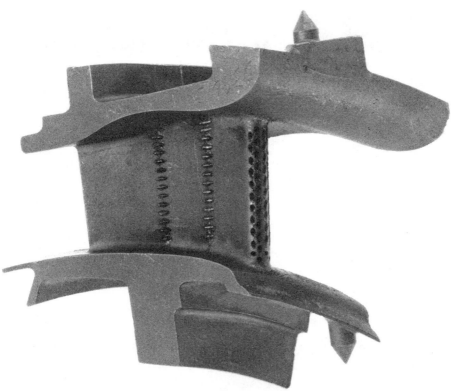

Fig. 6.20—A gas turbine blade containing cooling holes drilled with a pulsed laser beam

Table 6.2
Typical conditions for drilling with a pulsed ruby laser[a]

Material		Laser Variables		Hole diameter, in.	
Type	Thickness, in.	Pulse length, ms	Energy output, J	Entrance	Exit
Magnesium	0.062	1.8	2.1	0.014	0.008
Magnesium	0.062	2.0	3.3	0.016	0.012
Molybdenum	0.020	2.0	3.3	0.010	0.008
Molybdenum	0.020	2.25	4.9	0.010	0.008
Molybdenum	0.020	2.35	5.9	0.010	0.010
Copper	0.032	2.25	4.9	0.008	[b]
Type 304 stainless steel	0.036	2.35	5.9	0.020	0.010
Tantalum	0.062	2.35	5.9[c]	0.012	0.006
Tanalum	0.062	2.42[d]	8.0[d]	0.012	0.004
Ti-6%Al-4%V	0.096	2.35	5.9	0.016	0.006
Ti-6%Al-4%V	0.096	2.4	7.0	0.018	0.006
Ti-6%Al-4%V	0.096	2.4	7.0[c]	0.020	0.020
Tungsten	0.020	2.0	3.3	0.008	0.006
Tungsten	0.020	2.1	4.0	0.008	0.008
Tungsten	0.020	2.35	5.9	0.010	0.012

a A 94.8 mm objective lens was used in each case with the beam sharply focused at the surface. One laser pulse was used for each hole except where noted.   c Two laser pulses.
b Not measured.   d Extrapolated value.

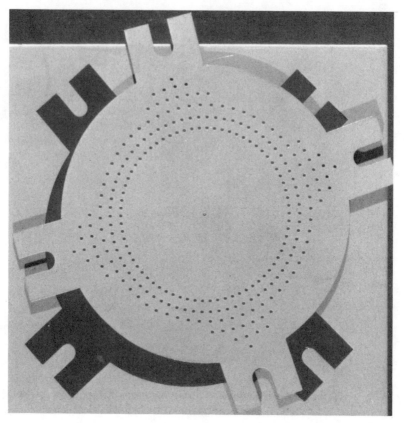

*Fig. 6.21—A 0.060-in. thick nickel alloy part cut and drilled with a pulsed
YAG laser beam and gas assist nozzle*

SAFETY

The basic hazards associated with laser operation are:

(1) Eye damage including burns of the cornea or retina, or both

(2) Skin burns

(3) Respiratory system damage due to evolution of hazardous materials during the beam-workpiece interaction

(4) Electrical shock

(5) Chemical hazards

(6) Cryogenic coolants

Of the above, eye damage is commonly associated with lasers. For laser beams operating at visible or near infrared wavelengths, even

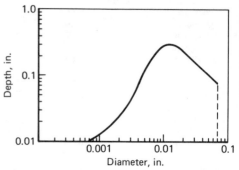

*Fig. 6.22—Relationship between diameter
and depth for pulsed laser drilled holes
in ferrous alloys*

a 5 milliwatt beam can cause retinal damage. Fortunately, safety glasses are available which are substantially transparent to visible light but are opaque to specific laser output. Selective filters for ruby, Nd-YAG, and other systems are available. Care must be taken to ensure that glasses appropriate to the specific laser system are used. At longer infrared wavelengths (such as 10.6 μm wavelengths of the CO_2 laser), ordinarily transparent materials such as glass are opaque. Clear, safety glasses with side shields may be used with these systems, so that the only light reaching the eye will be from incandescence of the workpiece. At high power, extreme brilliance will ensue on generation of a plasma, and shaded glasses will then be desirable.

Laser burns can be deep and very slow to heal. Exposure is to be avoided by appropriate enclosure of the beam. This is particularly important for nonvisible beams that provide no external evidence of their existence unless intercepted by a solid.

Since high voltages as well as large capacitive storage devices are associated with lasers, the possibility for lethal electrical shock is ever present. Electrical system enclosures should have appropriate interlocks on all access doors and provisions for discharging capacitor banks before entry. The equipment should be appropriately grounded.

The least obvious hazard is that of the potential products of the beam-workpiece interaction. For example, plastic materials used for "burn patterns" to identify beam shape and distribution in high power CO_2 systems can generate highly toxic vapors if irradiated in an oxygen-lean atmosphere. In deep penetration welding, fine metal fumes can be formed. With some metals, these may be carcinogenic. Also, severe plasma generation can produce ozone. Consequently, adequate ventilation and exhaust provisions for laser work areas is paramount.

Laser manufacturers are required to qualify their equipment with the U.S. Bureau of Radiological Health (BRH). Electrical components should be in compliance with NEMA standards. User action is governed by OSHA requirements. In all cases, American National Standard Z136.1, Safe Use of Lasers (latest edition), should be followed. Although hazards exist, they are well-defined and readily avoided with appropriate care.

Metric Conversion Factors

$1\ W/in.^2 = 1.55 \times 10^3\ W/m^2$
$1\ J = 1\ W \cdot s$
$1\ in. = 25.4\ mm$
$1\ in.^3 = 1.64 \times 10^4\ mm^3$
$1\ in./min. = 0.423\ mm/s$
$1\ ft/min = 5.1 \times 10^{-3}\ m/s$
$1\ \mu m = 3.94 \times 10^{-5}\ in.$ or 10^4 angstroms
$t_F = 1.8t_K - 460$

SUPPLEMENTARY READING LIST

Anderson, J. D., *Gas Dynamic Lasers,* New York: Academic Press, 1976.

Baardsen, E. L., et al., High speed welding of sheet steel with a CO_2 laser. *Welding Journal,* 52 (4); 227-9: 1973 Apr.

Banas, C. M., High power laser welding—1978. *Optical Engineering,* 17 (3), 210-16, 1978 May-June.

Bolin, S. R., Bright spot for pulsed lasers. *Welding Des. & Fab.,* 49 (8); 74-7: 1976 Aug.

Charschan, S. S., ed., *Lasers in Industry,* New York: Van Nostrand Reinhold, 1972.

Crafer, R. C., Improved welding performance from a 2kW axial flow CO_2 laser welding machine, *Advances in Welding Processes, 4th Int. Conf., Harrogate, England, 9-11 May 1978,* Cambridge, England: The Welding Institute, 1978.

Duley, W. W., CO_2 *Lasers,* New York: Academic Press, 1976.

Engel, S. L., How to make better laser welds. *Welding Des. & Fab.,* 51 (1): 1978

Estes, C. L. and Turner, P. W., Laser welding of a simulated nuclear reactor fuel assembly. *Welding Journal,* 53 (2); 66s-73s: 1974 Feb.

Gerry, E. T., Application of gas dynamic lasers. *Welding Journal,* 51 (5); 329-42: 1972 May.

Harry, J. E., *Industrial Lasers and Their Applications,* New York: McGraw-Hill, 1974.

Locke, E., et al., Deep penetration welding with high power CO_2 lasers. *Welding Journal,* 51 (5); 245s-49s: 1972 May.

Moorehead, A. J., Laser welding and drilling applications. *Welding Journal,* 50 (2); 97-106: 1971 Feb

Morgan-Warren, E. J., The application of laser welding to overcome joint asymmetry. *Welding Joural,* 58 (3); 76s-82s: 1979 Mar.

Schwartz, M. M., *Laser Welding and Cutting,* New York: Welding Research Council Bulletin, No. 167; 1971, Nov.

Seretsky, J. and Ryba, E. R., Laser welding dissimilar metals: titanium to nickel. *Welding Journal,* 55 (7); 208s-11s: 1976 July.

Sherwell, J. R., Design for laser beam welding. *Welding Des. & Fab.;* 50 (6); 106-10: 1977 June.

7

Friction Welding

Chapter Committee

F. J. WALLACE, *Chairman*
  Pratt & Whitney Aircraft Group,
  United Technologies

B. GRINSELL
  New Britain Machine Division,
  Litton Industries

F. R. HOCH
  Aluminum Company of America

D. E. SPINDLER
  Manufacturing Technology, Inc.

K. K. WANG
  Cornell University

C. D. WEISS
  Caterpillar Tractor Corporation

Welding Handbook Committee Member

I. G. Betz
  Department of the Army

7
Friction Welding

FUNDAMENTALS OF THE PROCESS

DEFINITION AND GENERAL DESCRIPTION

Friction welding is a solid-state joining process that produces coalescence by the heat developed between two surfaces by mechanically induced rubbing motion. The two surfaces are under pressure during the welding cycle. In most cases, the pressure is increased at the end of heating to facilitate forging or upsetting of the heated metal. Under normal conditions, the faying surfaces do not melt. Filler metal, flux, or shielding gas is not required with this process.

The basic steps in friction welding are shown in Fig. 7.1. First, one workpiece is rotated and the other is stationary as shown in Fig. 7.1(A). When the appropriate rotational speed is reached, the two workpieces are brought together and an axial force is applied, as in Fig. 7.1(B). Rubbing at the interface heats the workpieces locally and upsetting starts, as in Fig. 7.1(C). Finally, rotation of the one workpiece stops and upsetting is completed, as in Fig. 7.1(D).

A friction welded joint is similar in appearance to joints produced by flash and upset welding where heating is done by electrical resistance. Welding cycle times are similar in magnitude. Many applications are similar.

MODES OF OPERATION

With most friction welding applications, one of the two workpieces is rotated about an axis of symmetry with the faying surfaces normal to that axis. This means that in the normal case, one of the two workpieces must be circular in cross section at the joint location. Typical arrangements for single and multiple welding operations are shown in Fig. 7.2, (A) through (E). Figure 7.2(A) depicts the conventional and most commonly used mode in which one workpiece rotates while the other remains stationary. Figure 7.2(B) shows another mode in which both workpieces are rotated but in opposite directions. This procedure would be suitable for producing welds in small diameter workpieces where very high relative speeds are needed. Figure 7.2(C) shows still another mode where two stationary workpieces push against a rotating piece positioned between them. This setup might be desirable if the two end parts are long or are of such an awkward shape that rotation would be difficult or impossible by the other modes. A similar situation, shown in Fig. 7.2(D), involves two rotating pieces pushing against a long stationary piece at the middle. The same principle can be applied to the making of two welds back-to-back at the same time with one rotating spindle at the center, as shown in Fig. 7.2(E), for the purpose of improving productivity. This operation usually would be fully automated for mass production.

A special case is radial welding, where the force is perpendicular to the rotation axis. This concept is shown in Fig. 7.2(F). The sleeve or coupling is rotated and compressed as it is heated. An internal expanding mandrel supports the pipe

240

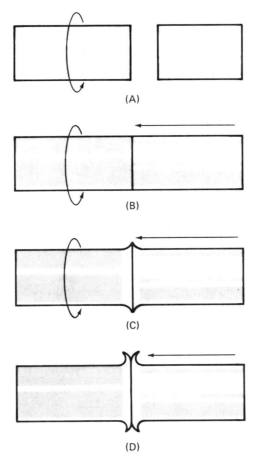

(A)

(B)

(C)

(D)

Fig. 7.1—Basic steps in friction welding

walls and prevents penetration of upset metal into the bore of the pipe.

The two pipes of Fig. 7.2(F) have beveled ends and are butted together and clamped securely in place. A solid ring with a composition similar to that of the pipe is positioned within the joint groove. The ring is beveled so that it will make initial contact at the apex of the joint groove. This is done to promote metal flow from this point outward and to reduce the initial high torque. After the necessary heat is developed from friction, ring rotation is terminated while the compressive load on the ring is maintained or increased to consolidate the joint. This method can also be used to join collars to solid shafts.

Orbital motion may be used to weld non-circular parts as shown in Fig. 7.2(G). The workpieces are aligned with the faying surfaces in contact and under pressure. One piece is then moved in a small circular motion relative to and in contact with the faying surface of the other piece, with neither piece rotating about its central axis. This variation provides uniform tangential velocity over the total interface area. When motion ceases, the workpieces must be rapidly aligned while the joint is still in a plastic state.

PROCESS VARIATIONS

There are two variations of friction welding known as continuous drive and inertia welding. In the continuous drive variation, one of the

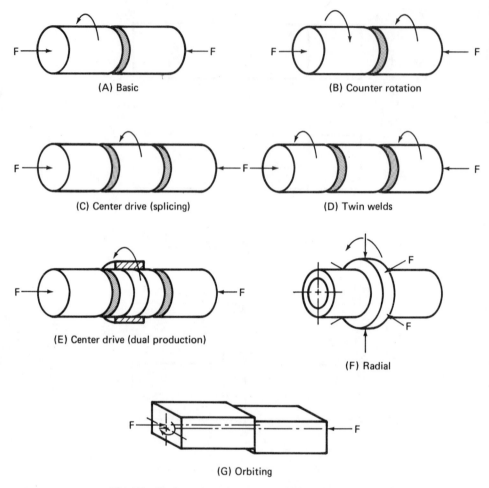

(A) Basic

(B) Counter rotation

(C) Center drive (splicing)

(D) Twin welds

(E) Center drive (dual production)

(F) Radial

(G) Orbiting

Fig. 7.2—Various arrangements of friction welding

workpieces is attached to a motor drive unit. The workpiece is rotated at a constant speed by the drive, and then an axial or radial force is applied. The workpieces rub together under pressure for a predetermined heating time or until a preset axial shortening takes place. The drive is disengaged and the rotating workpiece is braked to a stop. Welding force is maintained or increased to upset the metal, and held until the weld cools. Characteristics of typical continuous drive friction welding are shown in Fig. 7.3.

In the inertia welding variation, one of the workpieces is connected to a flywheel and the other is clamped in fixed position. The flywheel is accelerated to a particular speed, and it stores

mechanical energy as this is done. The drive is disengaged; the workpieces are brought together; and then a welding force is applied axially. This causes the faying surfaces to rub together under pressure. The flywheel energy is dissipated through friction at the interface; the weld zone heats as the flywheel speed decreases. At some point in the cycle, a speed is reached where less heat is being generated than is being dissipated. The torque peaks rapidly as the hot weld zone upsets under pressure, just before rotation stops. In some cases, a forging force may be applied as the torque peaks. Figure 7.4 shows the characteristics of inertia welding. In this variation, rotational speed decreases with time.

BONDING MECHANISM

The welding cycle can be divided into two stages: the rubbing or friction stage, and the upsetting or forging stage. The welding heat is developed during the first stage and the weld is consolidated and cooled during the second stage.

With identical workpieces, the bonding mechanism occurs in steps. When the pieces make contact, rubbing takes place between the faying surfaces, and strong adhesion takes place at various points of contact. The unit pressure is high. At some points, the adhesion is stronger than the metal on either side. Shearing takes place and metal is transferred from one surface to the other. As rubbing continues, the torque increases and a temperature is reached at which the transferring frágments become plastic.

The sizes of transferred fragments grow until they become a continuous layer of plasticized metal. During this period, the torque decreases to some minimum value, and then remains reasonably constant as metal is plasticized and forced from the interface. Axial shortening continues as the heat and pressure plasticize the metal.

With continuous drive, the drive is disconnected and a brake is usually applied to decelerate rapidly. As the speed decreases, the thickness of the very hot plasticized layer increases, and the torque drops to zero when rubbing stops.

With inertia welding, the speed decreases with time until the same behavior occurs at the interface. The thickness of the plasticized layer is related to the rubbing speed. As the speed decreases, the thickness of the hot plasticized layer increases, the generation of heat decreases, and the torque increases. The axial pressure forces the hot metal from the joint. During this time, the rate of axial shortening increases and then stops as the joint cools.

The bonding mechanism with dissimilar metals is more complex. A number of factors, including physical and mechanical properties, surface energy, crystal structure, mutual solubility, and intermetallic compounds, may play a role in the bonding mechanism. It is likely that some alloying will occur in a very narrow region at the interface as a result of mechanical mixing and diffusion. The properties of this layer may have a significant effect on overall joint properties. Mechanical mixing and interlocking may also contribute to bonding. The complexity makes prediction of weldability of dissimilar metals very difficult. It should be established by evaluation for a particular application with a series of tests designed for that purpose.

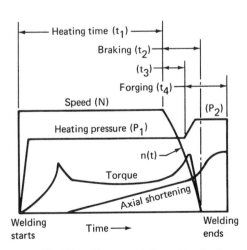

Fig. 7.3 — Characteristics of typical continuous drive friction welding

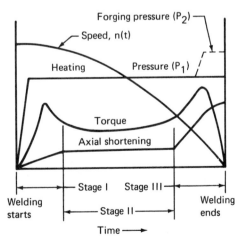

Fig. 7.4 — Characteristics of typical inertia welding

ADVANTAGES AND LIMITATIONS

Friction welding lends itself to joining workpieces of the same cross-section shape and area, as do flash and upset welding. With friction welding, electrical energy is first converted to mechanical energy and then to heat from rubbing action. No filler metal or flux is added with any of the processes. Normally, shielding gases are not required, but they are sometimes used with the reactive metals, such as titanium, to displace air from the joint during welding.

Advantages

(1) Surface cleanliness is not as significant as with upset welding because the rubbing tends to disrupt and displace surface films.

(2) The heat is very localized at the interface when compared to that of flash and upset welding, because all of the energy conversion occurs there. Consequently, the heat-affected zone is very narrow.

(3) The localized heating and the absence of melting make friction welding better suited for joining many dissimilar metal combinations than the other processes.

(4) In most cases, welded joints produced by this process are as strong as the base metal.

(5) The process is easily automated for mass production.

(6) The power requirements are much lower than for flash welding.

Limitations

(1) In general, one workpiece must have an axis of symmetry and be capable of being rotated about that axis.

(2) Preparation and alignment of the workpieces may be critical for developing uniform rubbing and heating, particularly with large diameters.

(3) Equipment costs may be higher than with flash and upset welding since precise mechanical design and construction equivalent to machine tools are required.

METALS WELDED

This process can be used to join a wide range of similar and dissimilar metals. Figure 7.5 indicates the combinations that have been joined according to the literature and equipment manufacturers' data. It should only be used as a guide. Specific weldability may depend upon a number of factors including specific alloy compositions, applicable process variation, component design, and service requirements.

In principle, almost any metal that can be hot forged and is unsuitable for dry bearing applications can be friction welded. Some metals may require postweld heat treatment to remove the effect of the severe deformation or quench hardening at the weld interface. Free-machining types of alloys should be welded with caution because redistribution of inclusions may create planes of weakness in the weld zone. Such welds exhibit low strength, ductility, and notch toughness.

There are a number of dissimilar metal combinations that have marginal weldability. These may involve combinations that have high and low thermal conductivities, a large difference in forging temperatures, or the tendency to form brittle intermetallic compounds. Examples are aluminum alloys to both copper and steel and titanium alloys to stainless steel.

The metallurgical structures produced by friction welding are generally those expected with short time at elevated temperature cycles. Time at temperature is short, and the temperatures achieved are generally below the melting point. With nonhardenable metals such as mild steel, changes in properties are negligible in the weld zone. On the other hand, with hardenable steels, structural changes may occur in the heat-affected zone. Consequently, they should be welded with a relatively long heating time to achieve a slower cooling rate and preserve toughness.

The interface structures of dissimilar metal combinations are significantly affected by the particular welding conditions employed. The longer the welding time, the greater the consideration that must be given to diffusion across the interface. Proper welding conditions will usually minimize undesired diffusion or intermetallic compound formation. Figure 7.6 shows the interface between aluminum and carbon steel. A very narrow diffusion zone is apparent.

In some cases, joints between dissimilar

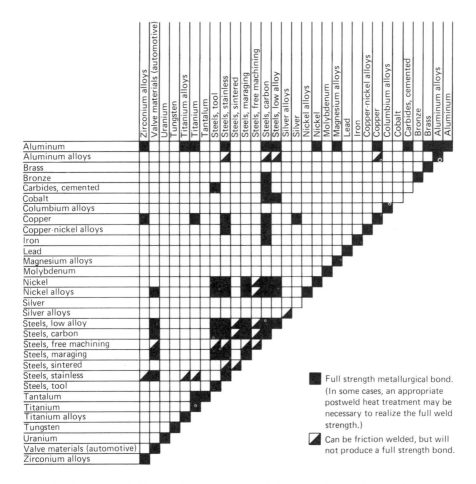

Note: This list was compiled from the literature and is provided as a generalization of weldability. It should be used only as a guide. Each case must be evaluated by appropriate tests.

Fig. 7.5—Friction weldability of similar and dissimilar metals

metals will show a mechanical mixing at the interface. This probably results as the plasticized metals are being mechanically worked during rotation under pressure. Figure 7.7 shows this action in a joint between Type 302 stainless steel and tantalum.

JOINT DESIGN

The nature of friction welding suggests that the joint face of at least one of the workpieces must be essentially round, except in the

case of orbital welding. The rotated workpiece should be somewhat concentric in shape because it is revolved at relatively high speed. Preparation of surfaces to be joined is not normally critical except in the case of alloys with distinct differences in mechanical or thermal properties, or both. Figure 7.8 illustrates the basic joint designs for combinations of bars, tubes, and plates. When bars or tubes are welded to plates, most of the flash comes from the bar or tube. This is true because there is less mass in the smaller section and the heat penetrates deeper into it.

Fig. 7.6—Interface of a friction weld between aluminum (top) and carbon steel (bottom) (×1000)

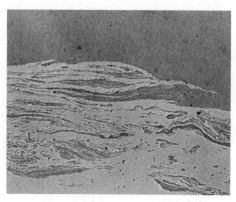

Fig. 7.7—Interface of a friction weld between tantalum (top) and Type 302 stainless steel (bottom) (×200 and reduced 66%)

Conical joints are usually designed with the faces at 45 to 60 degrees to the axis of rotation. For low-strength metals, large angles are preferred to support the axial thrust required to produce adequate heating pressure. One advantage of an angled joint over a 90 degree joint of equivalent interface area is that the change in tangential velocity across the interface is less.

This will provide more uniform heating across the interface.

Some joints between dissimilar metals with widely different mechanical or thermal properties are better produced with appropriate adjustment of faying surface area. The metal of lower strength or lower thermal conductivity may have a larger faying surface area. For applications where the flash cannot be removed conveniently, clearance for it can be provided in one or both workpieces.

FRICTION WELDING MACHINES

A typical friction welding machine consists of the following components:
(1) Head
(2) Base
(3) Clamping arrangements
(4) Rotating and upsetting mechanisms
(5) Power supply
(6) Controls
(7) Optional monitoring devices
This is true for both process variations. However, the machines for each variation differ somewhat in design and method of operation.

CONTINUOUS DRIVE

General Description

One of the workpieces to be welded is firmly clamped in a self-centering vise. The other workpiece is held in a centering chuck that is mounted on a rotatable spindle. The spindle is driven by a motor through a single or variable speed drive.

To make a weld, one workpiece is rotated and then thrust against the stationary work-

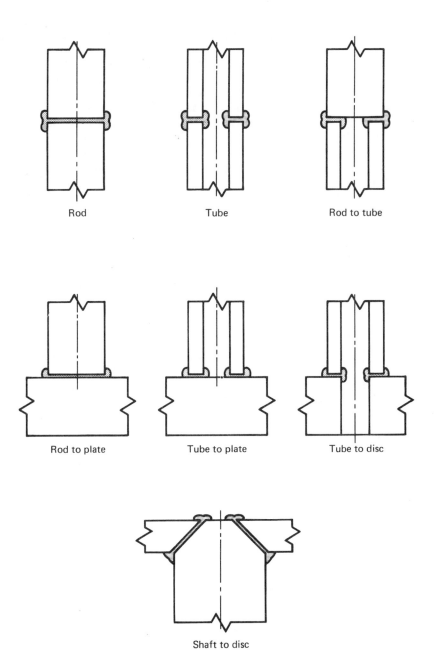

Rod

Tube

Rod to tube

Rod to plate

Tube to plate

Tube to disc

Shaft to disc

Fig. 7.8—Typical friction weld joint designs

piece by hydraulic force to produce frictional heat at the contact surfaces (see Fig. 7.1). The combination of speed and pressure raises the contact surfaces to a suitable temperature. Rotation is then stopped and the pressure is maintained or increased to upset the interface and complete the weld.

In its basic form, the machine spindle can be driven directly by a motor and allowed to stop under its natural deceleration characteristics and the retarding torque exerted by the weld. In practice, however, a clutch is normally used between the motor and spindle so that the motor can run continuously. The spindle can be engaged when required for the welding operation. This also conserves the starting energy that would be consumed if the motor is started for each weld.

It is also common practice, though not essential in all cases, to include a fast-acting brake on the spindle. The function of the brake is to rapidly terminate rotation at the end of a specified heating time or after a preset axial shortening of the weldment. This feature provides good control of overall weldment length and broadens the acceptable range of welding variables for critical applications.

The spindle can be rotated with force applied for whatever time period is required by each individual set of workpieces, depending upon their preweld length variations. Discreet differences between components, such as saw-cut ends or rough forging surfaces, can be corrected before braking and forging actions are initiated.

Alternatively, the spindle can be rotated under thrust for a preset time. This provides equal amounts of energy for each weld. It is usually best to machine the surfaces to be joined to ensure that they are flat and normal to the axis of rotation. This method of operation should provide good overall length tolerance after welding.

Generally, the former method is selected when friction welding components for critical applications. In these cases, some minimum loss in length must occur between the components to assure the removal of contaminants at the interface and a resulting sound weld.

There are two phases to the process: heating and upsetting. The basis of the process is to heat the faying surfaces to their forging temperature and then upset them together to produce a solid-phase bond. During the heating phase, the faying surfaces are rubbed together (by rotation) under pressure. A particular tangential speed and pressure are usually adequate for each similar metal combination.

Some alloys normally require preheating before welding. This can be done using a two-stage pressure application during the heating phase. First, a low pressure is applied during which heat is generated at the interface with very little axial shortening. The pressure is then increased to that necessary to accomplish welding. This operational technique, when used with hardenable steels, will produce a desirable condition after upsetting is complete. The preheat applied initially will reduce the cooling rate in the weld heat-affected zone. This should produce a more ductile joint and minimize the likelihood of cooling cracks.

The upsetting phase occurs at the end of the heating phase. Normally, a brake is applied to rapidly decrease the rotational speed to zero. At some time during deceleration, the pressure on the weld is increased to forge the components together. This forging pressure hot works the heat-affected zone and also forces plasticized metal from the interface into a "ram's horn"-shaped flash as shown in Fig. 7.1(D). This flash may be removed by machining in the welding machine or in a subsequent operation. In some cases, it may be left on the joint when it is not objectionable or when it serves to hold a part in place.

Welding Variables

There are a number of variables associated with this method as noted on Fig. 7.3. They are as follows:

(1) Rotational speed, N
(2) Heating pressure, P_1
(3) Forging pressure, P_2
(4) Heating time, t_1
(5) Braking time, t_2
(6) Forge delay time, t_3
(7) Forging time, t_4

In practice, some of these variables may be inherent in the machine design, such as braking time (t_2). If constant pressure is used, the forging variables do not apply. If forging does not start until rotation ceases, then t_2 and t_3 are equal. The three important variables with this equipment are speed, pressure, and heating time.

Speed. The function of rotation is to produce a relative velocity at the faying surfaces. For steels, the tangential velocity should be in the range of 250 to 350 ft/min. This is true for both solid and tubular workpieces. Tangential speeds below 250 ft/min produce very high torques that cause work clamping problems, non-uniform upset, and metal tearing. Production machines are designed usually to operate with speeds of 300 to 650 ft/min: for example, a spindle speed of 600 rpm can be used to weld steel products of 2 to 4 in. diameters (310 to 620 ft/min).

While high rotational speeds can be used, axial thrust and duration of heating must be carefully controlled to avoid overheating of the weld zone. This may be an advantage, however, when welding quench hardenable steels to control cooling rate and possible cracking.

For certain dissimilar metal combinations, low velocities can minimize the formation of brittle intermetallic compounds. From a weld quality standpoint, speed is not generally a critical variable.

Pressure. The effective pressure ranges are not necessarily narrow for heating and forging, although the selected pressures should be reproducible for any specific operation. The pressure controls the temperature gradient in the weld zone, the required drive power, and the axial shortening. The specific pressure depends upon the metals being joined and the joint geometry. Pressure can be used to compensate for heat loss to a large mass, as in the case of tube-to-plate welds.

Heating pressure must be high enough to hold the faying surfaces in intimate contact to avoid oxidation. Joint quality is improved in many metals, including steels, by applying a forging force at the end of the heating period.

For steels, a wide range of pressures is applicable for making sound welds. In the case of mild steel, heating pressures of 4500 to 8700 psi and forging pressures of 11 000 to 22 000 psi are acceptable. Commonly used values are 8000 and 20 000 psi, respectively. High-hot-strength alloys, such as stainless steels and nickel-base alloys, will require higher forging pressures.

If a "preheat" effect is desired to achieve a slower cooling rate, a pressure of about 3000 psi is applied for a brief period. The pressure is then increased to that required for welding.

Heating Time. For a particular application, heating time is determined during setup or from previous experience. Excessive time limits productivity and wastes material. Insufficient time may result in uneven heating as well as entrapped oxides and unbonded areas at the interface.

Heating time can be controlled in two ways. The first is with a suitable timing device that stops rotation at the end of the preset time. Preheat and forging functions could be incorporated with heating time using a sequence timer.

The second method is to stop rotation after a predetermined axial shortening. This method is set to consume a sufficient length to assure adequate heating prior to upsetting. Variations in surface condition can be accommodated without a sacrifice in weld quality.

Relationships Between Variables

For a set spindle speed, low pressure limits heating with little or no axial shortening. High pressure causes local heating to high temperature and rapid axial shortening. Figure 7.9 shows that with mild steel, axial shortening is approximately proportional to heating pressure. It also shows that for a given pressure during the heating phase, axial shortening is faster at low speed than at high speed. Tangential speeds above 1000 ft/min require long heating time without adding to weld quality.

For a given axial shortening with mild steel, the heating time will be significantly influenced by heating pressure and speed, as shown in Fig. 7.10. Heating time decreases at a decreasing rate as heating pressure is increased. It also decreases with speed at the same heating pressure.

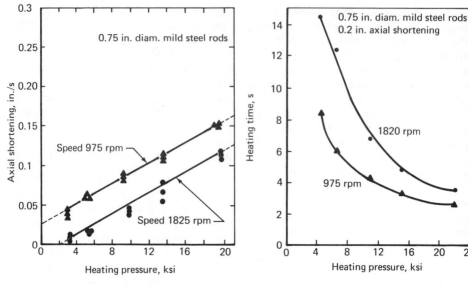

Fig. 7.9 — Relationship between axial shortening and heating pressure for mild steel with continuous drive friction welding

Fig. 7.10 — Relationship between heating time and heating pressure for mild steel with continuous drive friction welding

INERTIA DRIVE

General Description

With this type of machine, a flywheel is mounted on the spindle between the drive and the rotating chuck, as shown in Fig. 7.11. The flywheel, spindle, chuck, and work are accelerated to a selected speed corresponding to a specific energy level. When that speed is attained, driving is stopped and the flywheel and workpiece are allowed to spin freely. The two workpieces are then brought together and a specific axial thrust is applied with a hydraulic cylinder. The kinetic energy of the flywheel is transferred to the weld interface and converted to heat. As a result, the flywheel speed decreases and finally comes to rest. At the same time, the tangential velocity is decreasing with time to zero in an essentially parabolic mode. Heating time is only a matter of seconds.

In the majority of applications, inertia friction welding uses a single axial thrust to produce

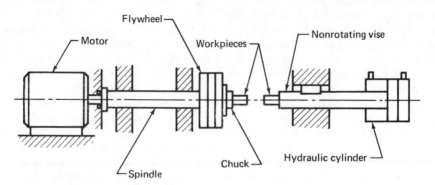

Fig. 7.11 — Basic arrangement of an inertia welding machine

heating pressure. However, the machines are normally equipped to apply two levels of thrust. When forging pressure is used, it is triggered at a selected speed setting near the end of the cycle (see Fig. 7.4). The technique can also be used to provide a preheat effect before welding occurs as with the conventional drive method.

An inertia welding cycle can be divided into three stages with each blending smoothly into the succeeding stage, as shown in Fig. 7.4. The first stage starts when the faying surfaces come in contact. It ends after the torque peaks and then drops to a nearly steady state, as in the continuous drive method. During this stage, the axial thrust is rapidly applied and rubbing raises the interface temperature to the plastic state. The rotational speed decreases as does the available energy.

During the second phase, the heat is conducted away from the interface at about the same rate that it is being generated. There is little axial shortening, but energy is being extracted from the flywheel and its speed continues to decrease.

When the speed decreases to some value, the energy being delivered by the flywheel becomes less than that being conducted from the interface. Torsional upsetting begins as the speed and temperature continue to drop. Torque again rises rapidly to a much higher level during this third stage as the metal stiffens. The speed and torque then decrease to zero as the weld cools.

Welding Variables

There are three welding variables with this method: moment of inertia of the flywheel, initial flywheel speed, and axial pressure. The first two variables determine the total kinetic energy available to accomplish welding. The amount of pressure is generally based on the material to be welded and the interface area. Axial shortening is controlled by adjusting initial flywheel speed.

The energy in the flywheel at any instant during the welding cycle is defined by the equation

$$E = \frac{Is^2}{5873} = \frac{Wk^2s^2}{5873}$$

where: E = Energy, ft · lb
I = Moment of inertia (Wk^2), lb ·ft^2

W = Weight of the flywheel system, lbs
k = Radius of gyration, ft
s = Instantaneous speed, rpm

With a particular flywheel system, the energy in the flywheel is determined by its rotational speed at any instant. If the mass of the flywheel is changed, the available energy at any particular speed changes. Therefore, the capacity of an inertia welding machine can be modified by changing the flywheel, within the limits of the machine capability.

During welding, energy is extracted from the flywheel and its speed will decrease. The total time for the wheel to come to rest will depend upon the average rate at which the energy is being removed and converted to heat.

The heating pattern at the interface can be adjusted to produce a bond over the entire area by varying the flywheel configuration, heating pressure, and speed. The conditions for welding a rod and a tube of the same alloy and equal diameters will likely be different because of the change in tangential velocity with radius. (The velocity at the center of a rod is zero and maximum at the periphery.) Also the heat input can be adjusted to control the width of the heat-affected zone and the cooling rate of the weldment. Figure 7.12 indicates the effect of flywheel energy, heating pressure, and tangential velocity on the heat pattern and flash formation of welds in steel.

Flywheel Configuration. The moment of inertia of the flywheel depends upon its section shape, diameter, and mass. For a specific application and initial speed, the energy available for welding can be increased by changing to a flywheel with a larger moment of inertia (I), and vice versa.

The amount of upset near the end of the welding cycle depends upon the remaining energy in the flywheel as well as the heating or forging pressure. For low carbon steel, upsetting usually starts at a peripheral velocity of about 200 ft/min. Large flywheels can prolong the forging or upsetting phase. If the flywheel is too small, the upset may be insufficient to consolidate the weld and eject impurities from the interface.

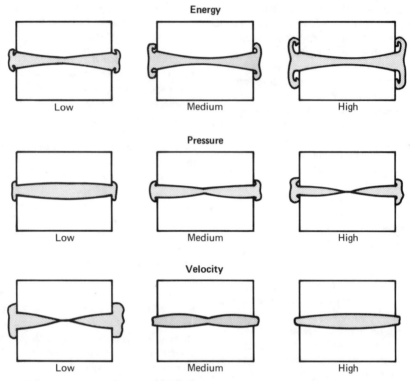

Energy

Low   Medium   High

Pressure

Low   Medium   High

Velocity

Low   Medium   High

Fig. 7.12—Effect of welding variables on the heat pattern at the interface and flash formation of inertia welds

The flywheel mass and its initial velocity may be varied inversely over a wide range for a given total energy requirement. This permits control of the size and shape of the heat-affected zone as well as the radial location of initial heat generation on the weld interface.

With a constant initial velocity and heating pressure, a larger flywheel will increase the available energy. The effect of this is shown in Fig. 7.12. As the available energy is increased, the amount of plasticized metal becomes greater as do the upset and flow of metal from the interface. The heating pattern remains fairly uniform but the excessive energy wastes metal in the form of flash.

Velocity. The instantaneous tangential velocity varies directly with the radius and rotational speed according to the following equation:

$$V_t = 0.52\, rs$$

where: V_t = tangential velocity, ft/min

    r = radius, in.

    s = instantaneous speed, rpm

With a solid rod, the velocity varies linearly from zero at the center to a maximum at the periphery.

For each metal, there is a range of peripheral velocities that produces the best weld properties. For welding solid bars of steel, the recommended initial peripheral velocity of the workpiece ranges from 500 to 1500 ft/min; however, welds can be made at velocities as low as 300 ft/min. If the velocity is too low, whether at the required energy level or not, the heating at the center will be too low to produce a bond across the entire interface and the flash will be rough and uneven. This is illustrated at the bottom of Fig. 7.12. At medium velocities of 300 to 800 ft/min, the heating pattern in steel has an hourglass shape at the lower end of the range and gradually flattens at the upper end of the range. At initial velocities above 1200 ft/min for steel, the weld becomes rounded and is thicker at the center than at the periphery.

Heating Pressure. The effect of varying heating pressure is generally opposite to that of velocity. As Fig. 7.12 shows, welds made at low heating pressure resemble welds made at high velocity with regard to the formation and appearance of weld upset and heat-affected zones. Excessive pressure produces a weld that lacks good bonding at the center and has a large amount of weld upset, similar to a weld made at low velocity. The effective heating pressure range for a solid bar of medium carbon steel is 22 000 to 30 000 psi.

WELDING PROCEDURES

SURFACE PREPARATION AND FIT-UP

As with any welding process, surface preparation can affect weld quality. Weld quality and consistency will be best when the faying surfaces are free of dirt, oxide or scale, grease, oil, or other foreign materials. In addition, the faying surfaces should mate together with very little gap.

In noncritical applications, some contamination and nonuniform contact of the faying surface may be tolerated. This is true if sufficient axial shortening is used to account for the gap and to extrude sufficient plasticized metal from the interface to carry away any contaminants. Sheared, flame cut, or sawed surfaces may be used with adequate axial shortening provided the surfaces are essentially perpendicular to the axis of rotation. If the surfaces are not perpendicular, joint mismatch could result. For best practice, the squareness should be within 0.010 in./in. of joint diameter.

Thick layers of mill scale should be removed from steel workpieces prior to welding to avoid unstable heating. A thin layer of rust may not be detrimental with adequate axial shortening.

Center projections left by cut-off tools are not harmful. However, pilot holes or concaved surfaces should be avoided since they may entrap air or impurities at the interface.

TOOLING AND FIXTURES

All gripping devices used for holding the workpieces must be reliable. Slippage of a workpiece in relation to the chuck results either in a poor weld or in damage to the gripping device or the workpiece.

The gripping mechanism of the chucking devices must be rigid and resist the applied thrust. The extension of the workpiece from the device should be as short as practical to minimize deflection, eccentricity, and misalignment. Grip diameter must be as large or larger than the diameter of the weld interface. Otherwise, the workpiece may shear at the grips. Serrated gripping jaws are recommended for maximum clamping reliability.

There are two basic types of tooling: rotating and nonrotating. The machine in Fig. 7.13 is equipped with both types. Each type, in turn, is either manual or power operated. As a rule, manually actuated tooling is used only for small quantity production.

Rotating tooling must be well-balanced, have high strength, and provide good gripping power. Collet chucks meet the above requirements and, therefore, are most frequently used. Typical collet tooling is shown in Fig. 7.14. Since total diametrical movement of a typical collet chuck is approximately 0.020 in., dimensional and ovality tolerances of the rotated workpiece must be within this limit. Specially designed collet chucks capable of larger movements are available, but they have a larger moment of inertia.

The most commonly used nonrotating gripping device is a self-centering, vice-like fixture with a provision for absorbing the thrust. This device permits reasonable diametrical tolerance

Fig. 7.13—A large inertia welding machine with associated tooling; above, (1) ways; below, (2) retracting cylinder, (3) rotary clamp actuator, (4) replaceable grips, (5) steady rest, (6) backstop

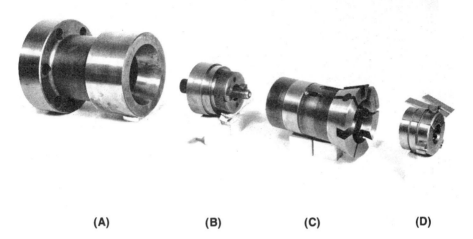

***Fig. 7.14—A collet type chuck frequently used to grip the rotating piece; (A) body,
(B) tie bar connector, (C) master collet, (D) insert collet***

in the stationary workpiece and yet maintains concentricity with the other piece in the collet chuck. More sophisticated devices may be used where concentricity is very critical.

Mating of faying surfaces and concentricity of the workpieces depend upon the accuracy of manufacture, projecting length from the clamping fixture, and the rigidity of the tooling.

HEAT TREATMENT

Prior heat treatment of the workpieces generally has little effect on the ability to friction weld specific alloys. However, it may affect the mechanical properties of the heat-affected zone and the gripping of the workpieces.

Postweld heat treatment is generally employed to produce the desired properties in the base metal or the welded joint, or both. The heat treatments should be those designed for the particular alloy being welded. In some cases, a postweld anneal is used to soften and stress relieve the joint for improved ductility.

In the case of dissimilar metal welds, a postweld heat treatment should not contribute to the formation or expansion of an intermetallic layer at the interface. An intermetallic layer might significantly lower joint ductility or strength. In any case, the postweld heat treatment should be evaluated for the application by destructive tests.

WELD QUALITY

JOINT DISCONTINUITIES

Weld quality is primarily dependent upon the proper selection of the welding variables. Fortunately, good welds can be made between like metals with a rather wide range of speeds, pressures, and times. Discontinuities such as gas pockets and slag inclusions are virtually

nonexistent. However, other types of discontinuities do occur. These are normally associated with improper surface preparation or welding conditions, or both.

Discontinuities at the center of a weld may occur for various reasons. The welding conditions used may not create sufficient heating at the center for coalescence. Figure 7.15 shows

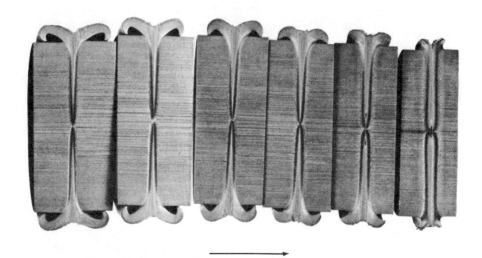

Excessive   Decreasing axial shortening   Insufficient

Fig. 7.15—The effect of axial shortening on the bonding and flash of friction welds

the effect of axial shortening on weld quality. These inertia welds were made with the same speed and inertial mass but with a decreasing heating pressure from left to right. Two welds exhibited center defects because insufficient axial shortening (pressure) was used, but the others did not. Lack of center bonding may also occur in continuous drive friction welds when inadequate speed, heating time, or pressure is used.

Concave faying surfaces that prevent uniform contact during the early stages of welding can limit center heating and entrap oxides. Figure 7.16 shows a weld with a discontinuity that resulted when a center hole for machining operations was not removed prior to welding.

INSPECTION AND TESTING

Process control is best performed by random destructive or proof testing of either production welds or specimens welded with the production machines. The frequency of testing is usually determined by service requirements, customer specifications, or company quality control procedures. Destructive inspection may include metallographic examination as well as tensile, impact, or bend tests.

Nondestructive testing methods are not generally acceptable for determining friction weld quality and soundness. However, ultrasonic techniques may be effective for the detection of subsurface discontinuities such as incomplete fusion at or near the center of the weld. A straight beam ultrasonic test conducted perpendicular to the weld interface gives the most reliable results. This test can be applied if one end of the weldment is accessible to the ultrasonic transducer. Lack of bonding at the inside surface of a tube-to-tube weld may be detected by ultrasonic inspection using the angle beam technique.

Discontinuities open to the surface may be detected by dye and fluorescent penetrant methods. The magnetic particle inspection method may detect discontinuities in magnetic materials when they are at or just below the surface. The flash should be removed from the joint before the inspection operation is performed.

PROCESS MONITORING

Automatic equipment is available for monitoring production friction welding operations. The purpose of monitoring equipment is to identify machine malfunction or change in operation due to part variations. It may identify the fault and signal the operator to stop the operation until appropriate adjustments are made. Process monitoring provides good production quality control.

Continuous drive friction welding equipment can be monitored for speed, heating and upsetting forces, time, and axial shortening. If used, forging force must be applied rapidly at the instant that drive power is disengaged. Excessive inconsistency in operation at this time in the cycle can produce welds of low quality.

Inertia welding has only two variables: the rotating speed at which axial thrust is applied, and the magnitude of axial thrust. These variables can be monitored with commercially available instrumentation.

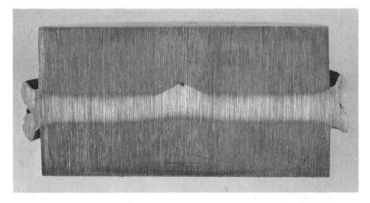

Fig. 7.16 — Discontinuity at the center of a friction weld caused by a prior center hole

APPLICATIONS

Friction welding is frequently being used instead of flash or upset welding for applications where one of the members to be joined has axial symmetry. Several typical applications follow.

An automotive transmission input shaft assembly and its two components are shown in Fig. 7.17. The assembly is an SAE 1010 steel stamping friction welded to the shoulder of an SAE 1141 steel rod. The weld was made in about two seconds.

The 0.060-inch-wall tubular container, shown in Fig. 7.18, was welded to a base of greater thickness in about one second. This assembly was formerly made by copper brazing. The process was changed to friction welding to reduce cost and to produce a more reliable and consistent part.

Figure 7.19 illustrates a tube-to-tube type weld with an interface area of approximately 7 in.² in a tractor track roller. This component of SAE 1046 steel is welded without preheat, but is heat treated to the desired hardness level.

A fabricated jet engine compressor wheel is shown in Fig. 7.20. It is composed of four 24-in. diameter sections of a high temperature, nickel-base alloy joined by three friction welds. Concentricity was held to within 0.015 in. and the axial shortening to within 0.010-in. tolerance. The electron photomicrograph of a cross section through one of these welds is shown in Fig. 7.21. The interface is free of any discontinuities.

Fig. 7.17—A friction welded automotive transmission input shaft

SAFETY

Friction welding machines are similar to machine tool lathes in that one workpiece is rotated by a drive system. They are also similar to hydraulic presses in that one workpiece is forced against the other with high loads. Therefore, safe practices for lathes and power presses should be used as guides for the design and operation of friction welding machines.

Machines should be equipped with appropriate mechanical guards and shields as well as two-hand operating switches and electrical interlocks. These devices should be designed to prevent operation of the machine when the work area, rotating drive, or force system are accessible to the operator or others.

Operating personnel should wear appropriate eye protection and safety apparel commonly used with machine tool operations. In any case, applicable OSHA.Standards should be strictly observed.

Metric Conversion Factors

1 ft/min = 5.08×10^{-3} m/s
1 psi   = 6.89 kPa
1 in.   = 25.4 mm
1 ft • lbf = 1.356 J
1 lb   = 0.45 kg
1 ft   = 305 mm
1 rpm   = 0.105 rad/s

Fig. 7.18—A thin-wall tubular container friction welded to a flange

Fig. 7.19—A friction welded track roller assembly

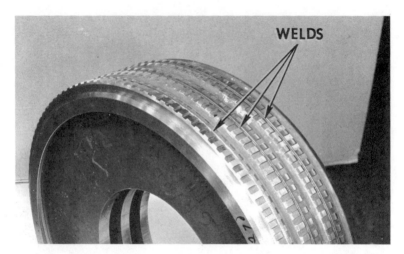

Fig. 7.20—A jet engine compressor wheel fabricated by friction welding

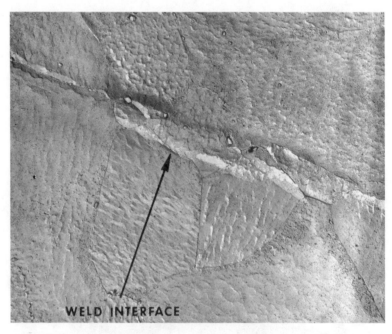

Fig. 7.21—Electron photomicrograph of a section through a friction weld in a high-temperature nickel base alloy (×5000)

SUPPLEMENTAL READING LIST

Adams, D. and Taylor, S., Detection of faults in friction welded studs by ultrasonics. *Weld. & Metal Fab.*, 37 (10): 412-21; 1969 Oct.

Dickson, G.R., et al., Experiments on friction welding some nonferrous metals, *International Conference on the Welding and Fabrication of Nonferrous Metals;* 1972 May 2-3; Eastbourne, Cambridge, England: The Welding Institute; 1972: 41-53.

Duffin, F.D. and Bahrani, A.S., The mechanics of friction welding mild steel. *Metal Construction*, 8 (10): 267-71; 1976 June.

Ellis, C. R. G., Continuous drive friction welding of mild steel. *Welding Journal*, 51 (4): 183s-197s; 1972 Apr.

Ellis, C.R.G. and Needham, J.C., Quality control in friction welding, Intl. Inst. of Welding; 1972; *IIW Document III—460-72*. Available from: *American Welding Society*, Miami, FL.

Ellis, C.R.G. and Nicholas, E.D., A quality monitor for friction welding, *Advances in Welding Processes*, 3rd International Conference; 1974 May 7-9; Harrogate, England. Cambridge, England: The Welding Institute; 1974: 14-18.

Forster, P.B., Heat under power (HUP) friction welding, *Advances in Welding Processes*, 3rd International Conference; 1974 May 7-9; Harrogate, England. Cambridge, England: The Welding Institute; 1974: 243-8.

Jessop, T.J., et al., Friction welding dissimilar metals, *Advances in Welding Processes*, 4th International Conference; 1978, May 9-11; Harrogate, England. Cambridge, England: The Welding Institute; 1978: 23-36.

Johnson, T.K. and Davies, J.W., A comparison of the capabilities of continuous drive friction and inertia welding, Dearborn, MI; Soc. of Manufacturing Engineers, 1973; Paper No. AD73-221.

Kuruzar, D.L., Joint design for the friction welding process. *Welding Journal*, 58(6): 31-5; 1979 June.

Needham, J.C. and Ellis, C.R.G., Automation and quality control in friction welding. *The Welding Institute Research Bulletin;* 12(12); 333-9 (Part 1); 1971 Dec. 13(2); 47-51 (Part 2); 1972 Feb.

Nessler, C.G., et al., Friction welding of titanium alloys. *Welding Journal*, 50(9): 379s-85s; 1971 Sept.

Nicholas, E.D., Radial friction welding, *Advances in Welding Processes*, 4th International Conference, 1978, May 9-11; Harrogate, England. Cambridge, England: The Welding Institute; 1978: 37-48.

Nicholas, F.D., Friction Welding: state of the art. *Welding Design and Fabrication*, 50 (7): 56-62; 1977 July.

Rich, T. and Roberts, R., The forge phase of friction welding. *Welding Journal*, 50 (3): 137s-45s; 1971 Mar.

Searl, J.C., The orbital friction welding process for noncircular components. *Engineers Digest*, 32: 33-36; 1971 Aug.

Wang, K.K., *Friction welding*, New York, Welding Research Council; 1975, Apr.; Bulletin 204.

Wang, K.K. and Linn, W., Flywheel friction welding research. *Welding Journal*, 53 (6): 233s-41s; 1974 June.

Wang, K.K. and Nagappan, P., Transient temperature distribution in inertia welding of steels. *Welding Journal*, 49 (9): 419s-26s; 1970 Sept.

Wang, K.K. and Rasmussen, G., Optimization of inertia welding process by response surface methodology. *Trans—ASME, Journal Engrg. Ind.*, 94, Series B (4): 999-1006; 1972 Nov.

8

Explosion Welding

Chapter Committee

V. D. LINSE, *Chairman*
Battelle Columbus Laboratories

H. E. OTTO
Southwest Laboratories

A. POCALYKO
E. I. du Pont de Nemours and Company

Welding Handbook Committee Member

W. L. WILCOX
Scott Paper Company

263

8

Explosion Welding

FUNDAMENTALS OF THE PROCESS

DEFINITION AND GENERAL DESCRIPTION

Explosion welding is a process that uses energy from the detonation of an explosive to join two pieces of metal. The explosion accelerates the pieces to a speed at which a metallic bond will form between them when they collide. The weld is produced in a fraction of a second without the addition of filler metal. This is essentially a room temperature process in that gross heating of the workpieces does not occur. The faying surfaces, however, are heated to some extent by the energy of the collision, and welding is accomplished through plastic flow of the metal on those surfaces. Welding takes place progressively as the explosion and the forces it creates advance from one end of the joint to the other. Deformation of the weldment varies with the type of joint. There may be no noticeable deformation at all in some weldments, and there is no loss of metal. Welding is invariably done in air, although it can be done in other atmospheres or in a vacuum. Most explosion welding is done on sections with relatively large surface areas, although there are applications for sections with small surface areas as well.

PRINCIPLES OF OPERATION

A typical arrangement of the components for explosion welding is shown in Fig. 8.1. Fundamentally, there are three:

(1) Base component
(2) Prime component
(3) Explosive

The base component remains stationary as the prime component is welded to it. This base component may be supported by a backer or an anvil. To do its job, the backer must have a large mass.

The prime component is moved toward the base component by the explosion during welding. This prime component usually is positioned parallel to the base component. However, for special applications they may be at some small angle to each other. In the parallel arrangement, the two are separated by some specified amount, referred to as the standoff distance. In the angular arrangement, a standoff distance (at the apex of the angle) may or may not be used. The explosion locally deforms and accelerates the prime component across the standoff distance so

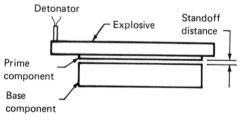

Fig. 8.1—Typical component arrangement for explosion welding

264

that it collides with and welds to the base component. Deformation and acceleration are accomplished increment by increment across the joint as the explosion takes place.

The explosive, usually in granular form, is distributed uniformly over the top surface of the prime component. The force which the explosion exerts on the prime component depends upon the detonation characteristics and the quantity of the explosive.

A buffer, usually of rubber material, may be required between the explosive and the prime component to protect the surface of that component from erosion by the detonating explosive.

Explosive Detonation

The action that occurs during welding is illustrated in Fig. 8.2. The manner in which the explosive is detonated is extremely important. Detonation must take place progressively across the surface of the prime component. The speed of the detonation front establishes the velocity at which the two components collide. This is known as the collision velocity and is one of the important variables of the process. The selection of an explosive that will produce the required detonation velocity is of utmost importance in consistently obtaining good welds. Moreover, the explosive must provide uniform detonation so that collision velocity will be uniform from the start to the finish of the weld.

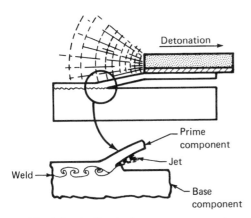

Fig. 8.2—Action between components during explosion welding

Prime Component Velocity And Angle

As the detonation front moves across the surface of the prime component, both the intense pressure in the front and the pressure generated by the expanding gases immediately behind the front accelerate the prime component, increment by increment, across the standoff distance. By the time an increment of the prime component located between the detonation front and the collision point reaches the base component, it has been rotated through a certain angle and accelerated to a certain velocity. This angle and velocity depend upon the type and amount of explosive; the thickness and mechanical properties of the prime component; and the standoff distance employed.

Collision, Jetting, and Welding

The important interrelated variables of the process are:

(1) Collision velocity
(2) Collision angle
(3) Prime component velocity

The intense pressure necessary to make a weld is generated at the collision point when any two of these variables are within certain well defined limits. These limits are determined by the properties of the particular metals to be joined. This pressure forces the surfaces of the two components into intimate contact and causes them to flow plastically. At the same time, a jet is formed at the collision point, as shown in Fig. 8.2. The jet sweeps away the original surface layer on each member, along with any contaminating film that might be present. This exposes clean underlying metal which is required to make a strong metallurgical bond. Residual pressures within the system are maintained long enough after collision to avoid springback of the metal and to complete the weld.

NATURE OF THE BOND

The interface between the two members of an explosion weld may be flat or wavy, depending upon the particular conditions employed in making the weld. A flat interface is formed

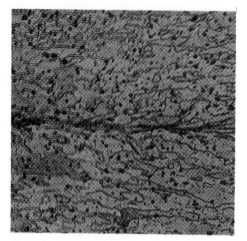

Fig. 8.3—Straight interface of an explosion weld in Type 2024 aluminum alloy

when the collision velocity is below some critical value for the particular combination of metals being welded. As shown in Fig. 8.3, little or no melting occurs in such a weld. Welds of this type usually possess satisfactory mechanical properties but as a rule are not sought in practice. Small variations in the collision conditions can result in lack of bonding.

Welds at collision velocities above the critical value have a wavy interface, as shown in Fig. 8.4. Welds with this interface generally have better mechanical properties than those with flat interfaces and can be made over a wider range of welding conditions.

Most welds with a wavy interface contain small pockets of jet material, as shown in Fig. 8.5. These pockets normally are located on the front and back slopes of the waves. The material is composed of some combination of the two parent metals, and partial or complete melting of this material generally occurs. The pockets will be ductile when the metal combinations can form solid solutions, but they may be brittle or may show discontinuities in those combinations that form intermetallic compounds. Pockets of the latter material may not be detrimental, if they are very small. Good welding practices will produce small pockets.

Large pockets, on the other hand, occur with excessive collision conditions (prime component velocity, collision velocity, and collision angle). Such conditions may even produce a continuous melted layer, as shown in Fig. 8.6. The large pockets and the continuous melted layer often contain a substantial number of shrinkage voids and other discontinuities that reduce strength and ductility. They are usually detri-

Fig. 8.4—Wavy interface of an explosion weld between Type 304 stainless steel and columbium

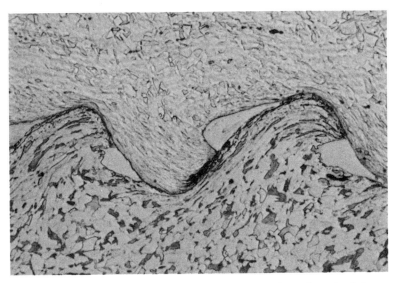

Fig. 8.5—Typical wavy interface with discrete melt pockets (stainless steel to mild steel)

mental to the soundness and serviceability of the weld. For this reason, welding practices that produce a continuous melted layer must be avoided.

CAPABILITIES AND LIMITATIONS

One attribute of the explosion welding process is its ability to join a wide variety of similar and dissimilar metals. The dissimilar metal combinations range from those that are commonly joined by other welding processes, such as carbon steel to stainless steel, to those that are metallurgically incompatible for fusion welding, such as aluminum or titanium to steel.

The process can be used to join components of a wide range of sizes. Surface areas ranging from less than 1 in.^{2} to over 300 ft^2 can be welded. Since the base component is stationary during welding, there is no upper limit on its thickness. The thickness of the prime component may range from 0.001 to 1.25 inches.

Geometric configurations that can be explosion welded are those which allow a uniform progression of the detonation front and, hence,

the collision front. These include flat plates as well as cylindrical and conical structures. Welds may also be made in certain complex configurations, but such work requires thorough understanding and precise control of the process.

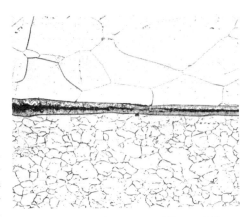

Fig. 8.6—Continuous melt layer at the interface of an explosion weld between a nickel base alloy and carbon steel

APPLICATIONS

METALS WELDED

As a general rule, any metal can be explosion welded if it possesses sufficient strength and ductility to withstand the deformation required at the high velocities associated with the process. Metals that will crack when exposed to the shock of detonation and collision with another component cannot be explosion welded. Metals with elongations of at least 5 percent (in a 2 in. gage length) and Charpy V-notch impact strengths of 10 ft • lbs or better can be welded by this process. In special cases, metals with low ductility can be welded by heating them to a temperature at which they will have adequate ductility. The commercially significant metals and alloys that can be joined by explosion welding are given in Fig. 8.7.

Explosion welding can produce significant changes in the mechanical properties and hardness of metals. This is indicated in Fig. 8.8. In general, strength and hardness increase and ductility decreases as a result of welding. This comes about from the severe plastic deformation encountered, particularly in the prime component (see Fig. 8.5). Explosion welding may increase the ductile-to-brittle transition temperature of carbon steel. Since the prime component undergoes severe plastic deformation during welding, these effects may be much more pronounced in that component than in the base component. Such effects may be erased by a postweld heat treatment (see Fig. 8.8). However, the particular heat treatment applied should be one that will not reduce the ductility of the weld by forming brittle intermetallic compounds at the interface.

The magnitude of any changes in mechanical properties should be established and taken into consideration in the design of load-bearing structures that are explosion welded. This is particularly true where impact loading

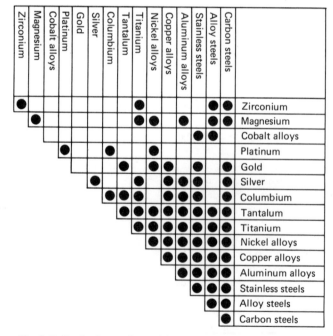

Fig. 8.7—Typical metal combinations that can be explosion welded

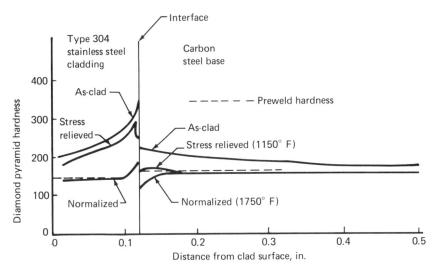

Fig. 8.8—Hardness profiles across stainless steel to carbon steel clad plate, as-welded and after heat treatment

may be expected in welds in carbon steel.

CLADDING

Plate

The cladding of flat plate constitutes the major commercial application of explosion welding. It is customary to supply explosion clad plate in the as-welded condition because the hardening which occurs immediately adjacent to the interface usually does not significantly affect the engineering properties of the plate. Despite this, some service requirements may demand postweld heat treatment. Clad plates usually are distorted somewhat during explosion welding and must be straightened to meet standard flatness specifications. Straightening is usually done with a press or a roller leveler.

Pressure vessel heads and other components can be made from explosion clad plates by conventional hot or cold forming techniques. Hot forming must take into account the metallurgical properties of the materials and the possibility that undesirable diffusion may occur at the interface. Compatible alloy combinations,

such as stainless and carbon steel, may be formed by methods traditionally used for clad materials. Incompatible combinations such as titanium and steel, on the other hand, may require special procedures to limit the formation of undesirable intermetallic compounds at the interface. Titanium clad steel, for instance, should be hot formed at temperatures no higher than 1400° F.

Reducing the thickness of clad plate by rolling provides a convenient means of producing bimetal sheets of proper thickness for subsequent processing.

Cylinders

Explosion welding can be used to clad cylinders on their inside or outside surfaces. One application of this is the internal cladding of steel forgings with stainless steel to make nozzles for connection to heavy-walled pressure vessels. Clad nozzles with inside diameters of 0.5 to 24 in. and lengths up to 3 ft have been made.

TRANSITION JOINTS

Fusion welded joints between two incompatible metals are difficult or impossible to make. Some of those that can be made exhibit low strength and ductility. Transition joints produced by explosion welding can provide a solution to that problem. Many such joints can be cut from a single large clad plate. Conventional fusion welding practices may then be used to attach the members of the transition joint to their respective similar metal components. Care must be taken, however, to limit the service temperature to a level suitable for the transition joint.

Electrical

Aluminum, copper, and steel are the metals most commonly used in electrical systems. Joints between them frequently are necessary to take advantage of the special properties of each. Such joints must be sound if they are to conduct high amperages efficiently, minimize power losses and avoid overheating of the member in service. Transition joints cut from

thick explosion welded plates of aluminum and copper or aluminum and steel provide efficient conductors of electricity. This concept is routinely used in the fabrication of anodes for the primary aluminum industry, an example of which is shown in Fig. 8.9.

When electrical connections must be disconnected periodically, mechanical joints between copper conductors serve well. Transition joints between copper and aluminum, in such cases, are installed by fusion welding the aluminum side of the joint to the aluminum member. The copper sides are then bolted together.

The temperature limit for transition joints between aluminum and steel should be not more than 500° F for long-term service. Copper-aluminum joints should be limited to 300° F. Transition welds are unaffected by thermal cycling below these temperatures. Short term exposure during attachment welding, for example, may reach 600° F and 500° F, respectively, without harm.

Fig. 8.9—Welded anode assembly using an aluminum to steel transition joint at the arrow

Structural

In the presence of an electrolyte such as salt water, aluminum and steel form a galvanic cell. In a mechanical connection, crevice corrosion can set in. A welded transition joint can minimize such corrosion. The transition joint is metallurgically bonded, and there is no crevice in which the electrolyte can act. Structural transition joints are used to attach aluminum superstructures to the steel decks of naval vessels and commercial ships.

Tubular

Tubular transition joints in various config-urations can be machined from thick clad plate. The interface of the explosion weld is perpendicular to the axis of the tube in this case.

Joints can also be fabricated in an overlapping or telescoping style similar to a cylindrical cladding operation. They offer the advantage of a long overlap and frequently require little or no machining after welding.

The majority of tubular transition joints involve the combination of aluminum to steel ranging in diameters from 2 to 12 inches. Other metal combinations for this type of joint include titanium to stainless steel, zirconium to stainless steel, zirconium to nickel-base alloys, and copper to aluminum. Typical tubular transition joints are shown in Fig. 8.10.

Fig. 8.10—Typical tubular transition joints

OTHER

Heat Exchangers

Explosion welding may be used to make tube-to-tube sheet joints in heat exchanger fabrication. The process is essentially a version of internal cylinder cladding. A small explosive charge is used to make the joint, as shown in Fig. 8.11. In most instances, the weld is located near the front of the tube sheet and has a length of approximately 1/2 in. or five times the thickness of the wall of the tube, whichever is greater. Points that must be considered in determining whether explosion welding is suitable for a particular tube-to-tube sheet application include the diameter of the tube, the ratio of the wall thickness to the diameter of the tube, the thickness of the ligament between the holes in the tube sheet, and the thickness of the tube sheet. Tubes may be welded individually or in groups. The number in any group is controlled by the quantity of explosive that can be detonated safely at any one time.

Most applications of explosion welding in tube-to-tube sheet joints involve tube diameters in the range of 0.5 to 1.5 inches. Metal combinations include steel to steel, stainless steel to stainless steel, copper alloy to copper alloy, nickel alloy to nickel alloy clad steel, and both aluminum and titanium to steel.

Buildup and Repair

Explosion welding may be used for buildup and repair of worn components. It is particularly applicable to the repair of inside and outside surfaces of cylindrical components. The worn area is simply clad with an appropriate thickness of metal and machined to the proper dimensions. In some instances, such as bearing surfaces, the repair can be made with a material that is superior to the original material.

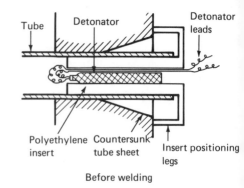

Before welding

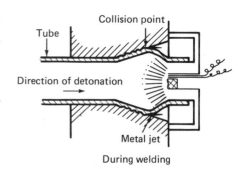

During welding

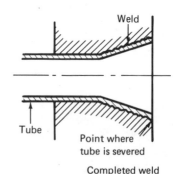

Completed weld

Fig. 8.11—Tube-to-tube sheet explosion welding

WELDING PROCEDURES

TYPES OF JOINTS

Explosion welding is limited to joints that overlap or have faying surfaces. In the case of cladding, the surfaces of both components have the same geometry and one component overlaps the other. The same is true of transition joints in pipe or tubing and tube-to-tube sheet joints. The weld in such joints should be long enough to insure that it will not fail in service by shear along the interface.

SURFACE PREPARATION

The surfaces to be joined should be clean and free of gross imperfections to produce welds of consistent soundness, strength, and ductility. The smoothness required depends upon the metals to be joined. In general, a surface finish of 150 microinches or better is required to obtain high quality welds.

FIXTURING AND BACKUP

For consistent quality, the welding conditions should be uniform over the entire area to be joined. These include standoff distance with parallel components or initial angle with inclined components and sufficient rigidity or support for the base component. For cladding, a relatively thick prime component may need support only at the edges, if deflection is not excessive. Where deflection is troublesome, the prime component may be placed on small blocks of rigid foam. These will be consumed during welding. Other supports may include sized metal balls and dimpled or corrugated metal strips. If the base component is relatively thin, it should be supported uniformly on a massive anvil to minimize deflection. It may even be necessary to straighten the base component before welding to ensure uniform contact with the anvil.

When joining pipe or tubing, an internal or external mandrel normally is required to back up the base component. An internal mandrel must be collapsible for easy removal after welding.

PROCESS VARIABLES

EXPLOSIVE

Two important properties of explosives for welding are detonation velocity and hazard sensitivity. The latter refers to the thermal stability, storage life, and shock sensitivity of the explosive. These affect handling safety.

The detonation velocity is proportional to the density of the explosive, but the pressure generated is proportional to density and the detonation velocity. The detonation velocity of an explosive may depend upon its thickness, the packing density, or the amount of inert material mixed with the explosive. Adjustments of the explosive for welding generally are made to decrease its detonation velocity.

Typical explosives which may be formulated to give suitable detonation rates are:

(1) Ammonium nitrate pellets with 6 to 12 percent diesel fuel

(2) Nitroguanidine plus inert material

(3) Amatol and sodatol with 30 to 55 percent rock salt

(4) Ammonium nitrate-TNT-atomized aluminum mixture

STANDOFF DISTANCE

Standoff distance will have some influence on the wave shape of the interface. Increasing the standoff distance increases the angle of approach between the prime and base components (see Fig. 8.2). This, in turn, increases the size of the wave until some maximum size is attained. Then the size of the wave will decrease as standoff distance is further increased.

Welding generally occurs when the angle of approach between the two components is between 5 and 25 degrees. When welding tubes to tube sheets, a suitable angle is achieved by tapering the hole in the tube sheet. A taper in the range of 5 to 10 degrees generally is suitable. In the case of parallel components, the angle of approach will depend upon other factors in addition to standoff distance. These are the detonation velocity and the thickness and mechanical properties of the prime component. A standoff distance between half and one times the thickness of the prime component normally is used. Those on the low side of the range are used with explosives having a high detonation velocity.

JOINT QUALITY

The quality of an explosion weld will depend upon the nature of the interface and the effect the process has on the properties of the metal components. The properties of the metal include strength, toughness, and ductility. The effect of welding on these properties can be determined by comparing the results of tension, impact, bending, and fatigue tests of welded and unwelded materials. Standard ASTM testing procedures may be used.

The quality of the bond can be determined by destructive and nondestructive tests. Since the size of test samples is limited by the thickness of the components, special destructive tests are used for evaluation of the bond. The tests should reflect the conditions the weld will have to endure in service.

NONDESTRUCTIVE INSPECTION

Due to the nature of explosion welds, nondestructive inspection is restricted almost totally to the ultrasonic method. Radiographic inspection is only applicable to welds between metals with significant differences in density and an interface with a large wavy pattern.

Ultrasonic Inspection

Ultrasonic inspection is the most widely used nondestructive method for the examination of explosion welds. It will not determine the strength of the weld, but it will indicate weld soundness. Pulse-echo techniques are used for clad steels in pressure vessels.[1] For such applications, the entire plate should be examined both before and after cladding. An ultrasonic frequency in the range of 3 to 4 MHz usually is adequate. Allowance needs to be made for the differences in acoustical impedance of various metals.

The ultrasonic instrument should be calibrated on standard samples containing both bonded and known unbonded areas which will provide a display signal amplitude of 50 to 75 percent of full scale for the bonded area. Unbonded areas reflect the signal before it can complete the circuit. This shows up in the height of the signal on the display scope. Recording devices can be used to give a permanent record of the results of the examination.

For large clad plates where scanning 100 percent of the surface area is not necessary, the examination can be carried out on a rectangular grid pattern laid out on the plate. Unbonded areas which are detected should be investigated to determine whether they are small enough to be acceptable or are so large or so numerous that they are unacceptable. The size and number of unbonded areas that can be permitted in a clad plate depend upon the application of the plate. Clad plates for heat exchangers sometimes require over 98 percent bond, and

---

1. See ANSI/ASTM A578, Standard Specification for Straight Beam Ultrasonic Inspection of Plain and Clad Steel Plates for Special Applications (latest edition).

limits are placed on the size and number of unbonded areas that are permitted.

Radiographic Inspection

Radiography can be used to inspect explosion welds in metals that have significantly different densities such as aluminum to steel. In such cases, all unbonded and flat bonded areas at the interface can be defined, regardless of their size and location.

Radiographs are marked to identify the plate and the precise location of the area they represent. The radiographs are taken perpendicular to the surface from the side with the high density metal. The film must be in intimate contact with the surface on the low density side. Radiographs can delineate a wavy interface as uniformly spaced light and dark lines. Unbonded areas do not show these light and dark lines, and such areas are interpreted as unbonded areas even though a flat bond may exist.

DESTRUCTIVE TESTING

Destructive testing is used to determine the strength of the weld and the effect of the process on the properties of the base metals. Standard testing techniques can be used, but specially designed tests sometimes are required to determine bond strength for some configurations.

Clad Plates

The requirements for carbon steel plates clad with copper, stainless steel, or nickel alloys are covered in appropriate ANSI/ASTM Standards.[2] These Standards use bend and shear tests to determine the strength of the bond and tension tests for the strength of the composite.

Chisel Test

The chisel test is widely used to quickly

2. See ANSI/ASTM Standard Specifications A263, A264, A265, and B432 (latest revisions).

Fig. 8.12—A chisel tested section from an aluminum clad steel plate

determine the integrity of the bond in an explosion weld. The test is performed by driving a chisel into and along the interface. The ability of the interface to resist separation by the force of the chisel provides an excellent qualitative measure of the bond strength. Figure 8.12 shows a section from an aluminum clad steel plate under evaluation in this manner. If the weld is not good, failure will occur along the interface in advance of the chisel point. If the weld is good, the chisel will cut through the weaker of the two parent metals.

Tension-Shear Test

This test is designed to determine the shear strength of the weld. The specimen is shown in Fig. 8.13. Equal thicknesses of the two components are preferred. The length of the shear zone, "d," should be selected so that little or no bending will occur in either component. Failure should occur by shear, parallel to the weld line. If failure occurs through one of the base metals, the shear strength of the weld is obviously greater than the strength of the base metal. In any event, the results are useful for comparison purposes only, using a common test specimen.

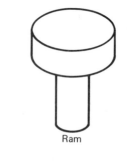

Ram

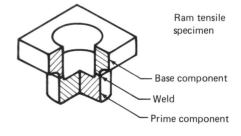

Ram tensile specimen

Base component

Weld

Prime component

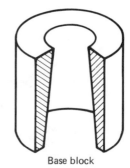

Base block

Fig. 8.14—Ram type tension test for bond strength

Tension Test

A special "ram" tension test can be used for evaluation of the tensile strength of explosion welds. As shown in Fig. 8.14, the specimen is designed to subject the weld interface to a tensile load. The cross-sectional area of the specimen is the annulus between the outside and inside diameters. The typical specimen has a short gage length which is intended to cause failure at the interface. The tensile strength of the weld can be obtained in this manner. If

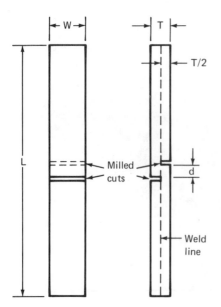

Fig. 8.13—Tension-shear test specimen

failure occurs in one of the base metals, the test shows that the weld is stronger than the metal.

The test is conducted by placing the specimen on the base block with the ram in the hole. A compressive load is then applied through the ram and base. Load at failure is recorded.

Metallographic Examination

Metallography can provide useful information about the quality of explosion welds. The section for metallographic examination should be taken so that the interface can be examined on a plane parallel to the direction of detonation and normal to the surfaces of the welded components. A well-formed, well-defined wave pattern generally is indicative of a good weld, and the amplitude of the wave can vary without

significant influence on the strength of the weld. Small pockets of melt can exist but these are not detrimental. A wave whose crest is bent over a large pocket containing voids in the swirls indicates that the collision angle and velocity were too high and the weld is poor. Excessive collision velocity with metals such as titanium and martensitic steels produces bands adjacent to the interface. These are a result of adiabatic shear. Proper welding conditions eliminate or minimize the occurrence of these bands.

Samples for metallographic examination should be taken from an area that is representative of the entire weld. Edge effects may result in local unbonded areas along the edges of a weld. Samples taken from such locations would not be representative of the rest of the weld.

SAFETY

Explosives and explosive devices are a part of explosion welding. Such materials and devices are inherently dangerous. Safe methods for handling them do exist. However, if the materials are misused, they can kill or injure anyone in the area and destroy or damage property.

Explosive materials should be handled and used by competent people who are experienced in that field. Handling and safety procedures must comply with all applicable federal, state, and local regulations. Federal jurisdiction on the sale, transport, storage, and use of explosives is through the U.S. Bureau of Alcohol, Tobacco,

and Firearms; the Hazardous Materials Regulation Board of the U.S. Department of Transportation; the Occupational Safety and Health Agency; and the Environmental Protection Agency. Many states and local governments require a blasting license or permit, and some cities have special explosive requirements.

The Institute of Makers of Explosives provides educational publications to promote the safe handling, storage, and use of explosives. The National Fire Protective Association provides recommendations for safe manufacture, storage, handling, and use of explosives.

Metric Conversion Factors

$$
\begin{aligned}
1 \text{ in.} &= 25.4 \text{ mm} \\
\text{ft/s} &= 0.305 \text{ m/s} \\
1 \text{ ft} \cdot \text{lb} &= 1.36 \text{ J} \\
t_C &= 0.556 \, (t_F - 32) \\
1 \text{ in.}^2 &= 645 \text{ mm}^2 \\
1 \text{ ft}^2 &= 0.093 \text{ m}^2
\end{aligned}
$$

SUPPLEMENTARY READING LIST

Berment, L. J., Small scale explosion seam welding. *Welding Journal,* 52 (3); 147-54; 1973 Mar.

Blazynski, T. Z., Implosive and explosive welding of mono- and bi-metallic duplex cylinders, *Intl. Conf. on the Welding and Fabrication of Non-ferrous Metals,* Eastbourne: 1972 May 2-3; 62-71. Cambridge, England; The Welding Institute, 1972.

Carpenter, S. H. and Wittman, R. H., Explosion welding, *Annual Review of Materials Science,* Vol. 5, 1975: 177-99.

Cowan, G. R., Bergmann, O. R., and Holtzman, A. H., Mechanism of bond zone wave formation in explosion clad metals. *Met. Trans.,* 2 (11): 3145-55; 1971 Nov.

DeMaris, J. L. and Pocalyko, A., *Mechanical Properties of Explosion Bonded Clad Metal Products,* Dearborn, MI; Society of Manufacturing Engineers, AD 66-113; 1966.

Explosive Welding: Cambridge, England: The Welding Institute 1976.

Explosive Welding, Proceedings of the Select Conference, Hove, 1968, Sept. 18-19, Cambridge, England: The Welding Institute, 1969.

Ezra, A. A., *Principles and Practices of Explosive Metal-Working,* Vol. 1, London: Industrial Newspapers Ltd., 1973: 1-31.

Hardwick, R., Methods for fabricating and plugging of tube-to-tube sheet joints by explosion welding. *Welding Journal,* 54 (4) 238-44; 1975 Apr.

Holtzman, A. H. and Cowan, G. R., *Bonding of Metals With Explosives,* New York: Welding Research Council, 1965 Apr; Bulletin 104.

Linse, V. D., *The Application of Explosive Welding to Turbine Components,* New York: Amer. Soc. of Mech. Engineers, 1974; Publication 74-GT-85.

Onzawa, T. and Ishii, Y., Fundamental studies on explosive welding. *Trans. of the Japanese Welding Society,* 6 (2); 1975 Sept.; Welding Research Abroad, 23 (3); 28-34; 1977 Mar. (Welding Research Council).

Otto, H. E. and Carpenter, S. H., Explosive cladding of large steel plates with lead. *Welding Journal,* 51 (7): 467-73; 1972 July.

Popoff, A. A., Explosion welding. *Mechanical Engineering,* 100 (5): 28-35; 1978 May.

Sahay, S. R. et al., Morphology and composition of bond zones in some Invar-based explosive welds. *Welding Journal* 51 (1): 23s-28s; 1972 Jan.

9
Ultrasonic Welding

Chapter Committee

P. C. KRAUSE, *Chairman*
  Sonobond Corporation
C. BIRO
  Bulova Systems and Instruments Company
G. K. DINGLE
  *Hughes Helicopters Division,
  Summa Corporation*

F. R. MEYER
  Christiana Metals Corporation
T. RENSHAW
  Fairchild Republic Company
A. SHOH
  Branson Sonic Power Company

R. G. VOLLMER
  *U. S. Army Aviation Research
  and Development Command*

Welding Handbook Committee Member

I. G. Betz
  Department of the Army

9

Ultrasonic Welding

FUNDAMENTALS

DEFINITION AND GENERAL DESCRIPTION

Ultrasonic welding is a solid-state welding process in which coalescence is produced at the faying surfaces by the local application of high-frequency vibratory energy while the workpieces are held together under moderately low static pressure. A sound metallurgical bond is produced without melting of the base metal. There is minor thickness deformation at the weld location.

Typical components of an ultrasonic welding system are illustrated in Fig. 9.1. The ultrasonic vibration is generated in the transducer. This vibration is transmitted through a coupling system or sonotrode[1], which is represented by the wedge and reed members in Fig. 9.1. The sonotrode tip is the component that directly contacts one of the workpieces and transmits the vibratory energy into it. The clamping force is applied through at least part of the sonotrode, which in this case is the reed member. The anvil supports the weldment and provides reaction to the clamping force.

Ultrasonic welding is used for applications involving both monometallic and bimetallic joints. The process produces lap joints between metal sheets or foils, joints of wire or ribbon to flat surfaces, cross wire or parallel wire joints, and other types of assemblies that can be supported on an anvil.

This process is being used as a production tool in the semiconductor, microcircuit, and electrical contact industries; in the fabrication of small motor armatures; in the manufacture of aluminum foil; and in the assembly of aluminum components. It is receiving acceptance as a structural joining technique in the automotive and aerospace industries. The process is uniquely useful for the encapsulation of materials, such as explosives, pyrotechnics, and reactive chemicals, that require hermetic sealing and cannot be processed by heat or electrical techniques.

PROCESS VARIATIONS

There are four variations in the process based on the type of weld produced. These are spot, ring, line, and continuous seam welding.

Spot Welding

In spot welding, individual weld spots are produced by the momentary introduction of vibratory energy into the workpieces as they are held together under pressure between the sonotrode tip and the anvil face. The tip vibrates in a plane essentially parallel to the plane of the weld interface, perpendicular to the axis of static force application. Spot welds between sheets are roughly elliptical in shape at the interface. They can be overlapped to produce an essentially continuous weld joint. This type of seam may contain as few as 5 to 10 welds per inch. Closer weld spacing may be necessary if a leaktight joint is required.

---

1. The sonotrode is the acoustical equivalent of the electrode used in resistance spot or seam welding.

Ring Welding

Ring welding produces a closed loop weld which is usually circular in form but may also be square, rectangular, or oval. Here, the sonotrode tip is hollow, and the tip face is contoured to the shape of the desired weld. The tip is vibrated torsionally in a plane parallel to the weld interface. The weld is completed in a single, brief weld cycle.

Line Welding

Line welding is a variation of spot welding in which the workpieces are clamped between an anvil and a linear sonotrode tip. The tip is oscillated parallel to the plane of the weld interface and perpendicular to both the weld line and the direction of applied static force. The result is a narrow linear weld, which can be up to 6 in. in length, produced in a single weld cycle.

Continuous Seam Welding

In continuous seam welding, joints are produced between workpieces that are passed between a rotating, disk-shaped sonotrode tip and a roller type or flat anvil. The tip may traverse the work while it is supported on a fixed anvil, or the work may be moved between the tip and a counter-rotating or traversing anvil. Area bonds may be produced by overlapping seam welds.

MECHANISM OF THE PROCESS

Ultrasonic welding involves a complex relationship between the static clamping force, the oscillating shear forces, and a moderate temperature rise in the weld zone. The magnitudes of these factors required to produce a weld are functions of the thickness, surface condition, and room temperature properties of the workpieces.

Stress Patterns

In all types of ultrasonic welding, static clamping force is applied perpendicular to the interface between the workpieces. The contacting sonotrode tip oscillates approximately parallel to this interface. The combined static and oscillating shear forces cause dynamic internal

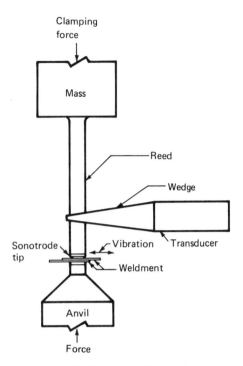

Fig. 9.1—Wedge-reed ultrasonic spot welding system

stresses in the workpieces at the faying surfaces that produce elastoplastic deformation.

Photoelastic stress models reveal significant aspects of these stress patterns. With applied static force only, the stress pattern is symmetrical about the axis of force application. With the superimposition of a lateral force, such as that occurring during one-half cycle of vibration, the force shifts in the direction of this lateral force, and shear stress is produced on that side of the axis. When the direction of the lateral force is reversed, as in the second half of the vibratory cycle, the shear stress shifts to the opposite side of the axis. During welding, shear stress shifts direction thousands of times per second.

As long as the stresses in the metal are below the elastic limit, the metal deforms only elastically. When the stresses exceed a threshold value, highly localized interfacial slip occurs, with no gross sliding. This action tends to break up and disperse surface films and permits metal-to-metal contact at many points. Continued oscillation breaks down surface asperities so that the contact area can grow until a physically continuous weld area is produced. At the same time, atomic diffusion occurs across the interface, and the metal recrystallizes to a very fine grain structure having the properties of moderately cold-worked metal.

Temperature Developed in Weld Zone

Ultrasonic welding of metals at room temperature produces a localized temperature rise from the combined effects of elastic hysteresis, localized interfacial slip, and plastic deformation. However, no melting of the weld zone metal occurs in a monometallic weld made under proper machine settings of clamping force, power, and weld time. Sections examined with both optical and electron microscopy have shown phase transformation, recrystallization, diffusion, and other metallurgical phenomena, but no evidence of melting.

Interfacial temperature studies made with very fine thermocouples and rapidly responding recorders show a high initial rise in temperature at the interface followed by a leveling off.

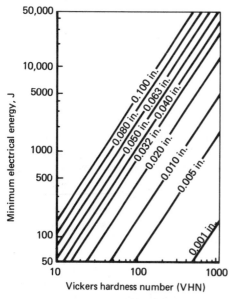

Fig. 9.2—The relationship between the minimum electrical energy required for ultrasonic (ceramic transducer) spot welding and metal hardness for several sheet thicknesses

The maximum temperature achieved is dependent upon the welding machine settings. Increasing the power causes an increase in the maximum temperature achieved. Increased clamping force increases the initial rate of temperature rise, but lowers the maximum temperature achieved. Thus, it is possible to control the temperature profile, within limits, by appropriate adjustment of machine settings.

The interface temperature rise is also associated with the thermal properties of the metal being welded. Generally, the temperature produced in a metal of low thermal conductivity, such as iron, is higher than that in a metal of high thermal conductivity, such as aluminum or copper.

Temperature measurements during welding of metals having a wide range of melting temperatures show that the maximum temperature in the weld is in the range of 35 to 50 percent of the absolute melting temperature of the metal when suitable welding machine settings are used.

Energy Delivered to the Weld Zone

The flow of energy through an ultrasonic welding system begins with the introduction of 60 Hz electrical power into a frequency converter. This device converts the frequency to that required for the welding system, which is usually in the range of from 10 to 75 kHz. The high-frequency electrical energy is conducted to one or more transducers in the welding system, where it is converted to mechanical vibratory energy of the same frequency. The vibratory energy is transmitted through the sonotrode and sonotrode tip into the workpiece. Some of the energy passes through the weld zone and dissipates in the anvil support structure.

Power losses occur throughout the system in the frequency converter, transducer, sonotrode, and the interfaces between these components. However, with a well-designed system, as much as 80 to 90 percent of the input power to the converter may be delivered into the weld zone.

For practical usage, the power required for welding is usually measured in terms of the high-frequency electrical power delivered to the transducer. This power can be monitored continuously and provides a reliable average value to associate with equipment performance as well as with weld quality. The product of the power in watts and welding time in seconds is the energy, in joules or watt-seconds, utilized in welding.

Power Requirements and Weldability

The energy required to make an ultrasonic weld can be related to the hardness of the workpieces and the thickness of the part in contact with the sonotrode tip. Analysis of data covering a wide range of materials and thicknesses has led to the following empirical relationship, which is accurate to a first approximation:

$$E = K(HT)^{3/2}$$

where: E = electrical energy, J (W•s)
K = a constant for a given welding system
H = Vickers hardness number
T = thickness of the sheet in contact with the sonotrode tip

The constant K is a complex function that appears to involve primarily the electromechanical conversion efficiency of the transducer, the impedance match into the weld, and other characteristics of the welding system. Different types of transducer systems should have substantially different K values.

Figure 9.2 shows the relationship between energy and hardness for various sheet thicknesses of any weldable metal based on the above equation. It provides a convenient first approximation of the minimum electrical input energy for a ceramic transducer type spot welding machine based on the Vickers hardness of the metal and the sheet thickness. Similar data

can be derived for ring, line, and seam welds. For seam welds, the energy would be expressed in terms of the unit length of seam.

ADVANTAGES AND DISADVANTAGES

Ultrasonic welding has advantages over resistance spot welding in that no heat is applied during joining and no melting of the metal occurs. Consequently, no cast nugget or brittle intermetallics are formed. There is no tendency to arc or expel molten metal from the joint as with resistance spot welding.

The process permits welding thin to thick sections as well as joining of a wide variety of dissimilar metals. Welds can be made through certain types of surface coatings and platings.

Ultrasonic welding of aluminum, copper, and other high thermal conductivity metals requires substantially less energy than does resistance welding.

The pressures used in ultrasonic welding are much lower, welding times are shorter, and thickness deformation is significantly lower than for cold welding.

A major disadvantage is that the thickness of the component adjacent to the sonotrode tip must not exceed relatively thin gages because of the power limitations of present ultrasonic welding equipment. The range of thicknesses of a particular metal that can be welded depends upon the properties of that metal.

In addition, the process is limited to lap joints. Butt welds can not be made because there is no effective means of supporting the workpieces and applying clamping force.

GENERAL APPLICATIONS

WELDABLE METALS

Most metals and their alloys can be ultrasonically welded. Figure 9.3 identifies some of the monometallic and bimetallic combinations that can be welded on a commercial basis. Blank spaces in the chart indicate combinations that have not been attempted or have not been successfully welded, so far as is known.

Various metals differ in weldability according to their composition and properties. The metals considered more difficult to weld are those that require either high power or long weld times, or both, and those that incur operational problems such as tip sticking or short tip life.

Aluminum Alloys

Any combination of aluminum alloys forms a weldable pair. This includes the high-strength structural alloys such as Types 2014, 2020, 2024, 6061, and 7075. They may be joined in any available form: cast, extruded, rolled, forged, or heat-treated. Soft aluminum cladding on the surface of these alloys facilitates bonding. Aluminum can be welded to most other metals as well as to germanium and silicon, two semiconductor materials.

Copper Alloys

Copper and its alloys, such as brass and gilding metal, are relatively easy to weld. High thermal conductivity does not appear to be a deterrent factor in ultrasonic welding as it is in the fusion welding processes.

Iron and Steel

Satisfactory welds can be produced in iron and steel of various types, such as ingot iron, low carbon steels, tool and die steels, austenitic stainless steels, and precipitation-hardening steels. The power requirements are somewhat higher than for aluminum and copper.

Precious Metals

The precious metals, such as gold, silver, platinum, palladium, and their alloys, can be ultrasonically welded without difficulty. Most precious metals can be satisfactorily welded to other metals and to germanium and silicon.

Refractory Metals

The refractory metals, including molybdenum, columbium, tantalum, tungsten, and some of their alloys, are among the most difficult metals to weld ultrasonically. Thin foil thicknesses of these metals can be joined if they are relatively free from contamination and surface or internal defects.

Other Metals

Nickel, titanium, zirconium, beryllium, magnesium, and many of their alloys can be ultrasonically welded in thin gages to themselves and to other metals. Metal foils and wires are readily joined to thermally sprayed metals on glass, ceramics, or silicon. Such welds are particularly useful in the semiconductor industry. Typical combinations are shown in Table 9.1.

Multiple-Layer Welding

Multiple-layer welding is feasible; for example, as many as 20 layers of 0.001-in. aluminum foil can be joined simultaneously with either spot welds or continuous seam welds. Several layers of dissimilar metals can also be welded together.

LIMITATIONS

Thickness

There is an upper limit to the thickness of any metal that can be ultrasonically welded effectively because the power output of available equipment is limited. For a readily weldable metal, such as Type 1100 aluminum, the maximum thickness in which reproducible high strength welds can be produced is approximately 0.10 in.; for some of the harder metals, the present upper limit may be in the range of 0.015 to 0.040 inch. This limitation applies only to the member of the weldment that is in contact with the welding tip; the other member may be of greater thickness.

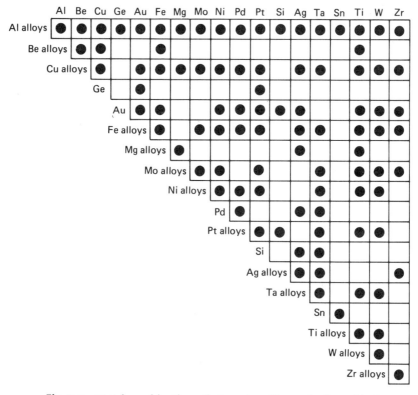

Fig. 9.3—Metal combinations that can be ultrasonically welded

On the other hand, very thin or small sections can be welded successfully. For example, fine wires of less than 0.0005-in. diameter and thin foils of 0.00017-in. thicknesses have been welded.

Where a weld is difficult to achieve with available power levels, good quality joints might be made by inserting a foil of another metal in between the two workpieces. Three examples of this are: (1) 0.0005-in. foil of nickel or platinum has been used between molybdenum components; (2) beryllium foil has been welded to AISI Type 310 stainless steel using an interleaf of thin Type 1100-H14 aluminum foil; and (3) the weldable range of Type 2014-T6 aluminum alloy has been extended by using a foil interleaf of Type 1100-O aluminum.

For some metals, use of abrasive or textured tips and anvils will decrease the required clamping force and the welding power. This may per-mit welding thicker sections with a particular machine size.

Geometry

Generally, ultrasonic welding can be used to join a metal within the thickness limitations of the process provided there is

(1) Adequate joint overlap

(2) Access for the sonotrode tip to contact the parts

(3) An avenue for anvil support and clamping force application

TYPICAL APPLICATIONS

Electronic Components

The greatest application of the process is in the electronics industry for the assembly of miniaturized components. Fine aluminum and gold lead wires are attached to transistors, diodes,

Table 9.1
Metal wire and ribbon leads that may be ultrasonically welded to thin metal surfaces on nonmetallic substrates

Substrate	Metal film	Material	Diameter or thickness range, in.
Glass	Aluminum	Aluminum wire	0.002-0.010
	Aluminum	Gold wire	0.003
	Nickel	Aluminum wire	0.002-0.020
	Nickel	Gold wire	0.002-0.010
	Copper	Aluminum wire	0.002-0.010
	Gold	Aluminum wire	0.002-0.010
	Gold	Gold wire	0.003
	Tantalum	Aluminum wire	0.002-0.020
	Chromel	Aluminum wire	0.002-0.010
	Chromel	Gold wire	0.003
	Nichrome	Aluminum wire	0.0025-0.020
	Gold	Aluminum wire	0.010
	Platinum	Aluminum wire	0.010
	Gold-platinum	Aluminum wire	0.010
	Palladium	Aluminum wire	0.010
	Silver	Aluminum wire	0.010
	Copper on silver	Copper ribbon	0.028
Alumina	Molybdenum	Aluminum ribbon	0.003-0.005
	Gold-platinum	Aluminum wire	0.010
	Gold on Molybdenum-lithium	Nickel ribbon	0.002
	Copper	Nickel ribbon	0.002
	Silver on Molybdenum-manganese	Nickel ribbon	0.002
Silicon	Aluminum	Aluminum wire	0.010-0.020
	Aluminum	Gold wire	0.002
Quartz	Silver	Aluminum wire	0.010
Ceramic	Silver	Aluminum wire	0.010

and other semiconductor devices. Wires and ribbons are bonded to thin films and microminiaturized circuits. Diode and transistor chips are mounted directly on substrates. Reliable joints with low electrical resistance are produced without contamination or thermal distortion of the components.

The encapsulation of microcircuits and other electronic components is effectively accomplished by ultrasonic ring welding. Containers such as transistor or diode cans are hermetically sealed without contamination of the high-purity internal components.

Ultrasonic welding is used to produce small weld-clad areas. One example is the weld cladding of aluminum or gold foil on integrated circuit frames to provide a surface for subsequent wire bonding. Another is the spot cladding of aluminum heat sinks with copper for subsequent attachment of semiconductor devices.

Electrical Connections

Electrical connections of various types are effectively made by ultrasonic welding. Both single and stranded wires can be joined to other wires and to terminals. The joints are frequently made through anodized coatings on aluminum or through certain types of electrical insulation.

Fig. 9.5—Starter motor armature with wires joined in commutator slots by ultrasonic welding

Fig. 9.4—Field coil assembled by ultrasonic welding

Other current-carrying devices, such as electric motors, field coils, harnesses, transformers, and capacitors, may be assembled with ultrasonically welded connections. A typical example is the field coil assembly for automotive starter motors shown in Fig. 9.4. Ultrasonic welds are used here for joining aluminum ribbon to itself, to copper ribbon, to consolidated stranded copper wire, and to copper terminals.

For the starter motor armature of Fig. 9.5, two wires are welded simultaneously into each slot of the barrel commutator. The entire process is accomplished automatically at rates up to 180 complete armatures per hour. Armatures for small motors in appliances, hand tools, fans, computers, and other electrical devices are assembled in a similar manner. If the motor has a flat-faced armature, multiple-wire joints are made simultaneously with a ring welding machine and a castellated sonotrode tip that contacts the part only in the desired weld areas.

Thermocouple junctions involving a wide variety of dissimilar metals can be produced by this means.

Foil and Sheet Splicing

The splicing of aluminum foil by continuous seam welding is an established production technique in foil rolling mills for joining broken and random lengths of thin sheet. Highly reliable splices, capable of withstanding annealing operations, are made rapidly in foils up to 0.005 in. thick and 72 in. wide. The splices are almost undetectable after subsequent working operations. Aluminum and copper sheet up to about 0.020 in. thick can be spliced together by ultrasonic welding using special processing and tooling.

Encapsulation and Packaging

Ultrasonic welding is used for a wide variety of packaging applications that range from soft foil packets to pressurized cans. Leaktight seals are produced by ring, seam, and line welding.

The process is useful for encapsulating materials that are sensitive to heat or electrical current. Such materials may be primary explosives; slow-burning propellants and pyrotechnics; high-energy fuels and oxidizers; and living tissue cultures.

Since ultrasonic welding can be accomplished in a protective atmosphere or vacuum, it permits sterile packaging of hospital supplies, precision instrument parts, ball bearings, and other items that must be protected from dust or contamination. This capability also permits encapsulation of chemicals that react with air.

Ring welds up to about 2.5-in. diameter can be produced, but these welds are limited to thin sections of aluminum or copper. Straight cylindrical containers are often welded with a flanging and re-forming technique such as that shown in Fig. 9.6. The cylinder ends are flared

Sequence:
1. As-received container
2. Flange formed
3. Cover welded to the flange
4. Cover trimmed
5. Cylinder redrawn

Fig. 9.6 — Cylinder closure by the flange-weld-redraw technique

Fig. 9.7 — Ultrasonically welded helicopter access door

to form a 90-degree flange. The covers are ultrasonically welded to the flange, and then the welded flange is subsequently re-formed to the original cylindrical geometry.

Line welding is used for packaging with one or more straight line seams, such as sealing the ends of squeeze tubes. Square or rectangular packets are produced by intersecting line welds on each of the four edges. Continuous seam welding is used to seal packages that can not be accommodated with ring or line welding.

Structural Welding

Ultrasonic welding provides joints of high integrity for structural applications within the limitations of weldable sheet thickness. The process is being used to assemble aircraft secondary structures, such as the helicopter access door in Fig. 9.7. This assembly consists of inner and outer skins of aluminum alloy joined by multiple ultrasonic spot welds. Individual ultrasonic welds had 2.5 times the minimum average strength requirements for resistance spot welds in the same metals and thicknesses. Assembled doors sustained loads of 5 to 10 times the design load without weld failure in air load tests. Significant savings in fabrication and energy costs were evident when compared to those of adhesive bonding.

In another application, small clips are attached to cylindrical reactor fuel elements with ultrasonic spot welds. Eight clips are attached to each element, and production rates of about 200 elements per hour are achieved in a semiautomatic setup.

Solar Energy Systems

Ultrasonic welding has reduced fabrication costs for some solar energy conversion and collection systems. Systems for converting solar heat to electricity frequently involve photovoltaic modules of silicon cells which are interconnected with aluminum connectors. An ultrasonic seam welding machine, operating at speeds up to 30 feet per minute, joins all connectors in a single row in a fraction of the time required for hand soldering or individual spot welding. After all connections are made on one side of the assembly, the same process is repeated on the opposite side.

Solar collectors for hot water heating systems may consist of copper or aluminum tubing attached to a collector plate. An automated ultrasonic system makes successive spot welds spaced on 1-in. centers between the plate and tubing as the assembly is passed beneath the welding tip. A 36-in. tube can be welded to a plate in about 2 minutes, at an energy cost of about 0.3 cent. The time and cost are much less than are required for soldering, resistance spot welding, or roll bonding these assemblies.

Other Applications

Applications in other areas have also been successful. Continuous seam welding was used to assemble components of corrugated heat exchangers. Strainer screens were welded without clogging the holes. Beryllium foil windows for space radiation counters were ultrasonic ring welded to stainless steel frames to provide a helium leaktight bond. Currently, pinch-off weld closures in copper and aluminum tubing, which are used with capillary tubes in refrigeration and air conditioning, are produced with special serrated bar tips and grooved anvils. Aluminum foil, surrounding fiber glass insulated ducting, is overlapped and welded with a traversing-head seam welding machine.

EQUIPMENT

GENERAL DESCRIPTION

An ultrasonic welding machine consists of the following components:

(1) A frequency converter to provide electrical power at the design frequency of the welding system

(2) A transducer-sonotrode system to convert this power into elastic vibratory energy and deliver it to the weld zone

(3) An anvil which serves as a support for the workpieces

(4) A force application mechanism

(5) Either a timing device to control the weld interval in spot, ring, and line welding or a rotating and translating mechanism for seam welding

(6) Appropriate electrical, electronic, and hydraulic or pneumatic controls

Vibratory Frequency

Ultrasonic welding can be accomplished over a broad frequency range from less than 0.1 to about 300 kHz. However, the frequencies used for welding machines are usually in the range of 10 to 75 kHz.

An ultrasonic welding machine is designed to operate at a single frequency. There is no critical frequency for welding of specific metals or thicknesses. Due to the practical fundamentals of transducer-sonotrode design, it is expedient to build both light, low power machines that operate at high frequencies and heavy, high power machines that operate at low frequencies. For example, welding machines in the power range of about 1200 to 8000 W operate with frequencies in the 10 to 20 kHz range. Conversely, small machines joining fine wires may have a power capacity of only a few watts and an operating frequency in the range of about 40 to 75 kHz.

A machine is designed to operate at some nominal frequency that may actually vary about 1 percent above or below the design frequency due to manufacturing variations. Adjustments are provided to tune the equipment to the optimum operating frequency.

Transducer-Sonotrode System

Both magnetostrictive and piezoelectric types of transducers are used in ultrasonic welding systems. Magnetostrictive materials have the property of changing length under the influence of varying magnetic flux density. Such transducers, which usually consist of a laminated stack of nickel or nickel alloy sheets, are rugged and serviceable for continuous duty operation but have low electromechanical conversion efficiency. Piezoelectric ceramic materials, such as lead zirconate titanate, are capable of changing dimensions under the influence of an electrical field. These materials have a conversion efficiency of more than twice that of magnetostrictive transducers. When they are operated at high duty cycles, both types of transducers must be cooled to prevent overheating and a loss of transduction characteristics.

The sonotrode system is designed to operate at the resonant frequency of the transducer and usually to provide gain in the amplitude of the delivered vibration. Sonotrode materials are selected to provide low energy losses and high fatigue strength under the applied static and vibratory stresses. A titanium alloy and stainless steel are the most commonly used sonotrode materials.

For high reliability, the various joints in the system must have high integrity and excellent fatigue life because of the high-frequency vibration. Brazed, welded, and mechanical junctions have been used, but most current welding machines use mechanical joints because of ease of interchangeability.

Transducer-sonotrode systems usually have acoustically designed mounting arrangements to ensure maximum efficiency of energy transmission when static force is applied through the system. These force-insensitive mounts prevent shift of the resonant frequency of the system and minimize loss of vibratory energy into the supporting structure.

Anvil

The anvil, in addition to supporting the workpiece, provides the necessary reaction to the applied clamping force. Its geometry is seldom

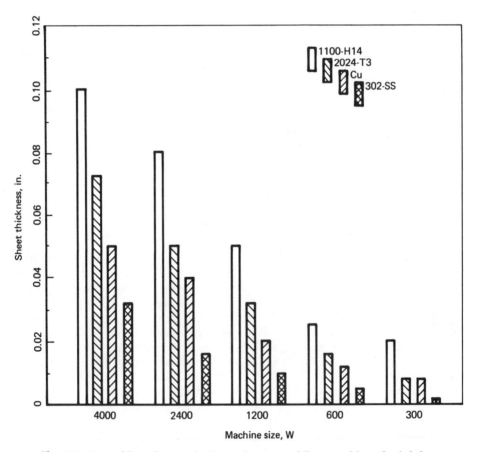

Fig. 9.8 — Capacities of several ultrasonic spot welding machines for joining selected metals

critical except that it must not permit the workpiece to vibrate in compliance with the applied frequency.

Clamping Mechanism

The static load is always applied normal to the plane of the weld interface. The means for applying this load depends upon the overall design of the welding machine. With larger units, hydraulic systems are satisfactory. Intermediate size units may incorporate pneumatically actuated or spring loaded systems. Miniature welding machines that require very small clamping loads may be spring actuated or dead-weight loaded. Such mechanical devices are suitable for production applications where frequent adjustments are not required.

Frequency Converter

The function of the frequency converter is to change electrical line power of 50 or 60 Hz to the design frequency of the welding system in an oscillator stage, and then to amplify the output power in an amplifier stage. The output

Fig. 9.9—Typical 4000 watt ultrasonic spot welding machine

of such a system is the high-frequency electrical power of the ultrasonic transducer in the welding head.

Most ultrasonic welding systems utilize a solid-state frequency converter of the silicon-controlled rectifier (SCR) or transistor type. Transistors can operate efficiently at high frequencies, but their power-handling capability is low. At high power levels, multiple units must be used, resulting in a more complicated and less reliable circuit. SCR's can handle more power per device and are normally useful at frequencies below 20 kHz. SCR's cannot be turned off by a control signal but require commutating circuitry, which adds to cost and inefficiency. The overall efficiency of SCR and transistor converters is approximately the same.

Automatic frequency control is a standard feature on most frequency converters. Both SCR and transistor circuits can be either controlled by free-running oscillators or operated in a self-excited mode through positive feedback derived from the load. The self-excited mode is preferable because such systems automatically track the mechanical resonance of the loaded transducer, which assures optimum load matching under all conditions.

Some ultrasonic welding machines also incorporate constant amplitude control. By keeping the transducer mechanical amplitude constant, regardless of load, transducer dissipation is held at a safe level for all loading conditions, and the sonotrode tip amplitude is constant.

In most ultrasonic welding machines, the frequency converter and the welding head are separate assemblies connected by lightweight cables. In some of the low power units, the converter is attached to the welding head.

TYPES OF EQUIPMENT

Spot Welding Machines

Basically, these machines all operate on the same principle. They deliver a single pulse of high-frequency vibration to the weld interface to produce a spot weld. Two different systems are used to apply the clamping force and transmit the ultrasonic energy to the workpiece: the wedge-reed system and the lateral drive system.

Spot welding machines range in size from about 10 W to 8000 W capacity. Figure 9.8 indicates the capabilities of various machines for welding some common metals.

Wedge-Reed System. In the wedge-reed system, the transducer drives a wedge-shaped coupler in longitudinal vibration, as shown in Fig. 9.1. The wedge is rigidly attached to and produces flexural vibration in the reed. The sonotrode at the end of the reed undergoes essentially lateral vibration in a plane parallel to that of the weld interface. For very high powers, two or more transducers may be used to drive the wedge.

Some systems have a movable head and a fixed anvil, with force being applied through the reed. Other systems have a fixed head and a movable anvil, and force is applied through the anvil. In both cases, the reed is acoustically designed for force insensitivity so that the vibration is not damped with force application. Figure 9.9 shows a typical machine of the wedge-reed type with a movable anvil.

Lateral Drive System. In this system, the sonotrode tip is attached to a lateral sonotrode which vibrates longitudinally to produce tip motion parallel to the weld interface. Two representative arrangements are shown in Fig. 9.10. A system used in low power welding machines is shown in Fig. 9.10(A). Clamping force at the tip results when a bending moment is applied to

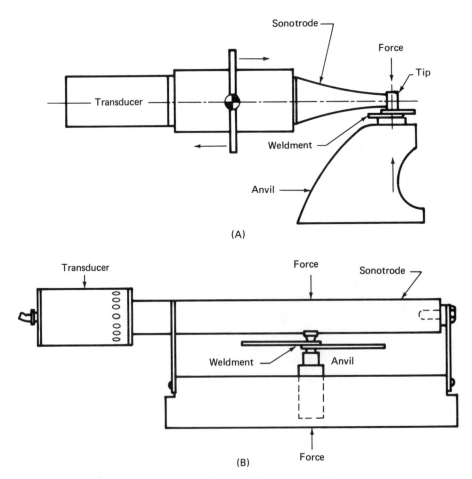

(A)

(B)

Fig. 9.10—Two types of lateral drive spot welding systems

the sonotrode, indicated by the arrows on the figure. This lateral sonotrode is surrounded by a metal sleeve which has a flange located at a vibratory node to isolate the acoustic system from the applied force. In the system shown in Fig. 9.10(B), the lateral sonotrode is mounted through resilient metal diaphragms located at vibratory antinodes. These diaphragms isolate the vibration from the force application system.

A typical 100 W welding machine incorporating a lateral drive system is shown in Fig. 9.11. Low power systems, such as those used for lead wire welding in semiconductor devices, may be incorporated in a micropositioner. They may also include automatic wire feed, work manipulation, positioning, and microscopic observation.

Ring Welding Machines

Ring welding machines utilize a torsionally driven coupling arrangement as shown in Fig. 9.12. The axial driving coupling members are tangent to the torsional coupler. They vibrate out of phase to produce a torsional vibratory displacement of the sonotrode tip in a plane parallel to the weld interface.

Ring welding machines are available in sizes ranging from 100 W to about 4000 W capacity. They are capable of making circular ring welds in diameters ranging from approximately 0.040 to approximately 2.5 inches.

Line Welding Machines

These machines usually incorporate an overhung, lateral drive, transducer-sonotrode system. The couplers extend beyond the sonotrode tip, and the clamping force is applied to the overhung portion of the sonotrode. This design eliminates high bending movements in the coupler when clamping forces are high.

A cluster of multiple transducer-sonotrode units attached to a single sonotrode tip is used for weld lengths greater than about 1 inch. An array for making 5-in. long line welds incorporates five such units with an interconnecting tip. Hydraulic cylinders apply force to each coupler, thereby equalizing the force for each increment of weld length.

Continuous Seam Welding Machines

Continuous seam welding machines usually incorporate lateral drive transducer-sonotrode systems and antifriction bearings. The entire transducer-sonotrode-tip assembly is rotated by a motor drive. The rotating tip rolls on the work along the desired path. Three arrangements are used for handling the work: a roller-roller system, a traversing anvil system, and a traversing head system.

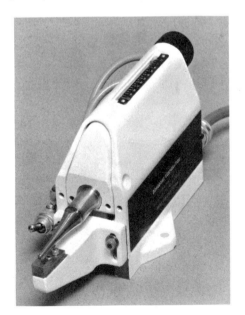

Fig. 9.11—Typical 100 watt spot welding machine with lateral drive system

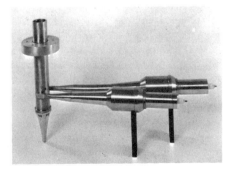

Fig. 9.12—Transducer-sonotrode system for ring welding machines

Roller-Roller System. In this arrangement, the workpiece is driven between a rotating disk tip and a counter-rotating anvil. A compact 100 W machine is used for thin foil materials. Heavier materials are welded with a 2000 W unit capable of welding rates up to 450 ft/min.

Traversing Anvil System. In this system, the rotating transducer-sonotrode system is mounted in a fixed position. The workpiece is located on an anvil that traverses laterally under the rotating coupler disk.

Traversing Head System. The predominant type of seam welding machine is the traversing head system in which the disk tip rotates as it traverses across the stationary workpiece. A typical unit is shown in Fig. 9.13. A 100 W system of this type is used in aluminum foil mills for splicing coils. Machines of higher power capacity are used for splicing thicker materials.

SONOTRODE TIPS AND ANVILS

Tip and Anvil Geometry

Welding is most effectively accomplished when the sonotrode tip and anvil are contoured to accommodate the specific geometry of the parts being joined.

For spot welding of flat sheets, the tip is contoured to a spherical radius of about 50 to 100 times the thickness of the sheet adjacent to the tip. The anvil face is usually flat. This provides a friction type drive in which slippage can occur between the tip and the top sheet or between the anvil and the bottom sheet as well as at the weld interface.

A positive type of drive is illustrated in Fig. 9.14, which shows the arrangement for welding a small rib to a tubular member. The sonotrode tip is contoured to mate with the rib so that they are locked together. With this drive, maximum energy is delivered at the weld interface.

When joining a wire to a flat surface, the tip is preferably grooved to match the wire so that the wire is not excessively deformed during welding. With small wires, such as those used for connections to semiconductor devices, the

tips must be precise in dimensions and finish. For joining two wires together, the anvil may be grooved to support and position both wires. The tip may fit into the groove and contact the upper wire.

Ring welding tips are usually hollow members having the shape of the desired welds: circular, elliptical, square, or rectangular. The wall thickness is determined by the desired weld width, and the edge of the tip contacting the work is convex. Anvils may be flat or appropriately contoured to mate with the workpiece. For example: in welding a lid to a cylindrical container, the anvil is usually recessed to accommodate the container with the flange in contact with the anvil surface.

Line welding tips have narrow, elongated shapes with the contacting surface any desired width up to about 0.1 inch. The anvil is designed to accommodate the workpiece. For line welding of can side seams, for example, the anvil consists of a cylindrical mandrel around which the can body blank is wrapped and supported with clamping jaws (Fig. 9.14).

Continuous seam welding tips are resonant disks. For welding flat surfaces, the disks are machined with a convex edge. The edge of the disk may also be contoured to mate with the workpiece; for example, the entire periphery of a disk can be grooved to permit continuous seam welding of a rib to a cylinder with an arrangement similar to that shown in Fig. 9.14.

Fig. 9.13—Typical traversing head, continuous seam ultrasonic welding machine

Tip Materials

As in resistance spot welding, wear of the sonotrode tip and anvil depends upon the properties and geometry of the parts being welded. Tips made of high-speed tool steel are generally satisfactory for welding relatively soft materials, such as aluminum, copper, iron, and low carbon steel. Tips of hardenable nickel-base alloys usually provide good service life with hard, high-strength metals and alloys. The material used for the sonotrode tip is also satisfactory for the anvil face.

Frequently, longer tip life and more effective welding are possible using tips and anvils with rough faces because they tend to prevent gross slippage between them and the workpieces. The roughening may be accomplished with electrical discharge machining (EDM) or sandblasting to a finish of about 200 microinch. Abrasive tips usually permit the use of lower powers and clamping forces than are required with smooth tips.

Tip Maintenance

When sonotrode tips begin to show wear, erosion, or material pickup, they may be reconditioned by cleaning and burnishing. Light sanding with 400-grit silicon carbide paper is usually sufficient. If the wear is excessive, the tips should be replaced. Most welding machines have mechanically attached tips to simplify replacement.

Occasionally, the tip may tend to stick to the weld surface, particularly if improper machine settings are used. The sticking may be alleviated by increasing the clamping force or decreasing the welding time. With some materials, it is effective to apply a lubricant, such as a faint trace of very dilute soap solution, to the surfaces being joined. If these measures are not adequate, tip sticking may usually be eliminated by welding with a tip having an insert of special grade tungsten carbide.

CONTROLS

The basic controls for an ultrasonic welding machine are relatively simple. They consist of a master switch for introducing line power and controls to adjust clamping force, power, welding time, and sometimes resonance. Appropriate readout is normally provided for all adjustments.

The welding cycle is generally controlled automatically and is usually actuated by dual palm buttons or a foot switch. The automatic cycle consists of lowering the sonotrode tip or raising the anvil, application of clamping force, introduction of the ultrasonic pulse, and retraction of the tip or anvil.

Other controls and indicators are included on some welding machines to monitor operation of the equipment or to provide flexibility in use. Means may be provided for adjustment of:

(1) Sonotrode stroke length

(2) Speed of sonotrode advance and retraction from the weldment

(3) Speed of traverse for continuous seam welding machines

(4) Height of anvil to provide clearance for the workpiece

(5) Anvil position, particularly on ring welding machines where precise alignment of tip and anvil are essential to ensure uniform

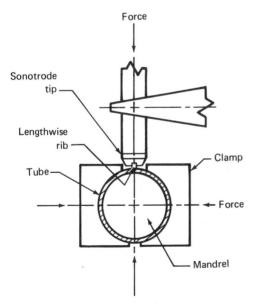

Fig. 9.14—Welding array for joining small longitudinal ribs to cylinders

contact around the periphery of the weld

Weld quality control monitors may also be included. One such device is a weld power meter which indicates the power delivered into the weld. A substantial change in load power indicates a faulty weld, which may be due to changes in part dimensions or surface finish, improper assembly of parts, or machine malfunction. On some machines, the high and low limits of acceptable power can be set, and deviations from this range can be used to trigger a visual or audible signal to alert the operator or to actuate a reject mechanism.

Another type of weld quality monitor is based on a constant energy principle. This system automatically adjusts welding time so that a predetermined amount of energy is delivered to each weld. When the energy can not be delivered within the preset time, an alarm is activated.

Automated equipment may also include frequency counters, weld counters, material-handling actuators, indexing mechanisms, and other devices to minimize operator functions and maximize production rates.

AUTOMATED PRODUCTION EQUIPMENT

Several features of ultrasonic welding equipment make it particularly adaptable to automated or semiautomated production lines: namely,

(1) The welding head can be readily interfaced with other automatic processing equipment. It can be mounted on any rigid structure and in any position with the tip contacting the work from any direction.

(2) The frequency converter may be located as far as 150 ft away from the welding head.

(3) Welding times are usually a fraction of a second, and production rates are limited primarily by the speed of the work-handling equipment.

(4) The process does not involve extensive heating of the equipment or the workpiece.

(5) In automatic filling and closing lines, accidental spillage of the contents on the weld interface usually will not significantly affect weld quality.

JOINING PROCEDURES

JOINT DESIGN

Joint designs for ultrasonic welding are less restrictive than for some other types of welding. Edge distance is not critical. The only restriction is that the sonotrode tip should not crush or gouge the sheet edge. Welds in structural aluminum alloys of several thicknesses have shown the same strength at both 1/8 in. and 3/4 in. from the edge. Weight and material savings are achieved by using the minimum acceptable overlap.

Ultrasonic welding places no restrictions on spot spacing or row spacing with any of the four types of welds. Consecutive or overlapped welds have no effect on the quality of previously made welds, except perhaps under resonance conditions described below.

Ring welding offers unique capabilities for hermetic sealing, as indicated by the joint designs in Fig. 9.15. Ring welds may also be preferred to spot welds for structural applications. The rings provide relatively uniform stress distribution with less stress concentration, less tendency toward cracking, and generally no parts resonance (see below).

CONTROL OF PARTS RESONANCE

Sometimes the workpiece may be excited to vibration by the ultrasonic welding system. When this occurs, inferior welds may result, previously made welds may fracture, or cracks may be generated in the workpiece. Any of several remedies may be applied singly or in combination.

Resonance vibration may be eliminated by altering the workpiece dimensions or the orientation of the workpiece in the welding machine. Damping of vibration in thin sections can frequently be accomplished by applying pressure-sensitive tape to the part. Clamping of masses

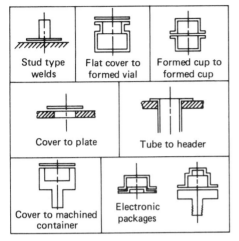

Fig. 9.15—Typical ultrasonic ring welding applications

to the workpiece or clamping into a comparatively massive fixture usually suffices in even the most difficult cases.

SURFACE PREPARATION

A good surface finish contributes to the ease with which ultrasonic welds are made. Some of the more readily weldable metals, such as aluminum, copper, or brass, can be welded in the mill finish condition if not heavily oxidized, and may require only the removal of surface lubricants with a detergent. Normally, thin oxide films do not inhibit welding since they are disrupted and dispersed during the process.

Metals that are heavily oxidized or contain surface scale require more careful surface preparation. Mechanical abrasion or descaling in a chemical etching solution is usually necessary to provide a clean surface for welding. Once the surface scale is removed, the elapsed time before welding is not critical as long as the materials are stored in a noncorrosive environment.

It is possible to weld metals through certain surface films, coatings, or insulations, but somewhat higher ultrasonic energy levels are required. Some types of films can not be penetrated and must always be removed prior to welding.

SPECIAL WELDING ATMOSPHERES

Ultrasonic welding usually does not require special atmospheres. With some metals, the process may produce discoloration of the surface in the vicinity of the weld. When such a surface is undesirable, it can be minimized by use of inert gas protection, such as small jets of argon impinging around the tip contact area. For packaging applications in which sensitive materials must be protected from contamination, welding can be accomplished in a chamber filled with inert gas.

PROCESS VARIABLES

The variables of ultrasonic power, clamping force, and welding time or speed are established experimentally for a specific application. Once determined, they usually require no adjustment unless there are alterations to the equipment, such as sonotrode tip changes or changes in the workpiece.

ULTRASONIC POWER

The power setting may be indicated in terms of the high-frequency power input to the transducer or the load power (the power dissipated by the transducer-sonotrode-workpiece assembly). As previously noted, the power requirement varies with the material and thickness of the workpiece adjacent to the sonotrode tip.

The minimum effective power for a given application can be established by a series of tests from which a threshold curve for welding is plotted. Details of this procedure are described later.

CLAMPING FORCE

An ultrasonic welding machine usually provides a fairly broad range of clamping forces. Table 9.2 shows typical ranges for machines of various power capacities. The clamping force range of machines with hydraulic or pneumatic force systems can be modified by changing the pressure cylinder.

The function of clamping force is to hold the workpieces intimately together. Excessive force produces needless surface deformation and increases the required welding power. Insufficient force permits tip slippage that may cause surface damage, excessive heating, or poor welds. Clamping force for a specific application is established in conjunction with ultrasonic power requirements.

WELDING TIME OR SPEED

The interval during which ultrasonic energy is transmitted to the workpieces in spot, ring, or line welding is usually within the range of 0.005 second for very fine wires to about 1 second for heavier sections. The need for a long welding time indicates insufficient power. High power and a short welding time will usually produce welds that are superior to those achieved with low power and a long welding time. Excessive welding time causes poor surface appearance, internal heating, and internal cracks.

The same factors of power and unit time are significant in continuous seam welding. With available equipment, the travel speed for hard, thick metals may be as low as 5 ft/min. Thin aluminum, 0.001-in. thick, can be welded at speeds up to about 500 ft/min.

FREQUENCY ADJUSTMENT

Adjustment of the frequency converter output to match the operating frequency of the welding system is necessary for good performance. A system has a given nominal frequency, but the best operating frequency may vary with changes in the sonotrode tip, the workpiece, or the clamping force. The method of frequency adjustment varies with different types of frequency converters. After the setting is established for a specific welding setup, usually no further adjustment is necessary.

INTERACTION OF WELDING VARIABLES

For a given application, there is an optimum clamping force at which minimum vibratory energy is required to produce acceptable welds. This condition can be established by plotting the threshold curve. This curve, illustrated in Fig. 9.16, defines the conditions of best dynamic coupling between the sonotrode tip and the workpiece and, thus, the minimum energy to produce strong welds.

The technique consists of making welds at selected power and clamping force settings and conducting cursory evaluation of weld quality as noted on Fig. 9.16. For ductile, thin sheets and fine wires, a useful criterion of successful bonding is the ability to pull a nugget from one of the workpieces when peel tested. Welds in hard or brittle metals may be evaluated on the basis of weld strength or evidence of material

Table 9.2

Typical clamping force ranges for ultrasonic welding machines of various power capacities

Machine power capacity, W	Approximate clamping force range, lbf
20	0.009-0.39
50-100	0.5-15
300	5-180
600	70-400
1200	60-600
4000	250-3200
8000	800-4000

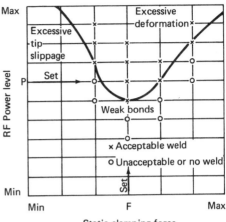

Optimum conditions for minimum power are clamping force F and power P

Fig. 9.16—Typical threshold curve relating RF power and clamping force

transfer when peel tested. The threshold curve is normally derived as follows:

(1) The welding time is set at a reasonable value. One-half second is a good starting point for most metals. For very thin metals, a shorter weld time is usually used.

(2) Welding is started at low values of clamping force and power, and a series of test welds is made with incrementally increasing values of clamping force at a fixed power level. The welds are evaluated and the results plotted as in Fig. 9.16, indicating acceptable and unacceptable welds.

(3) This procedure is then repeated at other values of ultrasonic power until an inverted bell-shaped curve is obtained.

These data will generate a curve separating the acceptable from the unacceptable welds. Welding is ordinarily done using the clamping force value for minimum acceptable power and a power level somewhat above the minimum. The product of the selected power and weld time is the total energy required. If welding time is decreased, then power must be increased ac-

cordingly. The threshold curve is a practical and efficient method for determining proper machine settings for all types of ultrasonic welds.

POWER-FORCE PROGRAMMING

Certain materials, such as the refractory metals and alloys, are more effectively welded when power-force programming is used. Weld strength is higher and cracking of the weld metal is minimized.

Power-force programming involves incremental variations in power and clamping force during the welding cycle. The cycle is initiated at low power and high clamping force. After a brief interval, the power is increased and the force reduced. The cycle is accomplished automatically with special logic circuitry.

WELD QUALITY

INFLUENCING FACTORS

Variations in weld quality may result from several factors which are generally associated with the workpieces and the welding machine or its settings. Weld quality is ordinarily not affected by normal manufacturing variations in metal parts. Metals that meet the specification requirements can usually be consistently welded without varying machine settings. Problems are sometimes encountered, however, if close tolerances are not held; for example, nickel, copper, and gold platings on metal surfaces frequently have thickness variations that affect weld quality. Surfaces for ring welding must be flat and parallel to ensure uniform welding around the periphery.

If there is any change in the workpieces during a production operation, the welding schedule must usually be adjusted to accommodate such change. Variations in weld quality during production runs have been traced to unauthorized changes in metal alloy, geometry, or surface finish. Wires such as magnet wires that are lubricated to facilitate coil winding may be ultrasonically welded without cleaning. A change in the type of lubricant may cause unacceptable welds unless machine settings are appropriately adjusted.

Uniform quality welding also depends upon the mechanical precision of the welding machine. Lateral deflection of the sonotrode or looseness of the anvil can produce unacceptable aberrations in the welds. Erratic weld quality may result from the use of a force-sensitive machine if power is lost and the frequency shifts as clamping force is applied. Sonotrode tips must be acoustically designed and precision ground to the desired contour. Their surfaces must be properly maintained to ensure reproducible welds.

PHYSICAL AND METALLURGICAL PROPERTIES

Ultrasonic welds have distinctive characteristics when examined both internally and externally.

Surface Appearance

The surface of the work at a weld location is usually roughened slightly by the combined compressive and shear forces. The roughness can be minimized with adjustments in the machine settings and with careful sonotrode tip maintenance. The surface contour depends primarily on tip geometry. Spot welds usually leave an elliptical impression because of the linear displacement of the tip. A weld spot impression is larger in soft, ductile metals, such as aluminum, than in hard metals of the same thickness with appropriate adjustment in machine settings. Spot size can be increased by using a larger tip radius.

The actual bond area does not necessarily duplicate the surface impression, except in thin sheet. Sometimes, spot welds have unbonded areas in the center. This condition can usually be eliminated by decreasing the tip radius or reducing the clamping force.

Thickness Deformation

A weld may show some thickness deformation because of the applied clamping force. Such deformation in sheet metals is usually less than 20 percent of the total joint thickness, even with soft metals. With contoured parts such as wires, deformation is somewhat greater unless the tip contour mates with the workpiece.

Microstructural Properties

Metallographic examination of ultrasonic welds in a wide variety of metals shows that a number of different phenomena occur as a result of the vibratory energy introduced into the weld zone. Three important ones are:

(1) Interfacial phenomena, such as interpenetration and surface film disruption

(2) Working effects, such as plastic flow, grain distortion, and edge extrusion

(3) Heat effects, such as recrystallization, precipitation, phase transformation, and diffusion

Ultrasonic welding is usually accompanied by local working along the faying surfaces, by interdiffusion or recrystallization at the inter-

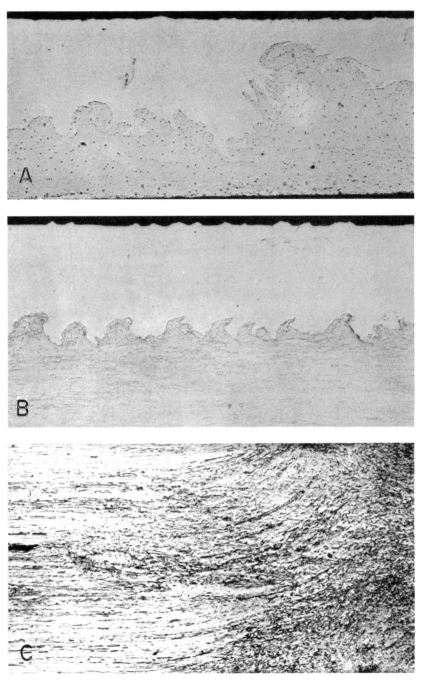

(A) 0.003-in. nickel foil (top) to 0.003-in. gold-plated Kovar foil (×150)
(B) 0.005-in. nickel foil (top) to 0.020-in. molybdenum sheet (×100)
(C) 0.008-in. arc-cast molybdenum sheet to itself (×70)

Fig. 9.17—Photomicrographs of typical ultrasonic welds

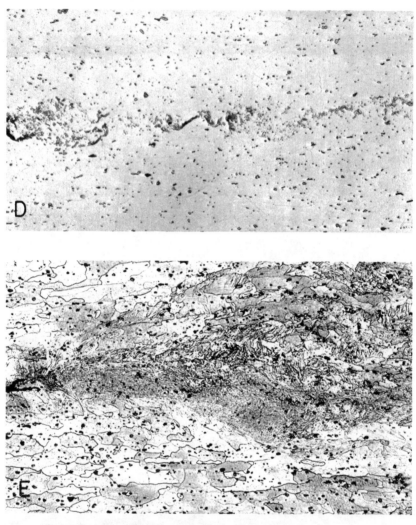

(D) 0.012-in. Type 1100-H14 aluminum sheet to itself (×250)
(E) 0.032-in. Type 2024-T3 aluminum alloy sheet to itself (×75)

Fig. 9.17 (cont.)—Photomicrographs of typical ultrasonic welds

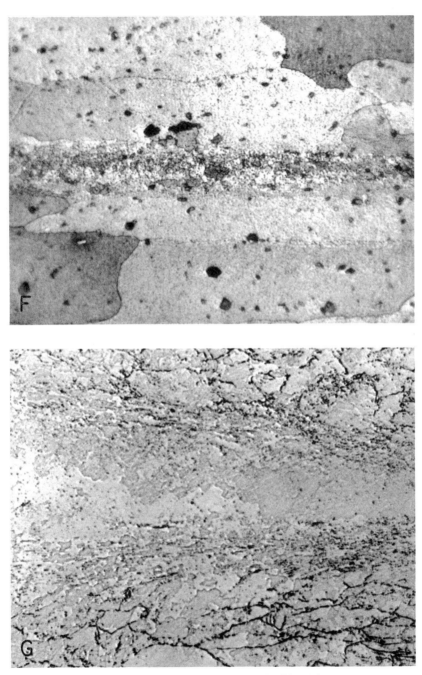

(F) 0.040-in. Type 2020 aluminum alloy sheet to itself (×375)
(G) 0.014-in. half-hard nickel sheet to itself (×250)

Fig. 9.17 (cont.)—Photomicrographs of typical ultrasonic welds

(H) 0.012-in. solution heat-treated and aged Inconel X-750 sheet to itself (×75)
(J) 0.032-in. die steel (0.9% C) (top) to 0.032-in. ingot iron (×500)

Fig. 9.17 (cont.)—Photomicrographs of typical ultrasonic welds

face, and by interruption and displacement of oxide or other barrier films. Surface films, which are broken up by the stress reversals and plastic deformations that occur along the interface, may be displaced in the vicinity of the interface or may simply be interrupted in continuity in random areas within the bond zone. The actual behavior of such films depends upon several factors, including the machine settings, the properties of the film and the base metal, and the temperature achieved at the interface.

The temperature effect is significant in welding certain metals. Recrystallization of the metal frequently occurs in the weld nugget. Sufficient heat may be generated in certain alloys that exhibit precipitation behavior or phase transformation to induce these effects. Although diffusion may occur across the interface, the extent of atom movements is limited by the short weld time.

More than one of the above effects may be apparent in the same weld, and different effects may occur in welds in the same metal produced at different machine settings.

Several typical examples are illustrated in Fig. 9.17. An extreme example of interpenetration across an interface is shown in Fig. 9.17(A), in which a Kovar[2] foil has intruded into as much as 75 percent of the thickness of a nickel foil. A gold plate on the surface of the Kovar has been dispersed throughout the highly worked region. Interfacial ripplets in a nickel-to-molybdenum weld, shown in Fig. 9.17(B), illustrate the plastic flow that occurs locally. Entrapped oxide is indicated by the dark patches on the extreme right of the figure. The weld between two sheets of arc-cast molybdenum, Fig. 9.17(C), shows very little interpenetration, and the bond line is thin. Figure 9.17(D) illustrates the surface oxide film dispersion that may occur during welding of aluminum sheet. General plastic flow along the interface is observed in the Type 2024-T3 aluminum alloy weld of Fig. 9.17(E) where the metal has recrystallized to a fine grain size.

Evidence of recrystallization has been observed in ultrasonic welds in several struc-

tural aluminum alloys, beryllium, low carbon steel, and other metals, even though they were not in the cold-worked condition prior to welding. For instance, in the Type 2020 aluminum alloy weld in Fig. 9.17(F), mutual deformation of the surfaces and subsequent recrystallization are evident. In Fig. 9.17(G), the elevated temperature during welding resulted in recrystallization of prior cold-worked nickel.

Still another effect of interfacial heating is illustrated in Fig. 9.17(H), which shows a weld in solution treated and aged Inconel X-750[3] alloy. In the aged condition, a precipitate normally appears throughout the grains and in the grain boundaries. In the vicinity of this interface, the oxide scale is dispersed and the grain boundaries appear to stop short of the interface, indicating that the precipitate was dissolved during welding. An example of alloying that may occur in the bond between ferrous metals of different carbon content is shown in Fig. 9.17(J).

MECHANICAL PROPERTIES

A variety of mechanical tests may be used to evaluate weld quality. The property most frequently tested is shear strength. In addition, data are reported on cross-tension strength, microhardness, corrosion resistance, and hermetic sealing properties. All available information indicates that the ultrasonic technique, properly applied, produces welds of acceptable strength and integrity.

Shear Strength

Shear tests are usually conducted on simple lap joints containing single- or multiple-spot welds or predetermined lengths of seam or line welds. For convenience, test specimen preparation and testing procedures essentially duplicate those used for resistance spot and seam welds.

Figure 9.18 shows the increase in shear strength with sheet thickness for single-spot specimens in two aluminum alloys. Usually in the thin gages of aluminum sheet, and often in the intermediate gages, failure under tension-shear load occurs by fracture of the base metal

2. A low-expansion iron-base alloy containing 29 percent nickel and 17 percent cobalt.

3. Tradename

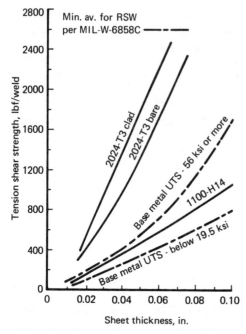

Fig. 9.18—Typical shear strengths of ultrasonic spot welds in aluminum alloy sheet

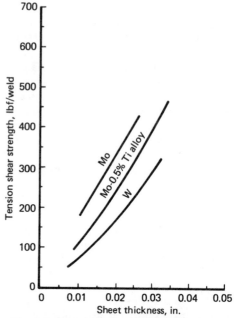

Fig. 9.20—Typical shear strengths of ultrasonic spot welds in several refractory metals and alloys

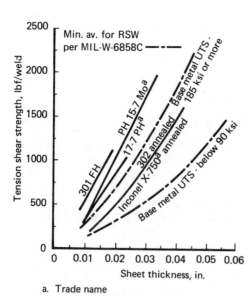

a. Trade name

Fig. 9.19—Typical shear strengths of ultrasonic spot welds in stainless steel and nickel-base alloys

or by tear-out of the weld button rather than by shear of the weld itself. Similar data for several ferrous and nickel alloys are shown in Fig. 9.19 and for several refractory metals and alloys in Fig. 9.20.

Typical spot weld strengths in a variety of metals are summarized in Table 9.3. Of particular interest is the low variability associated with the strength data. In most instances this is less than 10 percent.

Line welds and seam welds show approximately the same strength as the base metal, at least with thin gages. As an example, spot-seam welds in structural aluminum alloys have shown strengths equivalent to 85 to 95 percent of the ultimate tensile strength of the material under both shear and hydrostatic tests. Line welds in 0.001-in. type 5052-H16 aluminum alloy average 85 to 92 percent of the base metal strength. Continuous seam welds in thin gage 1100 aluminum show 88 to 100 percent joint efficiency.

Elevated temperature tests on welded specimens of several metals and alloys indicate that weld strength is no lower than that of the base material at the same temperature.

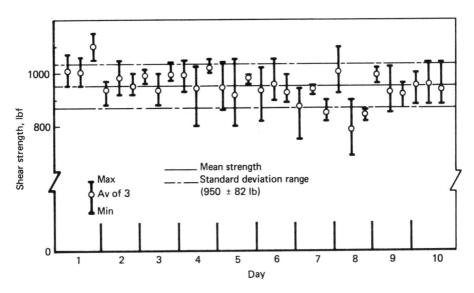

Fig. 9.21—Typical variance in ultrasonic weld shear strength in 0.040-in. Type 2024-T3 aluminum alloy

Table 9.3
Typical shear strengths of ultrasonic spot welds in several alloys

Metal	Alloy or type	Sheet thickness, in.	Mean shear strength with 90% confidence interval, lbf
Aluminum	2020-T6	0.040	1240 ± 50
	3003-H14	0.040	730 ± 40
	5052-H34	0.040	750 ± 30
	6061-T6	0.040	800 ± 40
	7075-T6	0.050	1540 ± 90
Copper	Electrolytic	0.045	850 ± 20
Nickel	Inconel X-750[a]	0.032	1520 ± 100
	Monel K-500[a]	0.032	900 ± 60
	René 41	0.020	380
	Thoria dispersed	0.025	910
Steel	AISI 1020	0.025	500 ± 20
	A-286	0.015	680 ± 70
	AM-350	0.008	310 ± 20
	AM-355	0.008	380 ± 40
Titanium	8% Mn	0.032	1730 ± 200
	5% Al-2.5% Sn	0.028	1950 ± 120
	6% Al-4% V	0.040	2260 ± 180

a. Trade names

Tensile Strength

Cross-tension tests on welds in selected metals indicate tensile strengths usually within the range of 20 to 40 percent of the tension-shear strength. With resistance welds, the ratio of direct tension to shear strength is usually taken as a criterion of weld ductility. The significance of this ratio for ultrasonic welds has not been established.

CORROSION RESISTANCE

The cast nugget of a resistance spot weld is frequently the site of localized corrosion attack when the weldment is exposed to an unfavorable environment. This is not true of ultrasonic welds. Weld specimens in aluminum alloys and stainless steels that have been exposed in boiling water, sodium chloride solutions, and other corrosive materials have shown no preferential attack in the weld.

QUALITY CONTROL

Nondestructive Evaluation

The weld power monitor, discussed previously, provides an effective means of monitoring weld quality at the time the weld is made. The operator can immediately detect an improper cycle and reject the part, or logic may be provided for automatic rejection of a part that is made at an unsatisfactory power level.

A number of postweld nondestructive techniques are also available. Ultrasound, radiography, and infrared radiation techniques may be used in specific applications. If hermetic sealing is the primary requirement of the weld, helium leak tests are effective.

Destructive Testing

An approach used for some applications involves destructive testing of randomly selected specimens during a production run. For relatively thin ductile sheet, a peel test will indicate adequate weld strength if failure occurs by nugget tear-out or fracture of the base metal. Metallographic sectioning for examination provides a reliable indication of weld quality, but it is slow and expensive.

For most applications, tension-shear testing is the most practical destructive test. Figure 9.21 shows typical variations in shear strength of random spot weld samples in 0.040-in. Type 2024-T3 aluminum alloy, produced with a specific machine setting for a number of days at different times of the day. The maximum, average, and minimum strength values for each set of weld samples are shown. The horizontal lines indicate the mean value and standard deviation range for the entire group. The process began to show poor control on the seventh and eighth days. Control was restored on the ninth day by making whatever adjustments were needed.

SAFETY

Ultrasonic welding equipment presents no unusual hazards to operating personnel. Most equipment is designed with interlocks and other safety devices to prevent operation under unsafe conditions. Nevertheless, consideration must be given to the health and safety of the operators, maintenance personnel, and other personnel in the area of the welding operations. Good engineering practice must be followed in the design, construction, installation, operation, and maintenance of equipment, controls, power supplies, and tooling to assure conformance to Federal safety laws (OSHA), State safety laws, and safety standards of the using company.

With high power equipment, high voltages are present in the frequency converter, the welding head, and the coaxial cable connecting these components. Consequently, the equipment should never be operated with the doors open or housing covers removed. Door interlocks are usually installed to prevent introduction of power to the equipment when the high voltage circuitry is exposed. The cables are shielded fully and present no hazard when properly connected and maintained.

Because of hazards associated with application of clamping force, the operator should never place hands or arms in the vicinity of the welding tip when the equipment is energized. For manual operation, the equipment is usually activated by dual palm buttons that meet the requirements of the Occupational Safety and Health Administration (OSHA). Both buttons must be pressed simultaneously to actuate a weld cycle, and both must be released before the next cycle is initiated. For automated systems in which the weld cycle is sequenced with other operations, guards should be installed for operator protection. Such hazards can be further minimized by setting the welding stroke to the minimum that is compatible with workpiece clearance.

Ring welding machines may be used for closure of containers filled with detonable materials. While no instance is known of premature ignition of such materials during ultrasonic welding, adequate provisions should always be made for remote operation by placing the welding machine either in a separate room from the control station or behind an explosion-proof barrier.

Metric Conversion Factors

1 in. = 25.4 mm = .025 m
1 ft/min = 5.08 $\times 10^{-3}$ m/s
1 microinch = 25 nm
1 ft = 0.305 m
1 lbf = 4.45 N
1 ksi = 6.89 MPa

SUPPLEMENTARY READING LIST

Alden, J. H., Ultrasonic sealing of foil. *Modern Packaging*, 34 (7): 129-133; 1961.

Avila, A. J., Metal bonding in semiconductor manufacturing—a survey. *Semiconductor Products and Solid-State Technology*, 7 (11): 22-26; 1964.

Chang, U. I. and Frisch, J., An optimization of some parameters in ultrasonic metal welding. *Welding Journal*, 53 (1): 24s-35s; 1974 Jan.

Estes, C. L. and Turner, P. W., Ultrasonic closure welding of small aluminum tubes. *Welding Journal*, 52 (8): 359s-369s; 1973 Aug.

Gencsoy, H. T., Adams, J. A., and Shin, S., On some fundamental problems in ultrasonic welding of dissimilar metals. *Welding Journal*, 46 (9): 145s-153s; 1967 Sept.

Hazlett, T. H. and Ambekar, S. M., Additional studies of interface temperature and bonding mechanisms of ultrasonic welds. *Welding Journal*, 49 (5): 196s-200s; 1970 May.

Jones, J. B., Ultrasonic welding. *Proceedings of the CIRP International Conference on Manufacturing Technology*, Ann Arbor, MI, 1387-1410; 1967 Sept.

Jones, J. B. and Meyer, F. R., Certain structural properties of ultrasonic welds in aluminum alloys. *Welding Journal*, 38 (7): 282s-288s; 1959 July.

Jones, J. B. and Powers, J. J., Ultrasonic welding. *Welding Journal*, 36 (8): 761-766; 1956 Aug.

Jones, J. B., et al., Phenomenological considerations in ultrasonic welding. *Welding Journal*, 40 (4): 289s-305s; 1961 Apr.

Koziarski, J., Ultrasonic welding: engineering, manufacturing and quality control problems. *Welding Journal*, 40(4): 349-358; 1961 Apr.

Littleford, F. E., Welding electronic devices by ultrasonics. *Industrial Electronics*, 6 (3): 123-126; 1967.

Meyer, F. R., Ultrasonic welding process for detonable materials. *National Defense*, 60 (334): 291-293; 1976.

Meyer, F. R., Assembling electronic devices by ultrasonic ring welding. *Electronic Packaging and Production*, 16 (7): 27-29; 1976.

Meyer, F. R., Ultrasonics produces strong oxide-free welds. *Assembly Engineering*, 20 (5): 26-29; 1977.

Rhines, F. N., Study of changes occurring in metal structure during ultrasonic welding. Metallurgical Research Laboratory, University of Florida, 1962; Summary Report TP 82-162.

Shin, S., and Gencsoy, H. T., Ultrasonic welding of metals to nonmetallic materials. *Welding Journal*, 47(9): 398s-403s; 1968 Sept.

Yeh, C. J., Libby, C. C., and McCauley, R. B., Ultrasonic longitudinal mode welding of aluminum wire. *Welding Journal*, 53 (6): 252-260; 1974 June.

10

Diffusion Welding and Brazing

Chapter Committee

M. M. SCHWARTZ, *Chairman*
Sikorsky Aircraft Division,
United Technologies

D. F. PAULONIS
Pratt and Whitney Aircraft Group,
United Technologies

Welding Handbook Committee Member

A. W. PENSE
Lehigh University

10

Diffusion Welding and Brazing

FUNDAMENTALS OF THE PROCESSES

DEFINITIONS AND GENERAL DESCRIPTIONS

Diffusion welding (DFW) is a joining process wherein the principal mechanism for joint formation is solid-state diffusion. Coalescence of the faying surfaces is accomplished through the application of pressure at elevated temperature. No melting and only limited macroscopic deformation or relative motion of the parts occur during welding. A solid filler metal (diffusion aid) may or may not be used between the faying surfaces. Terms which are sometimes used synonymously with diffusion welding include diffusion bonding, solid-state bonding, pressure bonding, isostatic bonding, and hot press bonding.

Several kinds of metal combinations can be joined by diffusion welding:

(1) Similar metals may be joined directly to form a solid-state weld. In this situation, required pressures, temperatures, and times are dependent only upon the characteristics of the metals to be joined and their surface preparation.

(2) Similar metals can be joined with a thin layer of a different metal between them. In this case, the layer may promote more rapid diffusion or permit increased microdeformation at the joint to provide more complete contact between the surfaces. This interface metal may be diffused into the base metal by suitable heat treatment until it no longer remains a separate layer.

(3) Two dissimilar metals may be joined directly where diffusion-controlled phenomena occur to form a bond. The mechanisms are similar to category (1) above with the added effects that dissimilar metals create.

(4) Dissimilar metals may be joined with a third metal between the faying surfaces to enhance weld formation either by accelerating diffusion or permitting more complete initial contact in a manner similar to category (2) above.

Diffusion brazing (DFB) is a process that produces coalescence of metals by heating them to a suitable temperature and by using a brazing filler metal or *in situ* liquid phase. The filler metal may be preplaced or formed between the faying surfaces or distributed by capillary action in the joint. Pressure may or may not be applied. The filler metal is diffused into the base metal to the extent that the joint properties approach those of the base metal. A distinct layer of brazing filler metal does not exist in the joint after the diffusion brazing cycle is completed. This characteristic distinguishes the process from brazing per se. The process is sometimes called liquid phase diffusion bonding, eutectic bonding, or activated diffusion bonding.

The distinction between diffusion welding and diffusion brazing may not be clear since a filler metal may be used with both processes. However, it is understood that melting actually takes place at the faying surfaces during the early stage of a diffusion brazing cycle. The filler metal layer itself may melt or a eutectic liquid may form from alloying between the filler metal and base metal. Diffusion at the interface continues with time at elevated temperature and any distinct layer of brazing alloy will finally disappear. Then the joint properties are nearly the same as those of the base metal.

If a filler metal is used and it does not melt or alloy with the base metal to form a liquid phase, the process is diffusion welding. The purpose of the filler metal is to aid bonding, particularly during the first stage of diffusion welding. It helps to eliminate voids at the interface that result when two rough surfaces are mated together. By proper selection, the filler metal will soften at welding temperature and flow under pressure to fill the interface voids. Also, it will diffuse into the base metal and produce a joint with acceptable properties for the application. The filler metal is considered to be a *diffusion aid,* not a brazing filler metal.

DIFFUSION WELDING PRINCIPLES

As illustrated in Fig. 10.1, metal surfaces have several general characteristics:

(1) Roughness

(2) An oxidized or otherwise chemically reacted and adherent layer

(3) Other randomly distributed solid or liquid products such as oil, grease, and dirt

(4) Adsorbed gas or moisture, or both

Because of these characteristics, two necessary conditions that must be met before a satisfactory diffusion weld can be made are:

(1) Mechanical intimacy of metal-to-metal contact must be achieved.

(2) Interfering surface contaminants must be disrupted and dispersed to permit metallic bonding to occur.

For conventional diffusion welding without a diffusion aid, a three-stage mechanistic model, shown in Fig. 10.2, adequately describes weld formation. In the first stage, deformation of the contacting asperities occurs primarily by yielding and by creep deformation mechanisms to produce intimate contact over a large fraction of the interfacial area. At the end of this stage, the joint is essentially a grain boundary at the areas of contact with voids between these areas. During the second stage, diffusion becomes more important than deformation, and many of the voids disappear as grain boundary diffusion

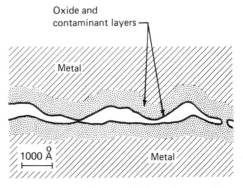

Fig. 10.1—Characteristics of a metal surface showing roughness and contaminants present

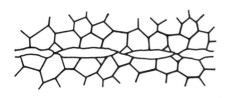

Initial asperity contact

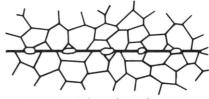

First stage deformation and interfacial boundary formation

Second stage grain boundary migration and pore elimination

Third stage volume diffusion pore elimination

Fig. 10.2—Three-stage mechanistic model of diffusion welding

of atoms continues. Simultaneously, the interfacial grain boundary migrates to an equilibrium configuration away from the original plane of the joint, leaving many of the remaining voids within the grains. In the third stage, the remaining voids are eliminated by volume diffusion of atoms to the void surface (equivalent to diffusion of vacancies away from the void). Of course, in a real system, these stages overlap, and mechanisms that may dominate one stage also operate to some extent during the other stages.

This model is consistent with several experimentally observed trends:

(1) Temperature is the most influential variable since it determines the extent of contact area during stage one and the rate of diffusion which governs void elimination during the second and third stages of welding.

(2) Pressure is necessary only during the first stage of welding to produce a large area of contact at the joining temperature. Removal of pressure after this stage does not significantly affect joint formation. However, premature removal of pressure before completion of the first stage is detrimental to the process.

(3) Rough initial surface finishes generally adversely affect welding by impeding the first stage and leaving large voids that must be eliminated during the later stages of welding.

(4) The time required to form a joint depends upon the temperature and pressure used; it is not an independent variable.

This model is not applicable to diffusion brazing or hot pressure welding processes where intimate contact is achieved through the use of molten filler metal and extensive (macro) deformation, respectively.

At the same time that intimate contact is being achieved as described above, various intervening films must be disrupted and dispersed so that metallic bonds can form. During initial mating surface contact (stage 1), the films are locally disrupted and metal-to-metal contact begins at places where the surfaces move together under shear. The subsequent steps in the process involve thermally activated diffusion mechanisms that complete film disruption and achieve intimate metal contact through void elimination (stages 2 and 3).

The barrier film is largely an oxide. Proper cleaning methods reduce the other components of film to negligible levels. Two actions tend to disrupt and disperse the oxide film. The first is solution of the oxide in the metal; the second is spheroidization or agglomeration of the film. Oxide films may be dissolved in titanium, tantalum, columbium, zirconium, and other metals in which interstitial elements are highly soluble. If the oxide is relatively insoluble in the metal, as is the case for aluminum, the disruption action for the trapped film is spheroidization. This leaves a few oxide particles along the weld line. However, if the weld is properly made, these oxide particles are no more detrimental than the inclusions normally present in most metals and alloys.

Both actions require diffusion. Solution occurs by diffusion of interstitial atoms into the metal and spheroidization by diffusion as a result of the excessive surface energy of the thin films. The time for solution of a film of thickness X is proportional to X^2/D, where D is the diffusion coefficient. The film must be kept very thin if diffusion welding times are to be within acceptable limits. Spheroidization occurs more rapidly if the oxide films are thin. Hence, control of the film thickness after cleaning and any increase in thickness during heating to welding temperature are critical factors in diffusion welding.

Once actual metal-to-metal contact is established, the atoms are within the attractive force fields of each other and a high strength joint is generated. At this time, the joint resembles a grain boundary because the metal lattices on each side of the line have different orientations. However, the joint may differ slightly from an internal grain boundary because it may contain more impurities, inclusions, and voids that will remain if full asperity deformation has not occurred. (Stage 2 in the model for achieving intimate contact is not yet complete.) As the process is carried to completion, this boundary migrates to a more stable non-planar configuration, and any remaining interfacial voids are eliminated through vacancy diffusion.

An intermediate metal (diffusion aid) is of significant practical importance in many systems, although the mechanisms so far described do not consider its use. When a diffusion aid is used or dissimilar alloys are welded, the additional factor of interdiffusion must be considered to develop a complete understanding of the DFW process.

DIFFUSION BRAZING PRINCIPLES

Diffusion brazing produces joint properties that are significantly different from those of conventional brazed joints. The main objective of the process is to produce joints having mechanical properties approaching those of the base metal in:

(1) Alloys that are not fusion weldable for their intended application, such as cast nickel-base superalloys for high temperature service and beryllium alloys

(2) Dissimilar alloy combinations that are not weldable

(3) Alloys where a combination joining and heat treating cycle is desirable to minimize distortion of the assembly during processing

(4) Alloys where conventional brazed joint properties are too low for the intended application, particularly at elevated temperature, such as in the application of high strength titanium alloys in aircraft

(5) Large, complicated assemblies where it is economical to produce many strong joints simultaneously and conventional brazing is unsuitable.

Two approaches to diffusion brazing are used. One utilizes a brazing filler metal that has a chemical composition approximately the same as the base metal but with a lower melting temperature. Melting temperature is suppressed by adding certain alloying elements to the base metal composition or to a similar alloy composition. For example, the melting temperature of a nickel-base high temperature alloy can be lowered by a small addition of silicon or boron. In this case, the brazing filler metal melts and wets the base metal faying surfaces during the brazing cycle. This approach is sometimes called activated diffusion bonding or transient liquid phase bonding.

The other approach is to braze with a metal that will alloy with the base metal to form one or more eutectic compositions. When the brazing temperature is slightly higher than the eutectic temperature, the filler metal and base metal will alloy together to produce a eutectic composition. The filler metal itself does not melt but an alloy (eutectic) is formed *in situ*. This method is also known as eutectic bonding. An example is the diffusion brazing of titanium alloys with copper.

With either approach, the assembly is held at temperature for a sufficient time for diffusion to produce a nearly uniform alloy composition across the joint. As this takes place, the melting temperature and the strength of the joint increase. The processing time depends upon the degree of homogeneity desired, the thickness of the initial filler metal layer, and the temperature. The relationship of heating rate to brazing temperature may also be important. A low heating rate will allow more solid-state diffusion to take place, and more filler metal will be required to provide sufficient liquid to fill the joint. Conversely, if a large quantity of filler metal and fast heating are used, the molten metal may run out of the joint and erode the base metal. The thick joint so formed will require a longer diffusion time to achieve a suitable composition gradient across it.

The composition gradient across the joint may be important with respect to response to subsequent heat treatment. This is particularly true for metals that undergo phase transformation during heating and cooling. Alloy composition will determine the transformation temperature and rate of transformation. Therefore, the phase morphology and mechanical properties of the joint can be controlled by the joint design and the brazing cycle.

ADVANTAGES AND LIMITATIONS

Diffusion welding and brazing have a number of advantages over the more commonly used welding processes and conventional brazing, as well as a number of distinct limitations on their applications.

Some of the advantages of the two processes are as follows:

(1) Joints can be produced with properties and microstructures very similar to those of the base metal. This is particularly important for lightweight fabrications.

(2) Components can be joined with minimum distortion and without subsequent machining or forming.

(3) Dissimilar alloys can be joined that are not weldable by fusion processes or by processes requiring axial symmetry.

(4) A large number of joints in an assembly can be made simultaneously.

(5) Components with limited access to the joints can be assembled by these processes.

(6) Large components of metals that require extensive preheat for fusion welding can be joined by these processes. An example is thick sections of copper.

(7) Defects normally associated with fusion welding are not encountered.

The two processes have limitations on their applications. The important ones follow:

(1) Generally, the duration of the thermal cycle is longer than that of conventional welding and brazing processes.

(2) Equipment costs are usually high, and this can limit the size of components that can be produced economically.

(3) The processes are not adaptable to high production applications, although a number of assemblies may be processed simultaneously.

(4) Adequate nondestructive inspection techniques for quality assurance are not available, particularly those that assure design properties in the joint.

(5) Suitable filler metals and procedures have not yet been developed for all structural alloys.

(6) The surfaces to be joined and the fit-up of mating parts generally require greater care in preparation than for conventional hot pressure welding or brazing processes. Smoothness and uniformity of joint clearance are important factors in quality control in the case of diffusion brazing.

(7) The need to simultaneously apply heat and a high compressive force in the restrictive environment of a vacuum or protective atmosphere is a major equipment problem with diffusion welding.

SURFACE PREPARATION

The surfaces of parts to be diffusion welded or diffusion brazed must be carefully prepared before assembly. Surface preparation involves more than cleanliness. It also includes (1) the generation of an acceptable finish or smoothness, (2) the removal of chemically combined films (oxides), and (3) the cleansing of gaseous, aqueous, or organic surface films. The primary surface finish is obtained ordinarily by machining, abrading, grinding, or polishing.

One property of a correctly prepared surface is its combined flatness and smoothness. A certain minimum degree of flatness and smoothness is required to assure uniform contact for diffusion welding or uniform joint gap for diffusion brazing. Conventional metal cutting, grinding, and abrasive polishing methods are usually adequate to produce the needed surface flatness and smoothness. A secondary effect of machining or abrading is the cold work introduced into the surface. Recrystallization of the cold worked surfaces tends to increase the diffusion rate in the weld region across the interface between them.

Chemical etching (pickling), commonly used as a form of preweld preparation, has two effects: The first is the favorable removal of nonmetallic surface films, usually oxides; the second is the removal of part or all of the cold worked layer that forms during machining, if that is done. The need for oxide removal is apparent because it prevents metal-to-metal contact.

Degreasing is a universal part of any procedure for surface cleaning. Alcohol, trichlorethylene, acetone, detergents, and many other cleaning agents may be used. Frequently, the recommended degreasing technique is intricate and may include multiple rinse-wash-etch cycles in several solutions. Since some of these cleaning solvents are toxic, proper safety precautions should be followed when they are used.

Heating in vacuum may also be used to obtain clean surfaces. The usefulness of this method depends to a large extent upon the type of metal and the nature of its surface films. Organic, aqueous, or gaseous adsorbed layers can be easily removed by vacuum heat treatment at elevated temperature. Most oxides do not dissociate during a vacuum heat treatment. It is possible to dissolve adherent oxides in some metals at elevated temperature. Typical metals are zirconium, titanium, tantalum, and columbium. Cleaning in vacuum usually requires subsequent vacuum or controlled atmosphere storage and careful handling to avoid the recurrence of surface adsorbed or chemisorbed layers.

Many factors enter into selecting the total surface treatment. In addition to those already mentioned, the specific welding or brazing conditions may affect the selection. With higher temperature or pressure, it becomes less important to obtain extremely clean surfaces. Increased atomic mobility, surface asperity deformation, and solubility for impurity elements all contribute to the dispersion of surface contaminants. As

a corollary, with lower temperature or pressure, better prepared and preserved surfaces are an asset.

Preservation of the clean surface is necessary following the surface preparation. One requirement is the effective use of a protective environment during the diffusion welding or brazing cycle. A vacuum environment provides continued protection from contamination. A pure hydrogen atmosphere will minimize the amount of oxide formed and it will reduce existing surface oxides of many metals at elevated temperature. However, it will form hydrides with zirconium, columbium, and tantalum that may be detrimental. Argon, helium, and sometimes nitrogen can be used to protect clean surfaces at elevated temperature. When these gases are used, their purity must be very high to avoid recontamination. Many of the precautions and principles applicable to brazing atmospheres can be applied directly to diffusion brazing or welding.[1]

DIFFUSION WELDING

PROCESS VARIABLES

Temperature

Temperature is an important diffusion welding process variable for a number of reasons:

(1) It is readily controlled and measured.

(2) In any thermally activated process, an incremental change in temperature will cause the greatest change in process kinetics when compared to most other process variables.

(3) Virtually all the mechanisms in diffusion welding are temperature-sensitive.

(4) Elevated temperature physical and mechanical properties, critical temperatures, and phase transformations are important reference points in the effective use of diffusion welding.

(5) Temperature must be controlled to promote or avoid certain metallurgical factors, such as allotropic transformation, recrystallization, and solution of precipitates.

Kinetic theory provides a means for understanding the quantitative effects of temperature in diffusion welding. Diffusivity can be expressed as a function of temperature as

$$D = D_0 e^{-Q/kT}$$

where

D = diffusion coefficient at T
D_0 = a constant of proportionality
Q = activation energy for diffusion
T = absolute temperature
k = Boltzmann's constant

From this, it is apparent that the diffusion controlled processes vary exponentially with

---

1. Brazing atmospheres are discussed in Vol. 2, *Welding Handbook,* 7th ed., 400-7.

temperature. Thus, relatively small changes in temperature produce significantly large changes in process kinetics.

In general, the temperature at which diffusion welding will take place is above $0.5\ T_M$, where T_M is the melting temperature of the metal. Many metals and alloys can best be diffusion welded at temperatures between 0.6 and $0.8\ T_M$. For any specific application the temperature, pressure, time at temperature, and surface preparation are interrelated.

Time

Time is closely related to temperature in that most diffusion controlled reactions vary with time. The diffusion length, x, is the average distance traveled by the average atom during a diffusion process. It can be expressed as

$$x = C(Dt)^{1/2}$$

where

x = diffusion length
D = diffusion coefficient at T
t = time
C = a constant

Thus, diffusion reactions progress with the square root of time (longer times become less and less effective), whereas they progress exponentially with temperature, as was previously shown.

Experience indicates that increasing both time at temperature and pressure increases joint strength up to a point. Beyond this point no further gains are achieved. This illustrates that time is not a quantitively simple variable. The simple relationship that describes the average distance traveled by an atom does not reflect the more complex changes in structure that result in the formation of a diffusion weld. Although atom motion continues indefinitely, structural changes tend to approach equilibrium. An example of similar behavior is the recrystallization of metals.

In a practical sense, time may vary over an extremely broad range, from seconds to hours. Production factors influence the overall practical time for diffusion welding. An example is the time necessary to provide the heat and pressure.

When the system has thermal and mechanical (or hydrostatic) inertia, welding times are long because of the impracticality of suddenly changing the variables. When there are no inertial problems, welding time may be as short as 0.3 min, as is the case when joining thoria-dispersed nickel to itself. On the other hand, it may be as long as 4 hours, as when joining columbium to itself with zirconium as a diffusion aid. For economic reasons, the time necessary for diffusion welding should be a minimum for best production rates.

Pressure

Pressure is an important variable. It is more difficult to deal with as a quantitative variable than either temperature or time. Pressure affects several aspects of the process. The initial phase of bond formation is certainly affected by the amount of deformation induced by the pressure applied. This is the most obvious single effect and probably the most frequently and thoroughly considered. Higher pressure invariably produces better joints when the other variables are fixed. The most apparent reason for this effect is the greater interface deformation and asperity breakdown. The greater deformation may also lower recrystallization temperature and accelerate the process of recrystallization at a given temperature.

The welding equipment and the joint geometry place practical limitations on the magnitude of welding pressure. The pressure needed to achieve a good weld is closely related to the temperature and time; there is some range of pressure in which good welds can be made. Pressure has additional significance when dissimilar metal combinations are considered. From economic and manufacturing aspects, low welding pressure is desirable. High pressure requires more costly apparatus, better control, and generally more complex part-handling procedures.

The pressures and temperatures employed are largely interdependent, but the pressure need not exceed the bulk yield stress of the material at the welding temperature. Thus, unless retaining dies are used, the pressure is usually kept slightly below the bulk yield stress at the temperature employed, which itself is selected to produce a weld in an acceptable time.

Metallurgical Factors

In addition to the process variables, there are a number of metallurgically important factors to be considered. Two factors of particular importance with similar metal welds are allotropic transformation and microstructural factors that tend to modify diffusion rates. Allotropic transformation (phase transformation) occurs in some metals and alloys. Heat treatable alloy steels are the most familiar of these, but titanium, zirconium, and cobalt also undergo allotropic transformation. The importance of the transformation is that the metal is very plastic during that time. This tends to permit rapid interface deformation at lower pressures, in much the same manner as does recrystallization. Diffusion rates are generally higher in plastically deformed metals as they recrystallize.

Another means of enhancing diffusion is alloying or, more specifically, introducing elements with high diffusivity into the systems at the interface. The function of a high diffusivity element is to accelerate void elimination. In addition to simple diffusion acceleration, the addition of these alloying elements may have secondary effects. The elements should have reasonable solubility in the metal to be joined, but not form stable compounds. Alloying must be controlled to avoid melting at the joint interface.

When using a diffusion-activated system, it is desirable to heat the assembly for some minimum time either during or after the welding process to disperse the high diffusivity element away from the interface. If this is not done, the high concentration of the element at the joint may produce metallurgically unstable structures. This is particularly important for joints that will be exposed to elevated temperature service.

Diffusion Aids

It is sometimes advantageous to use some form of diffusion aid or interlayer between the faying surfaces. One purpose of a diffusion aid is to provide a layer of soft metal between the surfaces. A soft metal layer permits plastic flow to take place at lower pressures than would be required without it during the first stage of welding (see Fig. 10.2). After the joint is formed, the diffusion of alloying elements from the base metal into the soft layer reduces the compositional gradient across the joint.

Intermediate diffusion aids may be necessary or advantageous in certain applications in order to

(1) Reduce welding temperature
(2) Reduce welding pressure
(3) Reduce process time
(4) Increase diffusivity
(5) Scavenge undesirable elements

Diffusion aids can be applied in many forms. They can be electroplated, evaporated, or sputtered onto the surface to be welded, or they can be in the form of foil inserts or powder. The thickness of the interlayer should not exceed 0.010 inch.

Generally, the diffusion aid is a more pure version of the base metal being joined. For example, unalloyed titanium often is used as an interlayer with titanium alloys, and nickel is sometimes used with nickel-base superalloys. Silver can be used with aluminum. Diffusion aids containing rapidly diffusable elements also can be used: for example, beryllium can be used with nickel alloys to decrease diffusion time. A properly selected diffusion aid will not melt at welding temperature or form a low melting eutectic with the base metal. An improperly chosen diffusion aid can

(1) Decrease the temperature capability of the joint
(2) Decrease the strength of the joint
(3) Cause microstructural degradation
(4) Result in corrosion problems at the joint

PROCESS VARIATIONS

Conventional

Conventional diffusion welding involves the application of pressure and heat to accomplish a weld along the entire length of one or more joints simultaneously. An interlayer may or may not be used. Pressure may be applied using gas pressure or a press (mechanical or hydraulic). Heat may be applied by any convenient means but electrical resistance heaters are the most common source. Forming parts to shape is done prior to or after welding using equipment designed for that purpose.

Continuous Seam

A relatively new process called continuous seam diffusion bonding (CSDB) joins components by "yield-controlled diffusion welding." With this process, the parts are positioned with tooling and then fed through a machine with four rollers. The top and bottom rollers are made of molybdenum and function much like resistance seam welding wheels. The two side rollers are used to maintain the shape of the components. The wheels and parts are heated by electrical resistance to the desired temperature. A special control system monitors part temperature. Welding temperature is usually between 1800 and 2000° F for titanium and between 2000 and 2200° F for nickel-base superalloys. The hot wheels apply pressure in the range of 1 to 20 ksi on the seam. The actual pressure depends upon the metal being joined, the joint design, the temperature, and the welding speed. An application of this process could be the joining of two flanges to a web to form a structural beam.

Combined Forming and Welding

Two process variations take advantage of the superplastic properties of certain metals or alloys. The alloys can deform or flow significantly at elevated temperatures under very small applied loads without necking or fracture. Titanium and its alloys exhibit this superplastic behavior in the temperature range of 1400 to 1700° F. Complex shapes can be formed using moderate gas pressures; then the shapes can be diffusion welded together, or vice versa.

One of these process variations is called creep isostatic pressing (CRISP). It is a two-step process combining creep or superplastic forming of titanium sheet structures with hot isostatic pressing to produce a diffusion welded one-piece structure.

Inherent in the CRISP process is the mating of two external skins. This is accomplished during the production of a structure. First, one skin is creep formed by gas pressure to the contour of a die. Then, shaped inserts are positioned on the skin and a second skin is creep formed

by gas pressure over the first skin and inserts. Diffusion welding of the formed sheets and inserts is achieved by hot isostatic pressing in an autoclave. This method eliminates the need for precision machined die sets and close dimensional tolerances in parts.

The other process variation takes advantage of the same properties of titanium and its alloys described previously; however, the welding is performed under low pressure conditions. This variation is called "superplastic forming diffusion bonding" (SPF/DB). Since superplastic forming and diffusion welding of selected titanium alloys can be accomplished using identical process temperatures, the two operations can be combined in a single fabrication cycle. The welding is accomplished under low pressure conditions.

The superplastic forming of the sheet can be done first, followed by welding, or the steps can be reversed. The order depends upon the design of the component. Forming is done first if this is required to bring the faying surfaces of the joint together for welding. If the faying surfaces are in contact, welding is done first and then the part is formed to shape in the die using inert gas pressure. A suitable nonmetallic agent can be used to prevent welding of areas where subsequent forming is to occur.

Superplastic forming of Ti-6%Al-4%V alloy sheet can be done by the application of low pressure argon at 1700° F in a sealed die. Gas pressure of about 150 psi is used for both forming and welding.

EQUIPMENT AND TOOLING

A wide variety of equipment and tooling is employed for diffusion welding. The only basic requirement is that pressure and temperature must be applied and maintained in a controlled environment. Various types of equipment have been developed, each with its special advantages and disadvantages. There are numerous variations of a given type of equipment or approach depending upon the specific application. A general description of four types of diffusion welding equipment follows.

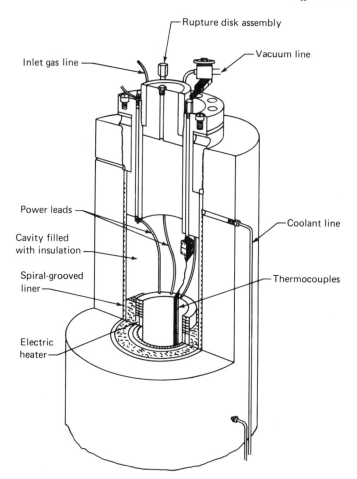

Rupture disk assembly

Inlet gas line

Vacuum line

Power leads

Cavity filled
with insulation

Spiral-grooved
liner

Coolant line

Thermocouples

Electric
heater

*Fig. 10.3—A typical high temperature, cold
wall autoclave*

Isostatic Gas Pressure

The pressure for welding can be applied
uniformly to all joints in an assembly using gas
pressure. One important requirement is that the
assembly to be welded be evacuated of air during
the welding cycle. The assembly itself may be
evacuated and sealed by fusion welding, if this is
possible. Otherwise, the assembly must be sealed
in a thin, gas-tight envelope which is evacuated
and sealed. Electron beam welding in vacuum is
a convenient process for evacuating and sealing
in one operation.

Gas pressure is applied externally against
the evacuated assembly at welding temperature.

Very high pressures can be applied using an
autoclave, but the assembly must be capable of
withstanding the applied pressure without macrodeformation. Some designs may require internal
support tooling with provisions for removing it
after welding.

The primary component of hot isostatic
equipment is a cold wall autoclave, which can
be designed for gas pressures up to 150 ksi and
for part temperature in excess of 3000° F. A
typical autoclave is shown in Fig. 10.3. Work to
be isostatically welded is placed in the heated
cavity. Internal water cooling is usually provided to maintain a low wall temperature. Closures on each end provide access to the vessel

cavity. Utilities and instrumentation are brought into the vessel through high pressure fittings located in the end closures. The high temperatures are produced with an internal heater. Resistance heaters of various designs are used. Alumina or silica insulation is used to reduce heat losses to the cold wall. Temperature is monitored and controlled by thermocouples located throughout the furnace and vessel. Pressurization is achieved by pumping inert gas into the autoclave with a multiple-stage piston type compressor. Temperature and pressure are controlled independently, and any combination of heating and pressurizing rates can be programmed. The autoclave is a pressure vessel, and it must be designed to meet applicable code requirements.

The most important consideration is the gastight envelope or can in which the specimen must be contained. If a leak develops in the can, pressure can not be applied to or maintained on the joint. Sufficient gas pressure is applied so that local plastic flow will occur at the joint interface and all void space will be filled as a result of local deformation. With proper conditions, essentially no macrodeformation and no change in part dimensions will occur during the operation.

The chief advantage of this technique is the ability to handle complex geometries. It is also well suited to batch operations where large quantities of relatively small assemblies can be processed simultaneously. The major drawbacks are the capital equipment costs and the size limitations imposed by the internal dimensions of the autoclave. Operational equipment ranges up to 36 in. inside diameter and inside lengths of up to 108 inches.

The gas pressure process variables are not uniform for different metals and combinations of metals. Usually joints are made at the highest possible pressure to minimize the temperature needed. This method is well suited for welding brittle metals or metals to ceramics and cermets because the isostatic pressure eliminates tensile stresses in the materials.

Presses

A very common approach for diffusion welding employs a mechanical or hydraulic press of some design. The basic requirements for the press are (1) sufficient load and size capacity, (2) an available means for heating, and (3) the maintenance of uniform pressure for the required time. It is often necessary to provide a protective atmosphere chamber around the weldment. Press equipment that can be adapted to diffusion welding applications is frequently available in a manufacturing or development organization.

There is no standard press design. Some units provide a vacuum or an inert atmosphere around the parts. Radiant, induction, and self-resistance types of heating are used. An advantage of a press setup is the ease of operation and the excellent process control available. A disadvantage is the practical limitation of press size when large component fabrication is considered. This approach does not lend itself to high production rates and rapid turn-around or batch operations.

Some of the limitations on size can be overcome by operating in a large forming or forging press without an inert atmosphere chamber. Heated platens are used to apply both heat and pressure to the components. The platens may be metallic or ceramic, depending upon the temperature and pressure employed. Castable ceramics are particularly useful because contours can easily be accommodated without extensive machining. Heating elements can be cast into a ceramic die to provide uniform heat during welding. Close tolerances must be maintained between the die and the part so that uniform pressure will be applied to the joint. This is a major problem with press type equipment. It is extremely difficult to maintain uniform pressure on the joint, and variations in weld quality can result.

Tooling requirements vary with application. If no lateral restraint is provided, upsetting may occur during the welding cycle. In such cases, lower pressure or temperature is usually required. Heated dies are required because of the time factor and die materials can be a problem. The die must be able to withstand both the temperature and the pressure and must be compatible with the metal to be welded. Interaction between the part and the die can be controlled by stopoff agents and sometimes by oxidizing the die surface. Atmosphere protection is often achieved by sealing the parts in evacuated metal cans that are designed to conform to the die shape.

Retorts can be used in conjunction with presses for diffusion welding of titanium parts. Tooling blocks and spacers of Type 22-4-9 stainless steel may be used to fill any voids between the titanium pieces to maintain their shapes. Presses with side and ,end restraining jacks can exert up to 2 ksi pressure on the retort in all directions. In actual production, the completed assembly pack (retort, heating pads, and insulation) is preheated before it is placed in the press. Large structures may require a preheat of as long as 40 hours. Several packs may be in assembly and preheat at one time. The actual time in the press will vary from 2 to 12 hours, depending upon the shape of the structure and the mass of titanium. The assembly pack is cooled to room temperature, dismantled, and the retort is then cut open. This approach is quite slow and may not be scaled readily to high production rates.

Resistance Welding Machines

Resistance welding equipment may be used to produce round diffusion spot welds between sheet metal parts. In general, modification of standard equipment is not necessary to achieve successful diffusion welds. The interface is resistance heated under pressure with this equipment. The cycle is designed to avoid melting of metal at the interface. Weld times are generally less than 1 second.

As in standard resistance welding, selection of a suitable electrode material is important. The electrodes must be electrical conductors, possess high strength at welding temperatures, resist thermal shock, and resist sticking to the parts. There is no universal electrode material because of potential interaction with the workpiece. Therefore, each system must be carefully evaluated from a metallurgical compatibility standpoint.

For some applications, a small chamber surrounding the electrodes is used to provide an inert atmosphere or vacuum during welding.

One advantage of this type of equipment is the speed at which diffusion welds can be made. Each weld is made in a very short time; however, only a small area is welded in each cycle and a number of welds are needed to join a large area.

Tooling

A number of important considerations must be observed when selecting tooling materials. The main criteria are
(1) Ease of operation
(2) Reproducibility of the welding cycle
(3) Operational maintenance
(4) Weld cycle time
(5) Initial cost of tooling
Furthermore, the materials must be capable of maintaining their proper positions and shapes throughout the heating cycle.

Suitable fixture materials may be limited when welding temperatures are above 2500° F. Only the refractory metals and certain nonmetallic materials have sufficient creep strength at such high temperatures: for example, tantalum and graphite may be suitable for fixturing tungsten. Ceramic materials are suitable for fixtures provided they are completely outgassed prior to welding.

Since pressure is required for diffusion welding, fixtures should be designed to take advantage of the difference in thermal expansion between the metals being joined and the fixture material. It is possible to generate at least part, if not all, of the pressure required for welding by appropriate selection of the fixture material and the clearances between the fixture and part. These principles have been used to join Type 2219 aluminum alloy tubing to Type 321 stainless steel: A precise method was devised to apply the correct welding pressure to a tubular assembly. Tooling was developed that provided a uniform and reproducible welding pressure by taking advantage of the difference in the thermal expansions of low alloy steel and stainless steel.

DIFFUSION BRAZING

PROCESS VARIABLES

Diffusion brazing operations are similar to those for conventional brazing.[2] Different methods of heating, atmospheres, joint designs, and equipment can generally be used interchangeably. With diffusion brazing, the brazing alloy, the processing temperature, and the time at temperature are selected to produce a joint with physical and mechanical properties almost identical to those of the base metal. To do this, it is necessary to essentially eradicate the braze layer by diffusion with the base metal.

Temperature and Heating Rate

The temperature cycle used for diffusion brazing depends upon the base metal and the design of the brazing system. When the brazing alloy composition is similar to that of the base metal, the assembly must be heated to the melting temperature of the brazing alloy as in conventional brazing. As the brazing alloy melts, it wets the base metal and fills the voids in the joint; then the temperature can be maintained or reduced to solidify the brazing alloy.

Some diffusion brazing systems form a filler metal *in situ* during the brazing cycle. The systems are generally designed to form a molten eutectic that flows and fills the voids in the joint at brazing temperature. The brazing temperature is somewhat higher than the eutectic temperature. For example, a plating of copper on a silver base metal faying surface will form a eutectic when heated to 1500° F. The eutectic melting temperature is 1435° F.

In systems where several eutectic and peritectic reactions take place at different temperatures, both the brazing temperature and the heating rate are important. Although a liquid phase can form at the lowest eutectic temperature, diffusion rates will be faster at higher temperatures. The heating rate will determine whether a molten eutectic is formed. If the heating rate is too low, solid-state diffusion will prevent

the formation of a molten eutectic. The voids at the faying surface will not be filled with "filler metal."

The maximum brazing temperature may be established by the characteristics of the base metal: for example, incipient melting in some nickel-base alloys. It may also be limited by the effect of temperature on the final metallurgical structure and by the heat treatment requirements for the weldment.

After brazing is accomplished, the part temperature may or may not be reduced to solidify the brazing layer. Temperature is then maintained while solid-state diffusion takes place.

Time

The duration of the diffusion brazing cycle will depend upon (1) the brazing temperature, (2) the diffusion rates between the filler metal and the base metal at temperature, and (3) the maximum concentration of interlayer metal permissable at the joint. The alloy composition at the joint may influence the response to heat treatment and the resulting mechanical properties of the joint. Therefore, the joint must be held at temperature for some minimum time to reduce the concentration of interlayer metal to an acceptable value.

Pressure

Normally, brazing is done with very little or no pressure on the joint. In some cases, fixturing may be necessary to avoid excessive pressure. This is particularly so when the molten filler metal is to flow into the joint by capillary action. When the filler metal is preplaced in the joint, excessive pressure may force low melting constituents to flow out of the joint before brazing temperature is achieved. In that case, the molten filler metal may not be sufficiently fluid to fill interface voids.

Metallurgical Factors

The metallurgical events that may transpire during diffusion brazing are similar to those that occur during diffusion welding. An additional factor is the variation in alloy content across the joint. It is well known that compo-

---

2. Conventional brazing is covered in Ch. 10, *Welding Handbook*, Vol. 2, 7th ed. 370-438.

sitional variations can significantly affect the response of a particular alloy to heat treatment. With metals that experience an allotropic transformation at elevated temperature, alloying can raise or lower the transformation temperature and the rate of transformation. Thus, the response to heat treatment across a diffusion brazed joint may vary with the concentration of the metal used to lower the melting temperature of the base metal: for example, copper stabilizes the beta phase in titanium and decreases the beta-to-alpha transition temperature.

Intermediate Metals

The intermediate metal may be either a brazing alloy or a metal that will alloy with the base metal at some elevated temperature to form a brazing alloy. In the latter case, a eutectic must form that melts at a temperature compatible with the metallurgy and design properties of the base metal. The metal or alloy may be in powder, foil, or wire form, or it may be plated onto the surface of the base metal. Close control of the amount of metal or alloy in the joint is essential for consistent joint properties in production.

Application of pure metals and simple alloys by electroplating or vapor deposition can be accurately controlled. Films of desired thickness can be deposited on the joint faces. However, these processes are not economical for all cases.

Preformed metal foil or wire shapes are better suited for many applications.

In the case of nickel- and cobalt-base heat-resistant alloys, elements commonly added to brazing alloys to depress the melting temperature also increase alloy hardness and brittleness. Consequently, these brazing alloys can only be produced in powder form. This presents a problem in diffusion brazing where precise amounts of filler metal are required. Boron in the range of 2.0 to 3.5 percent is used in nickel-base brazing alloys. This element can be diffused into the surfaces of nickel alloy foil or wire shapes to produce brazing alloy preforms. These preforms can provide good control of brazing alloy placement for diffusion brazing applications.

EQUIPMENT AND TOOLING

The equipment and tooling used for diffusion brazing are essentially the same as those used for conventional brazing. If furnace brazing is used, the entire cycle can be done in the same equipment. In some cases, it may be more economical and convenient to braze with one piece of equipment and then follow with a diffusion heat treatment with other equipment: for example, the brazing could be done with resistance welding or induction heating equipment, and the diffusion heat treatment could be performed in a furnace.

APPLICATIONS

A wide variety of similar and dissimilar material combinations can be successfully joined by diffusion welding and brazing. Most applications involve alloys of titanium, nickel, and aluminum, as well as dissimilar metal combinations. The ability to produce joint properties approaching those of the base metal depends largely upon the characteristics of the metals being joined: for example, excellent joint properties are readily produced in titanium alloys. The relatively low creep strength of titanium alloys and their ability to dissolve the oxide at the elevated temperature contribute to this.

On the other hand, nickel-base heat-resistant alloys are difficult to join, particularly in the solid state, because their creep strengths are high, requiring high pressures for diffusion welding. In addition, chromium oxides are stable at high temperatures and remain on the faying surfaces. Unless a proper interlayer metal is utilized, stable intermetallic precipitates can form at the interface and produce brittle joints.

The ability to achieve a joint with base metal properties depends upon the mechanism (cold work or heat treatment) used to develop the design strength of the base metal. If cold work is used, then the strength of the base metal will be irreversibly lowered by either process. On the other hand, heat treatable alloys can generally be strengthened with a thermal treatment combined with or subsequent to the joining operation.

TITANIUM ALLOYS

Many diffusion welding and brazing applications involve titanium alloy components, the majority of which are Ti-6%Al-4%V alloy.[3] The popularity of the processes with titanium alloys stems from the following factors:

(1) Titanium is readily joined by both methods without special surface preparation or unusual process controls.

(2) Diffusion welded or brazed joints may have better properties for some applications than conventional fusion welded joints.

(3) Most titanium structures or components are used principally in aerospace applications where weight savings or advanced designs, or both, are more important than manufacturing costs, within limits.

A number of well-established diffusion welding and brazing methods are available for joining titanium alloys. Welding can be accomplished using pressures in the range of several hundred to several thousand psi. High pressures are used in conjunction with low welding temperatures and when the assembly is welded in a closed container (retort). Inserts may be used to hold the structure to dimensions. When welding at higher temperatures without an enclosure, maximum pressure is usually limited by the allowable deformation in the part, and this pressure must be determined empirically. Pressures of 300 to 500 psi work well in many cases. In some applications, total assembly deformation and deformation rate, rather than pressure per se, are controlled during welding for process control.

Welding temperature is probably the most influential variable in determining weld quality; it is set as high as possible without causing irreversible damage to the base metal. For the commonly used alpha-beta type titanium alloys, this temperature is about 75 to 100° F below the

beta transus temperature: for example, Ti-6%Al-4%V alloy with a beta transus of approximately 1825° F is best diffusion welded between 1700 and 1750° F. The time required to achieve high weld strength can vary considerably with other factors, such as mating surface roughness, welding temperature, and pressure. Welding times of 30 to 60 minutes should be considered a practical minimum, with 2 to 4 hours being more desirable. Mating surface finish and pre-weld cleaning procedure are two other important considerations. Although the general rule that a smooth mating surface makes welding easier still applies, parts with relatively rough (milled or lathe-turned) mating surfaces can be successfully diffusion welded as long as welding temperature, time, and pressure are adjusted to accommodate such rough finishes. Freshly machined mating surfaces only need to be degreased with a suitable solvent prior to welding. Hydrocarbon and chlorinated solvents should not be used. A preferred cleaning method is acid cleaning in a HNO_3-HF solution. Any residue remaining from the cleaning operation must be removed by thorough rinsing.

A solid-state diffusion weld in Ti-6%Al-4%V alloy is shown in Fig. 10.4. The weld zone is virtually indistinguishable from the base metal in both appearance and properties. This weld was made at 1700° F for 4 hours using an applied pressure of about 500 psi. The mating surfaces were machined on a lathe and acid cleaned prior to welding. Extensive testing of welds of this quality has demonstrated that the mechanical properties of the welds are equivalent to the base metal.

Several industries have taken advantage of the benefits of the diffusion welding process, particularly the aerospace industry with its increased potential usage of titanium alloys. Two new vehicles, one for space travel and the other a long range bomber aircraft, are utilizing this process extensively. The engine mount of each space shuttle vehicle will have 28 diffusion welded titanium parts, ranging from large frames to interconnecting box tubes. This structure is capable of withstanding three million pounds of thrust. Square tubes of 0.75-in. thick wall and approximately 8-in. long sides were fabricated by diffusion welding in lengths up to 180 inches.

---

3. The weldability of titanium alloys is discussed in Ch. 73, Sec. 4, *Welding Handbook,* 6th ed. This chapter will be revised in Vol. 4, 7th ed.

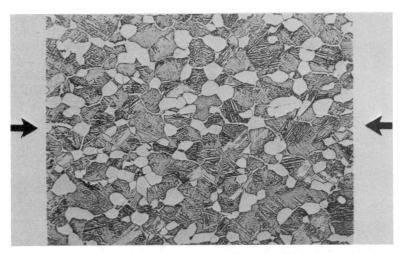

Fig. 10.4—A diffusion weld in Ti-6%Al-4%V alloy made at 1700° F for 4 hours with an applied pressure of 500 psi (×250)

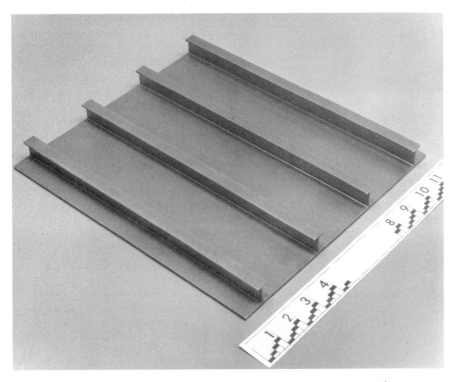

Fig. 10.5—A titanium alloy stiffened sheet structure fabricated by continuous seam diffusion bonding

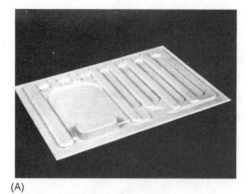

(A)

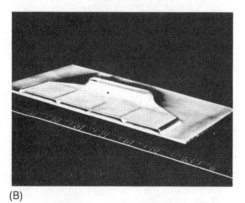

(B)

Fig. 10.6—Two titanium alloy parts fabricated by superplastic-forming diffusion bonding: (A) a door; (B) a windshield hot air nozzle

The advanced airplane has 66 diffusion welded parts of Ti-6%A1-4%V alloy. The wing carry-through structure is the largest diffusion welded composite structure in existence. It is the most critical structure in the vehicle.

The use of diffusion welding in the gas turbine industry reached a milestone with the production application of the Ti-6%A1-4%V component for an advanced, high thrust engine. This application marks the first production use of diffusion welding in a rotating engine component.

Continuous seam diffusion bonding has been used to produce stiffened skins fabricated as an integral one-piece structure. An example is shown in Fig. 10.5. One of the first applications of this method was the fabrication of curved Ti-6%A1-4%V alloy I-beams used as structural members to support boron-aluminum composite on a fighter airplane. These beams were made from 0.025-in. sheet.

The superplastic-forming diffusion bonding of titanium parts is also being investigated. Figure 10.6 shows two typical parts that were fabricated by the process.

Diffusion brazing techniques are also used for joining titanium alloys. Cycle times, temperatures, and preweld cleaning procedures are much the same as for diffusion welding. However, pressure is just sufficient to hold the parts in contact, and mating surface finish requirements are not stringent.

(A) Electroplated metal layers on the edge of a sheet

(B) A diffusion brazed joint

Fig. 10.7—A diffusion brazed T-joint between Ti-6%A1-4%V and Ti3%A1-2.5%V alloys

The mating surfaces are electrolytically plated with a thin film of either pure copper or a series of elements, such as copper and nickel. When heated to the brazing temperature of 1650 to 1700° F, the copper layer reacts with the titanium alloy to form a molten eutectic at the joint interface. The assembly is then held at temperature for at least 1.5 hours, or is given a subsequent heat treatment at this temperature for several hours, to reduce the composition gradient in the joint. Diffusion brazed joints made with a copper interlayer and a cycle of 1700° F for 4 hours had tensile, shear, smooth fatigue, and stress corrosion properties equal to those of the base metal. However, they had slightly lower notch fatigue and corrosion fatigue properties, and significantly lower fracture toughness. A typical photomicrograph of a diffusion brazed T-joint between Ti-6%Al-4%V and Ti-3%Al-2.5%V alloys is shown in Fig. 10.7. A Widmanstatten structure formed at the joint because the plated interlayer stabilized the beta phase.

Diffusion brazing is being used to fabricate light-weight cylindrical cases of titanium alloys for jet engines. In this application, the titanium core is plated with a very thin layer of selected metals that react with the titanium to form a eutectic. During the brazing cycle in a vacuum of 10^{-5} torr, a eutectic liquid forms at 1650° F. This liquid performs the function of a brazing alloy between the core and face sheets. The eutectic quickly solidifies due to rapid diffusion at the joints. The assemblies are held at temperature for one to four hours to reduce the composition gradient at the joint by diffusion. A typical diffusion brazed titanium alloy honeycomb structure is shown in Fig. 10.8.

NICKEL ALLOYS

Many nickel alloys, specifically the high strength heat resistant alloys, are more difficult to solid-state diffusion weld than most other metals. These alloys must be welded at temperatures close to their melting temperatures and with relatively high pressures because of their high temperature strengths. In addition, extra care must be taken in preparing the surfaces to be welded to ensure mutual conformity and cleanliness. Surface oxides that form on these alloys are stable at high temperatures and will not dissolve or diffuse into the alloy. During the welding operation, the ambient atmosphere must be carefully controlled to prevent interface contamination.

Pure nickel or nickel alloy interlayers, typically an electroplated layer or thin foil, are commonly used when diffusion welding nickel alloys. These interlayers, generally from 0.0001 to 0.001-in. thick, serve several functions. Their relatively low yield strength allows surface conformity to take place at relatively low welding pressures. More important, they are used during welding to prevent the formation of stable pre-

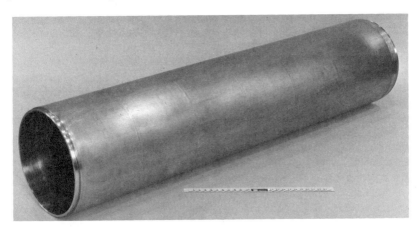

Fig. 10.8—Diffusion brazed titanium alloy honeycomb structure (81 in. long, 6 lb total weight)

Table 10.1
Typical diffusion welding conditions for some nickel-base alloys

Alloy[a]	Interlayer	Welding temp., °F	Pressure, psi	Time, h
Inconel 600	Ni	2000	100-500	0.5
Hastelloy X	Ni	2050	100-500	4
Wrought Udimet 700	Ni-35%Co	2140	1000	4
Cast Udimet 700	Ni-35%Co	2175	1200	4
René 41	Ni-Be	2150	1550	2
Mar-M 200	Ni-25%Co	2200	1000-2000	2

a. Tradenames

cipitates, such as oxides, carbides, or carbonitrides, at the joint interface. The diffusion welding time must be adequate to allow sufficient interdiffusion to occur in the joint region.

Some nickel base heat resistant alloys that have been diffusion welded are listed in Table 10.1, as are the welding conditions.

The pressure required for satisfactory welding is influenced strongly by the geometry of the parts being welded; therefore, the required pressure for each application must be determined empirically.

The significance of an interlayer and its composition was demonstrated by a series of diffusion welds in wrought and cast Udimet 700 alloy.[4] Welds were made without an interlayer and with 0.0002-in. interlayers of both pure nickel and Ni-35%Co alloy. The welding conditions were the same as those listed for this alloy in Table 10.1.

The microstructure of the welds in wrought Udimet 700 alloy are shown in Fig. 10.9. With no interlayer, fine Ti(C,N) and $NiTiO_3$ precipitates formed at the interface during welding and pinned the interfacial boundary, causing very poor joint properties. The nickel interlayer consisted of an electroplated layer on each surface. These layers probably welded together early in the cycle, and no precipitates were present to interfere with welding. Subsequent diffusion and grain boundary movement resulted in much improved mechanical properties. The pure nickel interlayer, however, resulted in preferential "up-

4. Tradename

hill" diffusion of aluminum and titanium into the nickel interlayer. This led to the formation of excessive amounts of the strengthening precipitate $Ni_3(Al, Ti)$ in the joint. The addition of 35% cobalt to the interlayer resulted in a completely homogeneous joint region because of the effect of cobalt on the formation of $Ni_3(Al, Ti)$

Nickel-base heat resistant alloys can be diffusion brazed using specially designed brazing filler metals and procedures. There are two reported variations that differ primarily in programming the thermal cycle to accomplish diffusion. Both utilize specially tailored interlayers that melt at brazing temperature. Then, through diffusion, either by extending the original thermal cycle or with a subsequent cycle, both methods produce high strength joints that resemble the base metal in both structure and mechanical properties.

With the first variation, a thin interlayer alloy of specific composition and melting point, 0.001 to 0.004-in. thick, is used as brazing filler metal. The parts are held together under slight compressive pressure (under 10 psi) and heated to the brazing temperature (typically 2000 to 2200° F) in vacuum or an argon atmosphere. At brazing temperature, the interlayer initially melts, filling the voids between the mating surfaces with a thin, molten layer. While the parts are held at temperature, rapid diffusion of alloying elements occurs between the interlayer and the base metal. This change of composition at the interface region causes the joint to isothermally solidify, thus forming a solid bond while still at temperature. After isothermal solidification occurs, the joint microstructure generally resembles that of the base metal except for some compositional and structural variations.

At this stage the joint has good properties, although not fully equivalent to those of the base metal. By permitting the part to remain at the brazing temperature for a longer time, the joint can be homogenized both in composition and structure until it is essentially equivalent to the base metal.

The second variation involves joining nickel-base components with a specially designed brazing filler metal that completely melts at some elevated temperature below the incipient melting point of the alloy or alloys being joined. Subsequent to this, the brazed component is given a diffusion heat treatment to homogenize the brazing filler with the base metal. This is followed by an appropriate aging heat treatment designed for the base metal.

Melting point depressants, such as silicon, boron, manganese, aluminum, titanium, and columbinum, are added to the base alloy to produce a brazing filler metal. The base alloy is simply "doped" with sufficient amounts of depressants so that the resultant alloy is molten at a temperature that does not impair the properties of the alloy to be joined. Ideally, brazing is accomplished at the normal solution heat treating temperature for a given alloy. Figure 10.10 shows a diffusion brazed joint made in wrought Udimet 700 using the first described procedure. An interlayer of 0.003-in. thick Ni-15%Cr-15%Co-5%Mo-3%Be filler metal was used in the joint with a processing cycle of 2140° F for 24 hours in vacuum. A microprobe chemical analysis across a joint showed a uniform chemical composition, essentially that of the base metal. Stress rupture tests at 1600° F and 1800° F showed that the diffusion brazed joints had essentially the same properties as the Udimet 700 base metal.

Diffusion brazed joints produced at lower temperatures and with shorter time cycles may not be uniform in composition. As a result, some elevated temperature mechanical properties of the joints may be lower than those of the base metal, particularly under stress-rupture conditions.

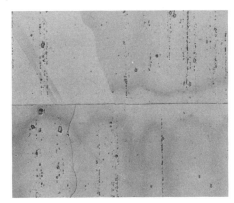

(A)

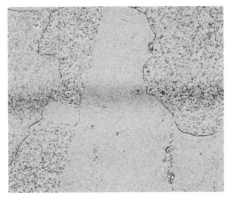

(B)

(C)

Fig. 10.9 — Diffusion welds in wrought Udimet 700 alloy with (A) no interlayer, (B) a nickel interlayer, and (C) a Ni-35%Co alloy interlayer (×250)

ALUMINUM ALLOYS

Aluminum alloys can be successfully diffusion welded as long as some means is employed to avoid, disrupt, or dissolve the tenacious surface oxide. A wide range of temperatures, pressures, and times may be utilized: for example, with Type 6061 aluminum alloy, welding conditions as divergent as 725° F and a pressure of 3800 psi for several hours or 1000° F and a pressure of 1000 psi for one hour have been satisfactory. However, the main boundary condition is the melting point of the specific alloy being joined. The welding operation is normally carried out in vacuum or inert gas although aluminum-boron fiber composites can be diffusion welded in air. For applications where little or no local deformation of the parts can be tolerated, the mating surfaces are first coated with a thin layer of silver or gold-copper alloy by electrolytic or vapor deposition. The coating prevents surface oxidation during welding.

Aluminum and aluminum-silicon alloys can be diffusion brazed using a copper interlayer. Sound, strong joints can be produced in aluminum by limiting the copper thickness to 20×10^{-6} in. and restricting the brazing temperature to between 1030 and 1060° F. The time at temperature should not exceed 15 minutes at the lower temperature limit or 7 minutes at the upper limit. Type A356.0 aluminum-7% silicon casting alloy can be diffusion brazed by electroplating one of the joint members with copper that will form a eutectic with the aluminum and silicon in the casting alloy when heated to 975° F.

To ensure optimum joint properties, copper thickness, brazing temperature, and time-at-temperature must be selected to promote isothermal solidification during brazing and thereby prevent the formation of the compound $CuAl_2$. Proper balancing of these variables results in strong joints that can withstand quenching from above the solution treating temperature which is required for heat treating Type A356.0 alloy to the T61 condition. Electroplating the cover sheets with 150 to 200 microinches of copper and holding between 980 and 1000° F for one hour are satisfactory conditions. After quenching and aging, the joint strength will equal that of the casting itself. Microstructurally, the brazed joint will be indistinguishable from the casting.

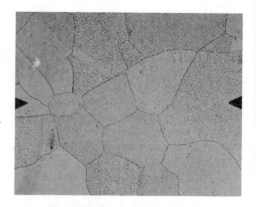

Fig. 10.10—A diffusion brazed joint in wrought Udimet 700 alloy (×100)

STEELS

Steels are not normally diffusion welded because they are more easily joined by conventional brazing or fusion welding methods, for most applications. Diffusion welding may be utilized successfully for specialized applications where high quality joints are required between large, flat surfaces: for example, plain carbon steels have been welded without an interlayer over a wide range of conditions. Two sets of variables that produced excellent welds in AISI 1020 steel are 1800 to 2200° F with a pressure of 1 ksi for 1 to 15 minutes and 2000 to 2200° F with a pressure of 5 psi for 2 hours. Welding can be accomplished either in a protective atmosphere or in air, provided the joint is first seal welded around the periphery to exclude air.

Stainless steels can be diffusion welded using conditions similar to those used for plain carbon steel; however, these steels are normally covered by a thin adherent oxide that must be removed prior to welding. This can be accomplished either by welding at high temperatures in dry hydrogen or by copper plating the faying surfaces after anodic cleaning. Copper oxide on the plating is relatively easy to reduce in hydrogen during heating to welding temperature. For illustrative purposes, sound welds were made in AMS 5630 martensitic stainless steel at 2000° F with a pressure of 100 psi for 1.5 hours using a 0.0001-in. thick copper interlayer.

Table 10.2
Diffusion welding conditions for some dissimilar metal combinations

Metal combinations	Interlayer	Temp., °F	Time, h	Pressure,[a] ksi	Atmosphere
Cu to Al	—	950	0.25	1	vacuum
Cu to 316 sst	Cu	1800	2	a	vacuum
Cu to Ti	—	1560	0.25	0.7	vacuum
Cu to Cb-1%Zr	Cb-1%Zr	1800	4	a	vacuum
Cu-10%Zn to					
Ti-6%Al-6%V-2%Sn	—	900	8	a	vacuum
4340 steel[c] to					
Inconel 718[b]	—	1730	4	29	vacuum
Nickel 200 to					
Inconel 600[b]	—	1700	3	1	not reported
Pyromet X-15[b] to					
T-111 Ta alloy	Au-Cu	1100	4	30	not reported
Cb-1%Zr to 316 sst	Cb-1%Zr	1800	4	a	vacuum
Zircaloy-2 to 304 sst[d]	—	1870-1900	0.5	a	vacuum

a. Pressure is applied with differential thermal expansion tooling.

b. Tradenames

c. Outgassing of the 4340 steel at 10^{-5} torr and 1850° F for 24 hours prior to welding was critical to the formation of satisfactory welds.

d. This combination formed a liquid eutectic phase at the welding temperature. Temperature and time must be controlled to minimize the thickness of the alloy zone.

DISSIMILAR METAL COMBINATIONS

Diffusion welding is particularly well suited for joining many dissimilar metal combinations, especially when the melting points of the two metals differ widely or when the materials are not metallurgically compatible. In such cases, conventional fusion welding is not practical because it would either result in excessive melting of one of the metals or in the formation of a brittle weld zone. Diffusion welding is also suitable when the high temperatures of fusion welding would cause an alloy to become brittle or lower its strength drastically, as is the case with some refractory metal alloys. Interlayer metals are sometimes used to prevent the formation of brittle intermetallic phases between certain metal combinations.

When determining conditions and interlayer requirements for diffusion welding a particular dissimilar metal combination, the effects of interdiffusion between the two metals must be considered. Interdiffusion can cause certain problems as a result of the following metallurgical phenomena:

(1) Intermediate phase or brittle intermetallic compound formation at the interface. Selection of an appropriate interlayer can usually prevent problems associated with this effect.

(2) Low melting eutectic phase formation. This effect can be used to advantages in some applications.

(3) Joint porosity due to dissimilar rates of metal transfer by diffusion in the region adjacent to the weld interface (Kirkendall porosity). Proper welding conditions or the use of an appropriate interlayer, or both, may prevent this problem.

One problem which often exists and which is not unique to diffusion welding is the difference in the thermal expansion characteristics of the two metals. Simply stated, any combination of dissimilar metals which is heated and cooled during welding or brazing will develop shear stresses in the joint if the thermal expansion characteristics of the dissimilar metals are not identical. The severity of the problem will vary depending upon the temperature span, the net difference between the rates of expansion, the size and shape of the parts, and the nature of the bond formed between them. This becomes a design problem, in part, since distortion can result. The most severe difficulty is cracking through the joint in cases where the bond strength or ductility, or both, are low and the shear stresses are high.

Representative conditions used for diffusion welding some dissimilar metal combinations are presented in Table 10.2. Often the temperature and the time used for a particular combination are selected as part of the necessary heat treatment for one of the alloys to develop design properties for the application.

INSPECTION

Establishing the quality of a diffusion welded or brazed joint is difficult with current nondestructive testing procedures. This is due to the nature of the joints. Usually, little or no porosity exists if the joint is made with properly developed procedures. The main defect in a diffusion weld is lack of grain growth across the original interface. Efforts to distinguish intimate contact without grain growth across the interface from a perfect bond have not been successful.

Radiography, eddy current, and thermal methods are relatively unsatisfactory for inspection of most applications. Dye penetrant methods are relatively successful for edge inspections.

Ultrasonic inspection has proved the most useful for internal inspection, especially if a hairline separation exists. The sensitivity varies with the metal being tested, the ultrasonic frequency, the skill of the operator, and the degree of sophistication of the equipment. In general, defects of less than 0.1 in. diameter are difficult to locate and a practical limit of about 0.04 in. exists. With special methods and very sophisticated equipment, it has been reported that defects equivalent to 0.005 in. diameter can be detected in some metals. These approaches cannot be considered routine and they only work under special conditions.

Ultrasonic inspection can not differentiate between complete intimate contact and an actual bond. The only method available to assure complete welding or brazing is metallographic examination. Since this is a destructive test, it cannot always be performed on the part in question. Fortunately, the processes are extremely reproducible if good process control is exercised. Random destructive sampling coupled with ultrasonic inspection will provide a high confidence level. This approach has been used in production with yields in excess of 95 percent, including parts deliberately cut up for evaluation.

Metric Conversion Factors

$$t_c = 0.556 \, (t_F - 32)$$
$$1 \, ksi = 6.895 \, MPa$$
$$1 \, psi = 6.895 \, kPa$$
$$1 \, lbf = 4.448 N$$
$$1 \, torr = 133.3 Pa$$
$$1 \, in. = 25.4 mm$$

SUPPLEMENTARY READING LIST

Bartle, P. M., Introduction to diffusion bonding. *Metal Construction and British Welding Journal,* 1 (5); May 1969.

Bartle, P. M., Diffusion bonding: a look at the future. *Welding Journal,* 54(11): 799-804; 1975 Nov.

Bartle, P. M. and Ellis, C. R. G., Diffusion bonding and friction welding, two newer processes for the dissimilar metal joint. *Metal Construction and British Welding Journal,* 1 (125): 88-95; 1969 Dec.

Bryant, W. A., A method for specifying hot isostatic pressure welding parameters. *Welding Journal,* 54(12): 433s-35s; 1975 Dec.

Doherty, P. E. and Harraden, D. R., New forms of filler metal for diffusion brazing. *Welding Journal,* 56 (10): 37-39; 1977

Duvall, D. S., Owczarski, W. A., and Paulonis, D. F., TLP*Bonding: a new method for joining heat resistant alloys. *Welding Journal,* 53(4): 302-14; 1974 Apr.

Duvall, D. S., Owczarski, W. A., and Paulonis, D. F., and King, W. H., Methods for diffusion welding the superalloy Udimet 700, *Welding Journal,* 51(2); 41s-49s; 1972 Feb.

Hoppin, G. S. and Berry, T. F., Activated diffusion bonding. *Welding Journal,* 49(11): 505s-9s; 1970 Nov.

King, W. H. and Owczarski, W. A., Additional studies on the diffusion welding of titanium. *Welding Journal,* 47(10): 444s-50s; 1968 Oct.

Mohamed, H. A. and Washburn, J., Mechanism of solid state pressure welding. *Welding Journal,* 54(9): 302s-10s; 1975 Sept.

Niemann, J. T. and Garrett, R. A., Eutectic bonding of boron-aluminum structural components. *Welding Journal,* Part 1, 53 (4); 175s-84s; 1974 Apr.; Part 2, 53 (8); 351s-9s: 1974 Aug.

Niemann, J. T. and Wille, G. W., Fluxless diffusion brazing of aluminum castings. *Welding Journal,* 57 (10): 285s-91s: 1978 Oct.

O'Brien, M., Rice, C. R., and Olson, D. L., High strength diffusion welding of silver coated base metals. *Welding Journal,* 55(1); 25-27; 1976 Jan.

Schwartz, M. M., Diffusion brazing titanium sandwich structures. *Welding Journal,* 57(9); 35-8: 1978 Sept.

Signes, E. G., Diffusion welding of steel in air. *Welding Journal,* 47(12); 571s-4s; 1968 Dec.

Tylecote, R. F., *The Solid State Welding of Metals;* New York: St. Martins Press, 1968: 301-18.

Weisert, E. D. and Stacher, G. W., Fabricating titanium parts with SPF/DB process. *Metal Progress,* 111 (3): 32-7: 1977 Mar.

Wells, R. R., Microstructural control of thin-film diffusion brazed titanium. *Welding Journal,* 55(1): 20s-8s; 1976 Jan.

Witherell, C. E., Diffusion welding multifilament superconducting composites. *Welding Journal,* 57 (6), 153s-60s; 1978 June

Wu, K. C., Resistance Nor-Ti-Bond joining of titanium shapes. *Welding Journal,* 50 (9): 386s-93s; 1971 Sept.

Zanner, F. J. and Fisher, R.W., Diffusion welding of commercial bronze to a titanium alloy. *Welding Journal,* 54 (4): 105s-12s; 1975 Apr.

11

Adhesive Bonding of Metals

Chapter Committee

C. R. RONAN, *Chairman*
Rohr Industries

N. J. DeLOLLIS
Sandia Corporation

M. J. HRABONZ
3M Company

J. P. McNALLY
Rohr Industries

J. D. MINFORD
Aluminum Company of America

D. K. RIDER
Bell Telephone Laboratories, Inc.

Welding Handbook Committee Member

R. L. FROHLICH
Westinghouse Electric Corporation

11

Adhesive Bonding of Metals

FUNDAMENTALS OF THE PROCESS

DEFINITIONS AND GENERAL DESCRIPTION

Adhesive bonding (ABD) is a materials joining process in which a nonmetallic adhesive material is placed between the faying surfaces of the parts or bodies, called adherends. The adhesive then solidifies or hardens by physical or chemical property changes to produce a bonded joint with useful strength between the adherends. During some stage of processing, the adhesive must become sufficiently fluid to wet the faying surfaces of the adherends.

Adhesive[1] is a general term that includes such materials as cement, glue, mucilage, and paste. Although natural organic and inorganic adhesives are available, synthetic organic polymers are usually used to adhesive bond metal assemblies. Various descriptive adjectives are applied to the term adhesive to indicate certain characteristics, as follows:

(1) Physical form: liquid adhesive, tape adhesive

(2) Chemical type: silicate adhesive, resin adhesive

(3) Materials bonded: paper adhesive, metal-plastic adhesive, can label adhesive.

(4) Application method: hot-setting adhesive, sprayable adhesive

Although adhesive bonding is used to join many nonmetallic materials, only the bonding of

metals to themselves or to nonmetallic structural materials is covered in this chapter.

Adhesive bonding is similar to soldering and brazing of metals in some respects but a metallurgical bond does not take place. The surfaces being joined are not melted, although they may be heated. An adhesive in the form of a liquid or tacky solid is placed between the faying surfaces of the joint. After the faying surfaces are mated with the adhesive in between, heat or pressure, or both, are applied to accomplish the bond.

In general, an adhesive system must have the following characteristics:

(1) At the time the bond is formed between an adhesive and the metal, the adhesive must be either a liquid or a readily deformed solid. This is required for the adhesive to flow or deform and then conform to the surfaces of the adherends. There must be adequate adhesive to completely fill the joint.

(2) In general, the adhesive cures, cools, dries, or otherwise hardens during the time the bond is formed or soon thereafter.

(3) The adhesive must have good mutual attraction with the metal surfaces, and have adequate strength and toughness to resist failure along the glue line under service conditions.

(4) As the adhesive cures, cools, or dries, it must not shrink excessively. Otherwise, undesirable internal stresses may develop in the joint.

(5) To develop a strong bond, the metal surfaces must be clean and free of dust, loose oxides, oil, grease, or other foreign materials.

---

1. Terms relating to adhesives are defined in ANSI/ASTM D907.

(6) Entrapped air, moisture, solvents, and other gases trapped at the interface between the adhesive and metal must have a way of escaping from the joint.

(7) The joint design and adhesive must be suitable to withstand the intended service.

PRINCIPLES OF OPERATION

At some stage in its application, an adhesive for metal bonding must become fluid. It may be a solid dissolved in a solvent, a chemically active liquid, or a solid transformed to the liquid state by heat only or by heat and pressure. The liquid must wet the surface of the adherends and come into close proximity so the forces of attraction will operate.

When an adhesive is placed on a metal surface, the adjacent adhesive molecules are being attracted by neighboring molecules as well as the metal atoms or foreign matter on the metal surface. If the surface energy of the adhesive is greater than that of the adherend surface, the adhesive will not wet. To accomplish wetting, the surface energy of the metal must be greater than that of the adhesive. Cleanliness of the metal surface is very important for providing the high surface energy required for wettability. Oil or grease on the surface will seriously lower the metal surface energy and impair bonding.

The spreading of the adhesive on a metal surface is also influenced by other surface conditions and by the mobility of the adhesive. In general, a metal surface contains a microscopic series of peaks and valleys which presents a larger exposed surface area than a perfectly flat surface. This gives a high adsorption force and good capillary action.

For the liquid adhesive to spread, it must have the proper viscosity. The higher its viscosity, the greater is the probability that the adhesive will not completely fill the valleys on the metal surface but will possibly entrap gases, liquid, or vapors in them. There is a tendency during bonding for the adhesive to absorb and then expel these foreign materials at the edge of the bondline. To assist in the desorption, the adhesive can be heated or subjected to pressure.

ADVANTAGES AND LIMITATIONS

Adhesive bonding has several advantages and limitations for metals joining when compared to resistance spot welding, brazing, soldering, or mechanical fasteners such as rivets or screws.

Advantages

Bonding Dissimilar Materials. It is possible to bond dissimilar metals with minimal galvanic corrosion in service provided the adhesive layer maintains electrical isolation between the metals. Many types of adhesive formulations are flexible enough to permit the bonding of dissimilar metals with widely different coefficients of thermal expansion. Such possibilities depend, of course, upon the size of the pieces and the degree of joint strength required. A single adhesive may be used for joining a number of dissimilar metal combinations in a single assembly. Adhesive bonding also makes it possible to join metals to nonmetallic materials such as various types of plastics. Figure 11.1 shows the bonding of rigid urethane foam to sheet steel with a free-flowing, room-temperature-curing, epoxy adhesive in the fabrication of insulating panels.

Bonding Thin Gage Metals. Very thin metal parts can be adhesive bonded in any combination of adherends. Three examples of this are: (1) multiple layers of thin metal sheets can be bonded together to form electric motor laminates; (2) various metal foils can be joined to themselves or to other materials; and (3) thin-gage metal sheets may be used as sandwich panel skins.

Low Processing Temperatures. The temperatures used for the heat curing of most adhesives are between 150 and 350° F, temperatures that are below the normal soldering range. Room-temperature-curing formulations are available that provide sturdy structural bonds for service temperatures up to 150° F under humidity conditions that do not exceed 70 percent relative humidity. They can be used to join heat-sensitive components without damage. Adhesive bonding should be considered when high temperature operations would cause metallurgical or struc-

Fig. 11.1— Rigid urethane foam bonded to sheet metal

tural damage to the parts.

Some adhesives can operate at temperatures somewhat higher than their curing temperatures, which is not the case with metal solders.

Combination Bonding and Sealing. The adhesive that bonds the components may also serve as a sealant or coating to provide protection from oils, chemicals, moisture, or a combination of these. In Fig. 11.2, a room-temperature-curing adhesive is being applied to seal the ends of an antenna circuit in the handle of a marine radio. The same adhesive is also used to bond neoprene tubing and plastic-coated lead-in wires to aluminum at three points within the radio housing.

Thermal and Electrical Insulation. Adhesives can provide thermal or electrical insulating layers between the two surfaces being joined: for example, almost all mass-produced printed circuits use adhesive bonding. In this application, the adhesive used to bond the copper conductor to the base material has electrical char-

acteristics similar to those of the base material. An adhesive may also serve as an insulator between adjacent conductors.

The addition of certain metallic or carbon fillers to adhesive formulations can make them electrically conductive. Some testing before acceptance should be done under simulated service conditions because corrosion may occur in some metal structures bonded with electrically conductive adhesives that are exposed to moisture. Metal powder additives can improve the thermal conductivity of adhesives.

Uniform Stress Distribution. Joints can be designed to distribute the load over a relatively large bonded area to minimize stress concentrations. In wall panel construction, for example, metal skin sheets are bonded to metal or paper honeycombs, foamed polystyrene, or other core materials.

Smooth Surface Appearance. Adhesives can ensure smooth, unbroken surfaces without protrusions, gaps, or holes. A typical example is the

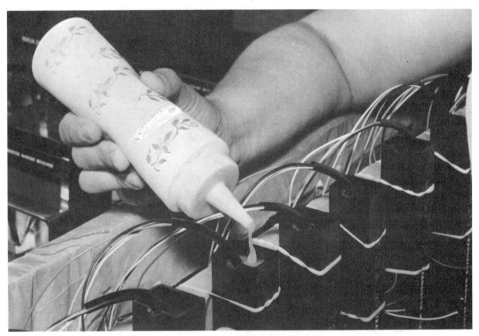

Fig. 11.2—Application of an epoxy adhesive to seal the end of a marine radio antenna

vinyl-to-metal laminate widely used in the production of television cabinets, housings for electronic equipment, and automotive trim. Figure 11.3 shows adhesive-bonded truck door panels where broad, smooth areas are required. Hood and roof stiffeners on automobiles are adhesive bonded rather than resistance spot welded to the panels to avoid marks that would be susceptible to rusting and might require filling, grinding, and polishing prior to painting.

Good Vibration and Sound Damping. Flexible adhesives can absorb shock and vibration which gives the joints good fatigue life and sound-dampening properties. The use of adhesives rather than rivets has increased joint fatigue life by a factor of ten or more, as was experienced with helicopter rotor blades. A combination of adhesives and rivets for joints in very large aircraft structures has increased the fatigue life of joints from 2×10^5 cycles for rivets alone to more than 1.5×10^6 cycles for the bonded and riveted joints. The large bonded area also dampens vibration and sound.

Weight Savings. Adhesive bonding may permit significant weight savings in the finished product by utilizing lightweight fabrications. Honeycomb panel assemblies, used extensively in the aircraft industry and the construction field, are excellent examples of lightweight fabrications. Typical panels are shown in Fig. 11.4. Not only is the honeycomb core material bonded to the metal face sheets, the honeycomb core itself is generally adhesive bonded. Although weight reduction can be important in the function of the product, it may also provide considerable cost savings in packing, shipping, and installation labor.

Simplification of Design. Adhesives often permit design simplification. In Fig. 11.5, aluminum die cast pump sections have been adhesive bonded to a steel core. Previously, the part was cast as one piece of steel, but blow holes in the casting resulted in an excessive number of rejects. Redesigning to an adhesive-bonded assembly reduced the number of rejects to near zero.

Fig. 11.3—Truck door panels of aluminum adhesive bonded to chipboard

Limitations

Adhesive bonding has certain limitations which should be considered in its applications, the most important of which are as follows:

(1) Adhesives will not support high peel loads above 250° F, not even elastomeric adhesives that have high tensile shear strengths at temperatures as high as 300° F. For applications where high peel strength is essential, some mechanical reinforcement may be necessary.

(2) Adhesives, including epoxy-phenolics that are designed for low creep at elevated temperatures, have an operational temperature ceiling of about 500° F. Some new high temperature adhesives derived from heat stable polyimides, polybenzimidazoles, and related compounds show promise for use at temperatures up to 900° F, but they are costly and difficult to process.

(3) Capital investment for equipment and tooling to process components may be high when large bonding areas and special service requirements are involved. The benefits of an adhesive-bonded joint must be balanced against the cost of autoclaves, presses, tooling, and other special equipment needed to perform the bonding operation.

(4) Process control costs may be higher than those for other joining processes. Surfaces must be properly cleaned, treated, and protected from contamination prior to bonding if the best bond durability is required. Surface preparation can range from a simple solvent wipe to multistep cleaning, etching, anodizing, rinsing, and

drying procedures which must be very carefully followed in critical structural bonding applications. Control of ambient temperature and humidity may also be necessary.

(5) To develop full strength, joints must be fixtured and cured at temperature for some time. On the other hand, mechanical fasteners provide design strength immediately and usually do not require extensive fixturing.

(6) Nondestructive inspection methods normally used for other joining methods are not generally applicable to the evaluation of adhesive bonds. Both destructive and nondestructive testing must be used with process controls to establish the quality and reliability of bonded joints.

(7) Service conditions may be restrictive. Many adhesive systems degrade rapidly when the joint is both highly stressed and exposed to a hot, humid environment.

Fig. 11.4—Adhesive bonded honeycomb panels

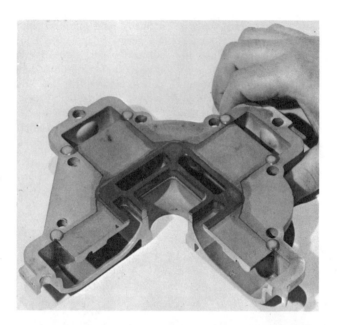

Fig. 11.5—Aluminum die cast pump sections adhesive bonded to a steel core

ADHESIVES

GENERAL DESCRIPTION

Adhesives may be either thermosetting or thermoplastic. The principal ingredients in most adhesive formulations are:

(1) A synthetic resin system
(2) An elastomer or flexibilizer
(3) Inorganic materials

Thermosetting resins are the most important materials on which metal adhesive formulations are based. Their properties can be modified for specific applications by the addition of modifying agents and fillers. Thermosetting adhesives harden or cure by chemical reactions such as polymerization, condensation, or vulcanization. These reactions occur with the addition of a hardener (curing agent) or catalyst. Heat, pressure, radiation, or other energy can accelerate the hardening or curing rate. Once they harden or cure, these adhesives cannot be remelted, and a broken joint cannot be rebonded by heating. Depending upon composition, thermosetting ad-

hesives may soften or weaken at high temperature and ultimately decompose.

Thermoplastic resins are long chain molecular compounds that soften upon heating and harden upon cooling. They undergo no chemical change upon heating, so the cycle can be repeated. However, they will oxidize and break down at excessively high temperatures. Many thermoplastic resins can also be softened at room temperature with organic solvents. They harden again as the solvent evaporates. Limited resistance to heat, solvents, and load-induced stresses makes thermoplastic resins generally unsuitable as structural adhesives. However, some thermoplastic resins or elastomers are combined with thermosetting resins such as epoxies and phenolics for improved flexibility, peel strength, and impact resistance.

Flexibilizers or elastomers are added to adhesive formulations to add resiliency, improve peel strength, and increase resistance to shock and vibration. Most nonvulcanizing, elastomeric-based adhesives may be considered as thermo-plastics and most vulcanizing formulations as thermosetting types.

Inorganic materials are added as fillers to improve the mechanical and physical properties of the adhesives. Fillers can add greatly to the stability of bonded joints by reducing shrinkage and thermal expansion and by increasing the modulus of elasticity of the adhesive.

TYPES

Solvent

The solvent type contact adhesives are predominantly elastomeric thermoplastics pro-duced as solutions. They achieve their bond strength upon removal of the solvent. The liquid adhesive is applied to the adherend surfaces and, for metal adherends, time is allowed for the solvent to evaporate. The adhesive-coated sur-faces are then joined under contact pressure. Sometimes heat is applied to fuse the coated surfaces after drying.

Hot Melt

The hot-melt adhesives are thermoplastics. After the adherends have been coated with ad-hesive and mated, heat and pressure are applied to the assembly. The joint is then cooled to so-lidify the adhesive and achieve a bond. These adhesives are not normally used for structural applications.

Pressure Sensitive

Pressure sensitive adhesives are formula-tions that instantly provide a relatively low-strength bond upon the brief application of pres-sure. They may be applied to any clean, dry surface. Since they are capable of sustaining only very light loads because of retention of their flow characteristics, they are not considered structural adhesives.

Chemically Reactive

The chemically reactive adhesives consist primarily of thermosetting resins in liquid and solid forms, including films and tapes. They are activated either by the addition of a catalyst or

hardener, or by the application of heat. Bond strength is achieved from the chemical reaction that takes place during the curing time. Catalysts or hardeners may be incorporated by the adhe-sive manufacturer or may be added by the user just prior to application. Such formulations gen-erally must be used within a prescribed period of time after mixing to avoid premature setting. Other chemically reactive, one-part adhesives contain a latent hardener that activates and cures the adhesive at elevated temperatures. Shelf life at room temperature for one-part adhesives may be limited to a few weeks unless it is prolonged by refrigerated storage. The time required for bonding is approximately 24 hours for room-temperature-curing adhesives and from a few seconds to several hours for heat-curing formu-lations.

Anaerobic

Anaerobic adhesives are unique in that they are shelf-stable, solventless, ready-to-use formu-lations that cure at room temperature. Their cure is inhibited by the presence of air (oxygen) in the package and during application. Once the joint is assembled and air excluded from the liquid adhesive, curing begins. Newer heat-curing anaerobics are capable of higher bond strengths than the room-temperature-curing products.

STRUCTURAL ADHESIVES

The ultimate objective of a structural adhe-sive is to create a bond that is as strong as the materials it joins. Since this goal is not always attainable, a structural adhesive can be defined as one that is used to transfer required loads between adherends in a structure for its life ex-pectancy when exposed to its service environ-ment. There are two general types of structural adhesives, both of which are thermosetting. They are the phenolic resin-base and the epoxy resin-base adhesives.

Phenolic resins are modified with thermo-plastics or elastomers for structural adhesive applications. These modified phenolics are avail-able as solutions in organic solvents and also as films, both supported and unsupported. Such adhesives feature high peel strengths, and tensile and shear strengths in the range of 3000 to 5000 psi.[2]

The epoxy resins combine the properties of excellent wetting action, low shrinkage, high tensile strength, toughness, and chemical inertness to produce adhesives noted for their strength and versatility. Unlike phenolic adhesives, epoxies do not form volatile products during curing. They can be applied in liquid form without a solvent carrier. Because of this, volatile entrapment is minimized. Only low pressure is necessary to maintain intimate contact between the adherends during bonding, resulting in greatly simplified equipment requirements.

Epoxy adhesives are available as free-flowing liquids, films, powders, stocks, pellets, and mastics. This variety of forms permits considerable latitude in the selection of application technique and equipment. Fillers or plasticizers may be added to minimize stresses that can develop when the adhesive and adherends have different coefficients of thermal expansion.

The wide choice of hardeners available for epoxy formulations offers curing cycles ranging from a few seconds at elevated temperatures to several minutes or hours at room temperature. However, the heat-resistant formulations require high temperature cures.

A large selection of hardeners makes it possible to produce an adhesive system having the best possible balance of properties for a given application. Figure 11.6 illustrates the effect of various hardeners on the bonding properties of an intermediate strength epoxy adhesive as compared to a specially formulated high-strength epoxy. This evaluation shows that different strength properties can be imparted to a single adhesive simply by changing the hardener. Other properties will also vary, depending upon the hardener selected.

The epoxy-based adhesives, as a general class, feature high shear and tensile strengths, but low creep and peel strengths. However, specially formulated epoxy adhesives provide improved peel strengths, especially those modified with nylon or carboxylic functional, nitrile copolymer rubber. Modified epoxy adhesives feature joint shear strengths in excess of 7000 psi and high peel strengths.

An epoxy-phenolic resin cured with dicyandiamide is an adhesive that performs well in the 400 to 500° F temperature range. It will produce a bond having good shear and creep properties at 500° F, but relatively poor peel resistance when cured in a representative cycle of 2 hours at 350° F and 150 psi pressure. Better "all-around" strength properties are often obtained by using this adhesive in conjunction with a specially formulated primer having a high peel strength. The elastomer-modified phenolics, which require curing at temperatures from 300 to 350° F and pressures up to 100 psi, offer good resistance to heat and water in service.

Another class of high-temperature-resistant resin systems, polybenzimidazole (PBI) and polyimide (PI), is available for structural adhesives for service temperatures in the range of −425 to 1000° F. These systems have been evaluated for bonding aluminum, stainless steel, titanium, beryllium, and other metals as well as reinforced plastics with excellent results. Shear strengths of over 2000 psi were obtained with stainless steels using a 1/2-in. lap joint. After exposure at 1000° F for 15 minutes, lap joints had shear strengths in excess of 1000 psi with stainless steel, titanium, and beryllium adherends. These adhesive systems have also been successfully utilized for fabricating stainless steel honeycomb sandwich constructions.

Although structural bonding has been successfully applied in aerospace applications for more than 25 years, stress corrosion problems were detected in bonded structures that were subjected to adverse service conditions. The combination of continuous or cyclic stress and a hot, humid atmosphere drastically affected the bond durability. Under this conditions, bonds have degraded and failed at stresses as low a 20 percent of the ultimate strength of the adhesive. The mechanism of this stress corrosion effect is not fully understood and is still under investigation. However, some generalizations can be made. Time to failure decreases as the applied

2. The mechanical properties stated for these adhesive systems pertain to bonded structures which have not been stressed prior to testing and are stored in a low or normal relative humidity condition (i.e., less than 70 percent relative humidity).

stress level increases. Initial failure occurs at the bond line between the adhesive and adherend when moisture permeates the adhesive. The moisture reacts with any oxide on the surface of the adherend by hydration. This weakens the oxide layer. Room-temperature-cured adhesives degrade more rapidly in a hostile service environment than heat-cured adhesives.

FORMS AND APPLICATION METHODS

Industrial adhesives are available in a number of forms:

(1) Liquids, ranging in viscosity from free-flowing to thick syrups

(2) Pastes
(3) Mastics
(4) Solids
(5) Powders
(6) Supported and unsupported films

The method used for application of a particular adhesive should be selected after careful consideration of these factors:

(1) Available forms of the selected adhesive
(2) Methods available for applying the various forms
(3) Joint designs and order of assembly
(4) Production rate requirements
(5) Equipment costs

Adhesives may be applied with rollers, brushes, caulking guns, trowels, spray guns, or

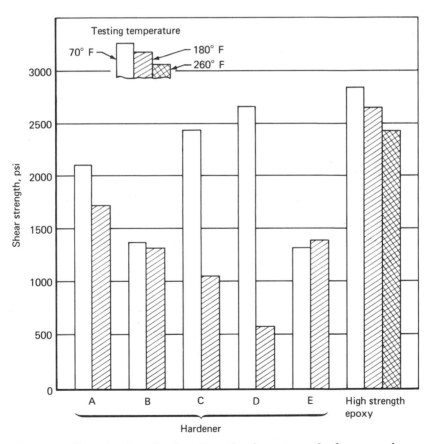

Fig. 11.6—Effect of various hardeners on the shear strength of an epoxy base adhesive compared to a high-strength epoxy adhesive

by dipping. The form of the adhesive and method of application must be compatible.

Liquid Adhesives

For small assembly work, liquid adhesives are commonly applied by brush, short-napped paint roller, or dipping. The more viscous liquids are often applied by trowel, extrusion gun, or plastic squeeze bottle. Polyethylene nozzles of bottle or tube containers should not be rubbed across prepared surfaces such as etched aluminum. This action may deposit a wax-like coating to which the adhesive will not adhere. Small applicators, resembling a ballpoint pen or a hypodermic needle, may be used to deposit very narrow glue lines for spot application. The silk screen process is also useful for applying an adhesive to selected areas. Automatic dispensing machines, which simplify the proportioning and the mixing of many two-component formulations, are also available.

For large areas such as curtain wall panels, liquid adhesive may be applied by spraying, flow coating, roller coating, or troweling. Depending upon the application method, consideration must be given to the viscosity and working (pot) life of the adhesive formulation.

Paste and Mastic Adhesives

Pastes and mastic adhesives may be applied with a smooth or serrated trowel, knife roller, or an extrusion device.

Some paste formulations contain a thixotropic additive that inhibits sag or flow during application and cure. This feature permits their use on vertical or overhead surfaces and may eliminate or significantly reduce the need for special clean-up operations.

Solid Adhesives

One method for applying a solid adhesive is to first heat the substrate to a temperature slightly above the melting point of the adhesive and then add the adhesive to the hot surface as it melts. Some rod and powder forms of epoxy adhesives are applied in this manner. Specially developed flame spray guns can also be used, but this method may require powders with a particle size within narrow tolerances. Also, care must be taken to avoid overheating the adhesive during application.

Film Adhesives

Adhesives in film or tape form are extremely simple to use and produce a bond line having relatively constant thickness and coating weight per unit area. These are important factors in most bonding applications. Films made from adhesives that are thermoplastic, thermosetting, or pressure sensitive are supplied in rolls or sheets that can be blanked or cut to the required shape with scissors. Film adhesives are particularly useful for bonding large areas such as honeycomb sandwich panels. Special films are available for use as adhesive backing on items such as nameplates and decals.

Generally, the film adhesive is placed in position between the adherends and activated with heat or solvent. In the case of pressure-sensitive film, pressure is applied to accomplish bonding. For applications demanding high strength at elevated temperatures, the films usually require both heat and pressure to create the bond.

Solvent reactive and pressure sensitive films are intended primarily for bonding large sheets where only nominal pressure is required. These types of films do not need special heating equipment. They are particularly useful for short-run production and for bonding parts at room temperature.

Duplex bonding films, which combine the properties of elastomeric adhesives and epoxies, are sometimes used for honeycomb sandwich constructions. The elastomeric side, usually a nitrile phenolic, is bonded to the facing to provide peel resistance. The epoxy side forms fillets around the cell walls, and this increases the effective bonding area and the resulting joint strength.

PRIMERS

Certain service conditions may require the use of primer for improved corrosion resistance, flexibility, shock resistance, or peel resistance of the adhesive bond. Primers may also be used to wet or penetrate the substrate or to protect a treated surface of a substrate prior to the application of adhesive. Most primers are low-viscosity solutions commonly applied by spraying. Brush applications may be satisfactory when relatively small areas are to be coated. In some cases, roll coating or dipping may be employed. Several primer coats may be required to build up the desired thickness, particularly if the adherends are porous. Air drying and full or partial curing are generally necessary prior to further processing.

ADHESIVE SELECTION

The selection of the proper adhesive for production depends basically upon the answers to four key questions:

(1) What materials are being joined?

(2) What are the service requirements?

(3) What method of adhesive application is most suitable?

(4) Are bonding costs competitive with other joining methods?

The service requirements of the completed assembly must be studied thoroughly. Several factors to be considered in describing the bonding application are:

(1) Type of loading

(2) Operating temperature range

(3) Chemical resistance

(4) Weather and climatic resistance

(5) Flexibility

(6) Differences in thermal expansion rates

(7) Odor or toxicity problems

(8) Color match

Once the service requirements are known, adhesive systems with good durability potential can be selected. The desired form of the adhesive and method of application can then be chosen based on availability of production equipment and scheduling requirements.

In adhesive selection, the tendency to overdesign must be avoided. Requirements for higher strength or greater heat resistance than is actually required for the specific application may exclude from consideration many formulations that are adequate for the job, and perhaps less costly or easier to handle in production.

In adhesive selection for a specific application, there are certain physical properties of the adhesive itself and mechanical properties of bonded joints that should be considered. These properties pertain to the behavior of an adhesive from the time that it is made until the bond is accomplished, as well as its performance in service. Table 11.1 lists some of the properties and the applicable ASTM Standards.

Table 11.1
ASTM standards for determining adhesive properties

Property	ASTM Designation(s)
Physical	
Aging	D1151, D1183, D3236
Chemical	D896
Corrosivity	D3310, D3482
Curing rate	D1144
Flow properties	D2183
Storage life	D1337
Viscosity	D1084
Volume resistivity	D2739
Mechanical	
Cleavage strength	D1062, D3433
Creep	D1780, D2293, D2294
Fatigue	D3166
Flexural strength	D1184
Impact strength	D950
Peel	D903, D1781, D1876, D2918
Shear	D1002, D2182, D2295, D2919, D3528
Tensile	D897, D1344, D2095

JOINT DESIGN

To incorporate as many of the advantages of adhesive bonding as possible, joint design should be a part of the early stages of product planning. If bonding is being considered as part of a redesign program, structural adhesives should not be substituted directly for other joining methods. The joint should be redesigned to take advantage of adhesive bonding.

Although the primary objective is a strong assembly capable of meeting service requirements, proper joint design can often lead to other cost-saving benefits. Through good design, it may be possible to achieve satisfactory results with an economical adhesive formulation, to utilize a simple bonding process, and to minimize the quality control steps needed to ensure reliability.

Joint design often influences the form and characteristics of the bond line. The design must provide space for sufficient adhesive and a means of getting the adhesive into the joint area. It must also play an important role in determining the thickness of the bond line. Thin bond lines, in the range of 0.003 to 0.005 in., are better than thick lines for high-strength adhesives for the following reasons:

(1) Less likelihood that the adhesive will creep or flow

(2) Smaller deformation under pressure

(3) More resistance to cracking, especially with hard adhesives

(4) Less danger of internal stresses and gross porosity in the adhesive

(5) Greater saving in adhesive

When considering a new design or a redesign for adhesive bonding, three rules that should be observed are:

(1) The design loading should produce shear or tensile loading on the joint; cleavage or peel loading should be minimized.

(2) The joint design should ensure that the static loads do not exceed the adhesive plastic strain capacity.

(3) If the joint will or might be exposed to low cyclic loads, the joint overlap should be increased sufficiently to minimize the possibility of creep in the adhesive.

These rules may be difficult to achieve in practice. Some stress concentrations are unavoidable, and it is difficult to design a joint that will be stressed in one mode only.

The four main types of loading are illustrated in Fig. 11.7. An adhesive bonded joint performs best when loaded in shear: that is, when the direction of loading is parallel to the plane of the faying surfaces. With thin-gage metal bonds, joint designs can provide large bond areas in relation to the metal cross-sectional area. This makes it possible to produce joints that are as strong as the metal adherends.

The relationship between joint strength and overlap length for a double-shear lap joint is shown in Fig. 11.8. The joint strength and overlap distance are proportional up to some limit (point A on Fig. 11.8). Then, the unit increase in strength decreases as the overlap distance increases. Beyond some overlap (point B), the failure load does not change significantly with overlap distance.

Figure 11.9 indicates the shear stress distribution across lap joints caused by load P with short, medium, and long overlaps. With short overlap, Fig. 11.9 (A), the shear stress is uniform along the joint. In this case, the joint can creep

Tensile

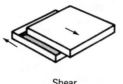
Shear

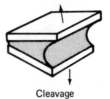
Cleavage

Peel

Fig. 11.7—Four principal types of loading

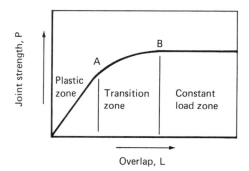

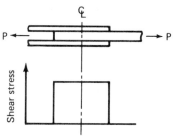

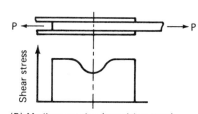

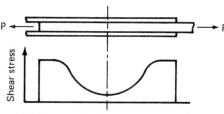

Fig. 11.8—Relationship between joint strength and overlap in shear

to a slight bending moment that creates a cleavage load. Another example is a single lap joint that is designed to withstand expected shear stress but sees cleavage or peel loads when the joint rotates slightly as the load forces tend to align, as shown in Fig. 11.10. These problems can usually be minimized by selecting an adhesive designed to carry the type of loading expected and by employing the proper joint design.

Several of the more common types of joints for sheet metal are shown in Fig. 11.11. A butt joint design, shown in Fig. 11.11(A), is not

under load with time and failure may occur prematurely. When the overlap exceeds some value, the adhesive at the ends of the joint carries a larger portion of the load than the adhesive at the center. Therefore, the shear stress at the center is lower, as shown in Fig. 11.9 (B), and the likelihood of creep is decreased. With long overlap, Fig. 11.9 (C), the portion of the joint overlap that sees low shear stress is a greater percentage of the total, and creep potential is minimized. The joint overlap for minimum creep will depend upon the mechanical properties of the base metal, the adhesive properties and thickness, the type of loading, and the service environment.

Difficulties may arise when cleavage or peel type loading is present. Cleavage loading produces nonuniform stress across the joint, and this causes failure to initiate at the edge of the adhesive. Obviously, such a joint is considerably weaker than the same bonded area under uniform shear or tensile stress.

The situation is even more critical when the adhesive is subject to peel type loading. A very narrow line of adhesive at one edge of the joint must withstand the load. Peel loading produces failure at only a fraction of the tensile load needed to rupture a bond of the same area.

As noted earlier, unidirectional loading is rarely accomplished. Most joints are subjected to loads which combine cleavage or peel loading with tension or shear stress in the bond. One example is a straight butt joint that is designed to be stressed strictly in tension but is subjected

(A) Short overlap (plastic zone)

(B) Medium overlap (transition zone)

(C) Long overlap (constant load zone)

Fig. 11.9—Change in shear stress distribution with overlap for constant load, P

P

P

Fig. 11.10 — Lap joint rotation as a result of loading

recommended. Cleavage loading may develop if the applied loading is eccentric. A scarf joint (Fig. 11.11B) is a better design because the bonded area can be greater than with a butt joint. Cleavage stress concentrations at the edges are minimized by the tapered edges of the adherends. Although widely used in wood bonding, this configuration is difficult to make in metals with regard to alignment and pressure application during curing.

The single lap joint, Fig. 11.11(G) is probably the most commonly used type and is adequate for many applications. The beveled lap joint, Fig. 11.11(H), has less stress concentration at the edges of the bond because of the beveled edges. The thin edges of the adherends deform as the joint rotates under load, and this minimizes peel action.

If joint strength is critical and the components are thin enough to bend under load, a joggle lap joint, Fig. 11.11(I), is better. The load is aligned across the joint and parallel with the bond plane, thus minimizing the possibility of cleavage loading.

If sections to be bonded are too thin to permit edge tapering, a double-strap joint, Fig. 11.11(E), will give good results. The best design is the beveled double-strap joint, Fig. 11.11(F), with its tapered straps.

Adhesive bonding can be used to advantage on extruded, cast, or machined components. The butt joints shown in Fig. 11.12 can easily be incorporated into machined or extruded shapes

that are to be assembled by adhesive bonding. The tongue and groove joint not only aligns the load-bearing interfaces with the plane of shear stress but also provides good resistance to bending. The landed-scarf tongue and groove joint offers production advantages. Its configuration automatically aligns the parts to be mated, controls the length of joint, and establishes the thickness of the glue line. It is a good design for an assembly that will see high compressive forces, and it offers a clean appearance.

Corner and T-joint designs are shown in Fig. 11.13. The use of beveled or tapered reinforcing members requires a cost analysis to determine if the improved joint properties are justifiable. Joints requiring machined slots or complex corner fittings are seldom of interest in sheet metal designs.

Adhesive bonding is also useful for tube joints, a number of examples of which are shown in Fig. 11.14. Large bonded areas give very strong joints with clean appearance. Processing may be complicated with some designs. During assembly of designs (A) and (B), the adhesive may be pushed out of the joint. Design (C) partially overcomes this problem. Adhesive in the corners is forced into the joint by a positive pressure filling action during assembly. Tapered or scarfed tubular joint designs (D), (E), and (F) will produce a positive pressure on the adhesive during assembly to completely fill the gap, but they are costly to produce. Design (G) shows a tubular sleeve joint that can be filled by injecting the adhesive under a positive pressure through a hole in the sleeve. This technique results in completely filled and bonded joints at reasonable fabrication cost.

Adhesive bonded assemblies comprise over 50 percent of the total area of at least one modern airplane. They include about 400 major assemblies including sections measuring 3 by 13 in., tapered spar caps 408 in. long, and panels measuring up to 52 by 192 inches. Bonded stiffeners are used on all single curvature panels forming the fuselage skin. Use of bonding resulted in an appreciable reduction of weight because areas of high stress concentration can be reinforced locally with built-up laminations of material. Extensive testing has

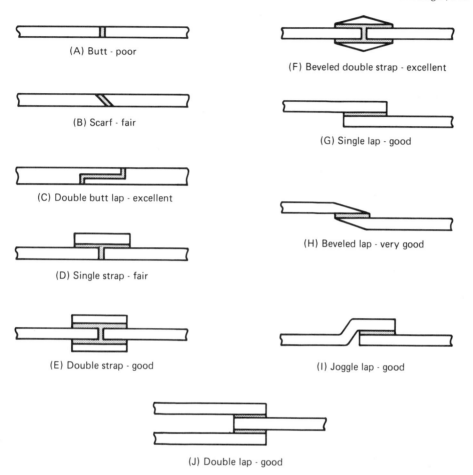

Fig. 11.11—Adhesive bonded joint designs for sheet metal

indicated that adhesive bonded joints may have up to 20 times the fatigue life of conventional riveted joints.

The automotive industry is a large user of adhesives. Nonstructural uses are common; load-bearing bonded assemblies are designed to be reliable throughout the expected life of the vehicle. Automatic transmission bands and brake linings are major applications of adhesive bonding. Double shell panels are bonded with a high-strength vinyl plastisol adhesive. Adhesive bonding reduces the number of subassembly details in a panel by about 50 percent, provides a smooth exterior surface, reduces noise level, and improves corrosion resistance.

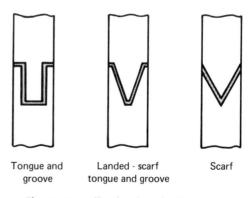

Tongue and groove   Landed - scarf tongue and groove   Scarf

Fig. 11.12—Adhesive bonded butt joint designs for machined or extruded shapes

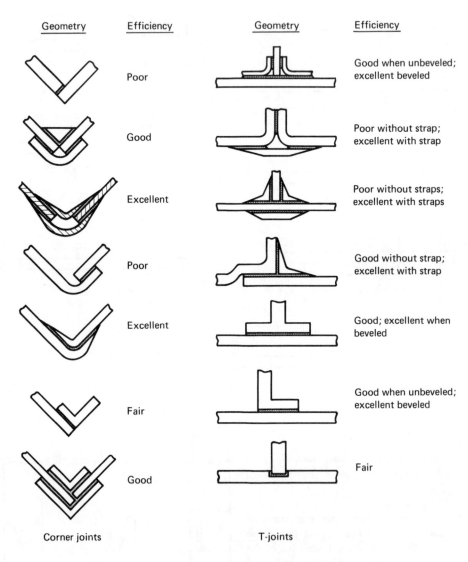

Geometry	Efficiency	Geometry	Efficiency
	Poor		Good when unbeveled; excellent beveled
	Good		Poor without strap; excellent with strap
	Excellent		Poor without straps; excellent with straps
	Poor		Good without strap; excellent with strap
	Excellent		Good; excellent when beveled
	Fair		Good when unbeveled; excellent beveled
	Good		Fair
Corner joints		T-joints	

Fig. 11.13—Corner and T-joint designs for adhesive bonding

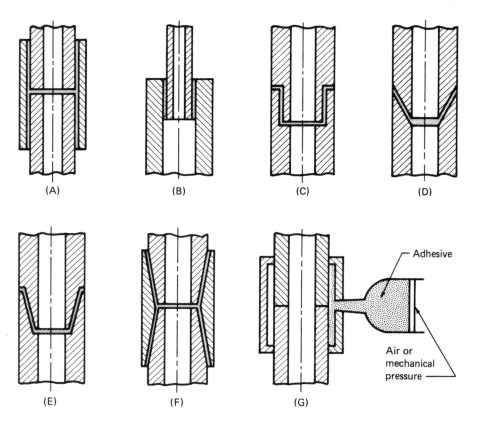

Fig. 11.14—Tubular joint designs for adhesive bonding

SANDWICH CONSTRUCTION

The greatest advantages of structural adhesive bonding are probably realized in sandwich construction. First used in the aircraft industry to meet the demand for a high stiffness-to-weight ratio, sandwich panels are now widely used throughout industry. Their characteristics make them equally valuable in the fabrication of walls, truck bodies, refrigerators, cargo pallets, and a great many other commercial applications.

The sandwich skins, or facings, behave approximately as membranes stabilized by a lightweight core material that transmits shear between the skins. Basically, a high stiffness-to-weight ratio is achieved by placing the two load carrying skins as far from the neutral axis as possible with the lightweight core bonded to them. In a sense, the face sheets perform the same functions as the flanges of an I-beam, and the core performs as the beam web.

The selection of skin and core materials for sandwich construction is dictated by service requirements and economic considerations, including the cost of fabrication and materials. Many different skin and core materials may be used as sandwich components. Sheets of metal, plastic, wood, and fire-resistant inorganics are all commonly used as skin materials. Core materials are of the following three basic types:

(1) Solid types such as hardwood, balsa, cement-asbestos board, or metal

(2) Honeycomb or corrugated types made of various materials, typically metal foils, resin-impregnated paper, or reinforced plastics

(3) Open- or closed-cell foamed materials such as polystyrene, polyurethane, polyisocyanurate, and glass

Metal foil honeycomb cores are available with and without perforations in the cell walls. The perforations permit equalization of pressure in the completed panel and also vent gases produced by some types of adhesives during bonding. Cores are also available in truss and waffle configurations that are produced by corrugating and folding or by pressing the material to shape between matched dies.

Expanded plastic cores may be supplied in a preformed state that requires a separate bonding step, or the plastic may be foamed in place and thus simultaneously formed and bonded to the skins of the sandwich.

Special consideration should be given to selecting the material combination that will produce the optimum composite structure for a specific application: for example, a honeycomb core between thin metal skins can provide high strength and low weight; a foamed plastic core can be used to assure high thermal insulation with almost any skin material.

Several basic types of adhesives with very similar properties are used in sandwich manufacture. These basic formulations are modified to meet specific environmental conditions. A careful study of the design, environmental factors, and material properties should be made before selecting the adhesive.

SURFACE PREPARATION

Frequently, the weakest link in an adhesive-bonded joint is the interfacial bond between the adhesive and the adherends. Exactly the right adhesive may have been selected for the job, the joints may have been designed properly, and correct application and cure procedures and equipment used. However, if the adherends are not properly cleaned and prepared to receive the adhesive, the bond will fail. Surfaces should be cleaned by procedures that ensure that the bond between the adhesive and metal surfaces is as strong as the adhesive itself. Failure should occur in the adhesive rather than at the bond line when the joint is tested under simulated service conditions.

The degree of surface preparation depends chiefly upon the nature of the material and also, to some extent, on the service requirements, bonding cycle, and probable nature of contaminants. For some less critical applications, a solvent wipe or washing in a detergent solution may prove to be adequate. Care should be taken,

however, to remove all cleaning agents from the surface by rinsing and drying thoroughly prior to the application of the adhesive. However, for the best joint performance, the surfaces must be prepared using procedures that will provide the best bond between the adhesive and adherend.

It is equally important to avoid recontamination of the clean surfaces during processing. Components should be handled with clean gloves, tongs, or hooks, and all contact with the bonding area should be avoided. Priming, bonding, or a combination of these should be accomplished as soon after surface cleaning as possible. In the interim, parts should be stored in a clean, dry place.

METAL PREPARATION

Metal surfaces may be cleaned by chemical means or mechanical abrasion. These are the two basic methods, but some variation in procedures is advised by various adhesive manufacturers. Metal faying surfaces should be free from oxide scale, deep scratches, burrs, and other irregularities. Cleaning should be done after machining, heat treating, welding, sandblasting, abrading, deburring, polishing, or similar treatment that might leave foreign material on the metal surface.

The surface preparation procedure should be reproducible, easily controlled, and production oriented if it is to be economical. In addition, it should satisfy the following requirements:

(1) Remove all contaminants from the surface

(2) Make the surface chemically receptive to the adhesive or primer and give satisfactory wetting characteristics

(3) Prevent poorly adhering or low-strength compounds from forming on the adherends

(4) Remove minimum amounts of metal

(5) Not cause embrittlement or corrosion, nor form a surface prone to environmental attack

Surface preparation methods for aluminum alloys, stainless steels, carbon steels, magnesium alloys, titanium alloys, and copper alloys are given in ASTM D2651, *Standard Recommended Practice for Preparation of Metal Surfaces for Adhesive Bonding.* The best corrosion-resistant surface preparation for aluminum alloys is stated in SAE Aerospace Recommended Practice (ARP) No. 1524.

To obtain optimum strength characteristics on aluminum, surface preparation is usually done in steps that include:

(1) Vapor degreasing

(2) Drying

(3) Chemical cleaning such as an alkaline precleaning and rinse followed by an acid etch

(4) Anodizing

(5) Careful rinsing in clean water

(6) Air drying, forced drying, or a combination of these

The vapor degreasing solution should be checked periodically for oily contaminants and decomposition products, and the solvent changed when necessary. Solvent degreasing is rarely recommended, but if it is used, the adherends should only be wiped with clean cloths or disposable tissues.

The recommended composition of the bath for chemical cleaning varies from one manufacturer to another. It depends upon the kind of metal being treated.

Forced drying in an oven after water rinsing is preferred since this reduces the possibility of recontamination from dust and impurities when air drying at room temperature. Forced drying temperatures must be selected judiciously. Temperatures that are too high can adversely affect the surface condition.

Mechanical abrading is sometimes used to clean metal surfaces and increase the effective bonding area by roughening. Grinding, filing, wire brushing, sanding, and abrasive blasting are some methods. Abrasive cleaning is not usually as effective as chemical treatment, but it may be adequate in certain applications. It is best to follow abrading with degreasing in case the abrading material was contaminated.

For magnesium, one of the corrosion-inhibiting pretreatments developed specifically for this metal should be used. It is possible to clean magnesium mechanically, but care must be taken to prevent the exposure of any magnesium dust to an open flame. Mechanical pretreatments should not be used for structural magnesium joints or for joints that will be subjected to severe environments. Some treatments for magnesium have been found to be effective only with certain adhesives. Certain corrosion-inhibiting pretreatments produce a weak surface layer that fails before the adhesive; these are not satisfactory for structural applications. If such pretreatments are used, the joints should be reinforced with mechanical fasteners.

A thorough check of treating solution composition should be made periodically. Special consideration should be given to the rate of metal processing over a given period of time. Failure to control the concentration of strong acid or alkali solutions may result in excessive metal loss. Aluminum and magnesium alloys are more reactive than stainless steel and titanium alloys, and exposure of these metals to high concentrations of acid or alkali may affect adhesion characteristics with certain kinds of adhesives. The time between composition checks depends upon the rate of exhaustion of the treating solution.

PREPARATION OF OTHER MATERIALS

Rigid plastics can be lightly sanded to reduce gloss and remove mold release compounds. Then they should be wiped and flushed with an oil-free solvent. Certain types of plastics, such as fluorocarbon isomer and polyethylene, are difficult to bond and may require chemical treatment.

Glass is easily cleaned by wiping with a suitable solvent. Joint durability can be greatly enhanced, particularly in moist environments, by first cleaning with a laboratory glassware cleaning solution or with a 30 percent hydrogen peroxide solution and then priming the surfaces with a silane finish.

INSPECTION OF PREPARED SURFACES

The affinity of a clean metal surface for water is the most common test for a chemically clean surface. It is called the water-break test. If clean water from the rinse bath spreads smoothly over the metal surface as it runs off, it indicates that the surface is clean. If it collects in droplets, there is probably a thin film of oil on the surface. The oil should be removed completely, and the water-break test repeated.

If water drops placed on a flat, dry, treated surface spread rapidly and uniformly, this indicates that the surface is free of oil or grease. If the contact angle of the water drop is low (10 degrees or less), the surface has been cleaned adequately to remove greases, oils, and other nonpolar contaminants. When contact angle measurement is used as a quality control device for cleanliness, inspection should be made immediately after the metal has been dried. If the water remains in droplet form, the surface has not been suitably prepared.

ASSEMBLY AND CURE

The procedures and equipment used for assembly and cure of bonded components depend upon:

(1) The type of adhesive used

(2) The type, size, and configuration of the assembly

(3) The service requirements of the completed assembly

ASSEMBLY

Drying time is an important factor when solvent-dispersed adhesives are used. Since this time varies with different formulations, it is essential that the adhesive manufacturer's recommendations be followed. Solvent evaporation rate may be increased by moderate heating with infrared lamps, a hot air oven, or other methods.

If there is sufficient porosity in one component to allow the solvent to escape, the parts may be mated during the drying time. In any case, the assembly must not be heated for curing until the solvent has evaporated. It is also essential that coated parts be mated before the tack range of the adhesive has expired.

Parts may be mated immediately after they are coated with chemically reactive adhesives. Mating may be delayed, but it must be done before the adhesive starts to "body" or thicken excessively.

FIXTURING

Provision should be made for positioning the components for mating and holding them in place while the adhesive cures or sets. Assembly fixtures are frequently used for positioning. They may be simple jigs or self-contained equipment with provision for applying pressure or heat, or both. The fixture design depends upon the amount of heat and pressure needed to cure the adhesive and the size and configuration of the assembly.

Fixturing is particularly important when a

contact adhesive is used. Care should be taken to align the parts accurately before they are mated since a strong bond is created instantly upon contact of the two coated surfaces. The use of release sheets, untreated kraft paper for example, is often helpful to avoid premature contact. Positioning may not be so critical with some formulations of less aggressive tack if the assembly can be slightly adjusted after mating without damage to the bond.

The fixture should properly position the parts to meet assembly tolerances and glue line thickness requirements. It should be lightweight for ease of handling and heat transfer. A heavy fixture presents a large heat sink which may retard heating and cooling rates. This may be detrimental for some adhesive systems. Nevertheless, the fixture must be strong enough to maintain dimensions under the curing conditions for the assembly. The fixture material also affects heating and cooling rates: for example, a material of low thermal conductivity will retard heating and cooling. The expansion rate of the fixture material should nearly match that of the assembly to minimize part distortion and subsequent stressing of the adhesive.

Pressure-sensitive tape may be used to hold parts in position if it can withstand the curing temperature. Tapes are particularly useful with epoxy formulations that cure at room temperature or slightly warmer and require only moderate pressures.

Adhesive bonding may be combined with resistance welding or mechanical fasteners to improve the load carrying capacity of the joint. The adhesive is applied to the adherends first. Then the components are joined together with spot welds or mechanical fasteners to hold the joints rigid while the adhesive cures. Figure 11.15 illustrates typical design combinations. These techniques significantly reduce or eliminate fixturing requirements and decrease assembly time when compared to conventional adhesive bonding methods.

PRESSURE APPLICATION

With certain adhesive formulations, it is necessary to apply and maintain adequate pressure during cure to:

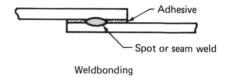

Weldbonding

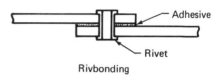

Rivbonding

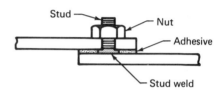

Fig. 11.15—Adhesive bonding in combination with resistance welds and mechanical fasteners

(1) Produce a uniformly thin glue line over the entire bonded area for optimum strength characteristics

(2) Facilitate flow or spreading of viscous adhesives

(3) Counteract any internal pressure caused by the release of volatiles

(4) Overcome minor imperfections in the faying surfaces

(5) Compensate for solvent loss and dimensional changes

Pressure may be applied to the joint by several methods which include the following:

(1) Dead weights such as bags of sand or shot

(2) Mechanical devices such as clamps, wedges, bolts, springs, and rollers

(3) Inflated tubes

(4) Air pressure bearing on the assembly located in a flexible, evacuated bag

(5) Mechanical or hydraulic presses

(6) Autoclaves

Inflated tubes are used in conjunction with a rigid backing fixture. When inflated, the tube presses uniformly along the bond line. Ambient air pressure is adequate for some applications.

It is applied by enclosing the assembly in a thin, air-tight bag and then evacuating the bag. Autoclaves are used in a similar fashion in that the assembly is placed in a thin, gas-tight bag vented to ambient pressure. The bagged assembly is placed in an air-tight chamber which then is pressurized to several atmospheres. The pressure forces the bag to conform to the part and transmits the pressure to the assembly.

Phenolic-based adhesives generally require curing pressures in the 300 psi range, although adequate bonds may be attained with pressures as low as 50 psi.

Flat panels coated with a neoprene contact bond adhesive are generally mated by passing them through rollers under as much pressure as the components will withstand without crushing. A weighted hand roller or other pressure device can also be used.

For sandwich panel fabrication, the upper pressure limit is governed by the compressive strength of the core material. The lower limit depends upon the minimum requirements of the adhesive formulation. For sandwich panels containing solid inserts or edgings, special fixtures may be used to apply higher pressure at the specific locations.

Throughout the curing cycle, pressure should be as uniform and constant as possible over the entire bond area. If necessary, irregular surfaces can be built up with pads of compressible material. In some cases, soft rubber pads are used to compensate for variations in the dimensions of sheet material and fixtures. The mass of such materials should be minimized to avoid heat sink and insulating effects. Matched tooling is not often used for curved panels because of the high cost. A better method is to use a male or female tool in conjunction with the vacuum bag or autoclave technique.

CURING TEMPERATURE

Since variations in the thermal conductivity of the components influence the amount of heat transmitted to the adhesive layer, curing temperature should be measured at the glue line. Otherwise, the adhesive may not develop the desired properties for the application because of improper curing temperature.

Most phenolic-based structural adhesives require curing at elevated temperatures, generally from about 300 to 400° F, for periods ranging from 0.5 to 2 hours. Many one-part epoxy adhesives can be cured at temperatures as low as 250° F. A great number of two-component epoxy systems cure at room temperature; however, their properties are generally better when they are cured at elevated temperatures.

When neoprene contact-bond adhesives are used, the adhesive-coated surfaces are frequently heated during the drying cycle and mated under pressure while still warm. When design requirements are not so stringent, the adhesive may be dried and the components mated at room temperatures. Joint properties tend to be more variable, however, than those obtained when the hot contact bonding procedure is used.

As a general rule, curing time decreases as the curing temperature is increased within limits. Even the epoxies designed to cure at room temperature will cure faster when heated to moderately elevated temperatures. Curing time may be reduced from a number of hours to several minutes by heating. On the other hand, some room-temperature-curing adhesives will not cure properly below 60° F. This may be an important factor in field applications.

In some instances, longer curing times at elevated temperatures will improve the bond strength of the joint for service above room temperature. Post-curing of the bonded joint without pressure can also improve the heat resistance of the bond.

OVENS

Ovens are a widely used, inexpensive method for heat curing when only moderate pressure or simple positioning of the parts is required. They can be heated by gas, electricity, or steam. Adhesives that give off flammable vapor or solvent during cure should not be exposed to open flame or electrical elements. Ovens should be vented, temperature-controlled, and fitted with an air-circulating fan for uniformity of heat throughout. Infrared lamps and ovens are commonly used for contact-bond rapid drying neoprene formulations.

HEATED PRESSES

Hydraulic platen presses are frequently used for applying heat and pressure to flat assemblies. These are usually heated by electric heating elements, high-pressure steam, hot water, or some other heat-exchanging fluid.

When the work is placed in the press at temperatures below about 150° F, it is called "cold entry." Entry at the adhesive curing temperature is known as "hot entry." In general, adhesives that release volatiles perform better when cold entry is employed. Certain adhesives are also affected by the rate of temperature rise or heat input. These factors influence the chemical reactions, the flow, and the density of cured adhesives of the volatile releasing types. For example, cold entry and a rate of temperature rise

of less than 10° F per minute result in better shear strength at elevated temperatures for certain nitrile-phenolic film adhesives. Other adhesives, such as epoxy-phenolics, require either stepped heat input or release of pressure (breathing) at specific temperatures to allow volatiles to escape. Nonvolatile adhesives, such as epoxies, are not affected to any great extent by entry temperature or by the rate of heat input.

Large autoclaves are used for bonding aircraft assemblies and other extremely large parts. The typical operating range of such autoclaves is 200 psi maximum pressure and 350° F maximum temperature. Pressure is generally provided by compressed air, and curing temperature is achieved with steam heated tubes or electrical elements.

QUALITY CONTROL

TESTING

Adhesive bonded joints are inspected and tested to determine their quality and performance under the specific loading and environmental conditions they will see in service. Based on the test results, quality requirements can be established. Inspection methods and procedures can be specified to assure that quality. The advantages and limitations of inspection and testing procedures must be understood in order to apply them successfully. There are a number of military and industry standard specifications for testing adhesive bonded joints (Table 11.1).

Testing may also consist of accelerated, simulated, or actual use tests of the end product devised by the individual manufacturer or an industry group. For this reason, industry associations may be good sources of information on testing procedures.

If an adhesive is to be used with a metal for which no performance data exist, or if it is to be used in an unusual environment, it should be subjected to some testing. Single overlap shear

specimens can be used to evaluate the compatibility of a metal surface condition with an adhesive system and to evaluate the effect of any unusual environmental exposure. If an adhesive is to be used in a structural joint under stress in a certain environment, test joints should be simultaneously subjected to both the stress and the environmental conditions expected in service.

PROCESS CONTROL AND QUALITY ASSURANCE

Good process control usually requires inspection of all cleaning and processing equipment, evaluation of all materials, and control of storage time and conditions.

Adhesives and primers should be evaluated to assure conformance to the requirements of the design and the user's or adhesive manufacturer's specifications. Certified test reports from the manufacturer may be acceptable in lieu of actual performance tests.

Periodic tests should be performed to determine that cleaning, mixing, and bonding procedures are adequately controlled. Lap shear tests are generally satisfactory for control of mixing, priming, and bonding. Peel tests should be performed to ascertain the adequacy of cleaning procedures. The climbing drum method, described in ASTM D1781, as well as a recently developed crack extension (wedge) test, may also be used for this purpose.

The crack extension (wedge) test is designed for rapid screening of adhesive joint durability in a controlled humidity and temperature environment. The test specimen design for aluminum alloys is shown in Fig. 11.16. One or more specimens are cut from an adhesive bonded panel. The wedge is forced between the adherends and bends them apart. This separates the adhesive and produces cleavage loading at the apex of the separation. The location of the apex of the sheet separation is recorded.

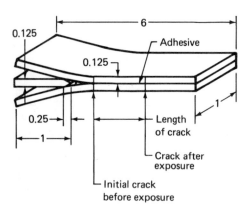

Dimensions are in inches

Fig. 11.16 — Crack extension (wedge) specimen

The wedged specimens are then exposed at 120° F to an air environment of 95 to 100 percent relative humidity for 60 to 75 minutes. The water should contain less than 200 parts per million total solids. The distance that the apex moved during exposure is measured within two hours after exposure.

The test is used for surface preparation process control by comparing test results with a maximum acceptable increase in adhesive crack length. It is also used for adhesive durability characteristics and surface preparation procedures. The test was originally designed for adhesive bonded aluminum. However, it may be suitable for other metals with design modifications to account for differences in stiffness and yield strength.

The frequency of testing will depend upon the volume of parts produced and the requirements for the application. However, many manufacturers who employ adhesive bonding in critical applications perform suitable quality control tests at least daily to ensure that the process is within specifications. Any production parts rejected for dimensional reasons or structural damage should be destructively inspected for joint quality.

EVALUATION OF FABRICATED PARTS

After the mechanical and processing properties of an adhesive system have been determined through destructive laboratory testing, the ability of manufacturing departments to duplicate these properties should be established. Therefore, rather complete testing of the first part or the first few parts of the production run is recommended.

Test loads should be applied in the same manner in which the part will be loaded in use. However, actual loading conditions are often difficult to simulate. In cases involving multidirectional loads, design loads may be applied in each plane individually. The part can then be loaded to failure in the most critical load path to determine if it meets the minimum design strength.

When it is impractical to load a completed part for test because of geometry or difficulty in mounting, many companies fabricate test specimens that are either an integral part of the assembly or separate panels processed in the same manner as the part. Mechanical properties of such specimens closely represent the actual strength of the part. This procedure can provide close control over materials and processing equipment.

NONDESTRUCTIVE INSPECTION

There are several nondestructive inspection methods other than visual that may be applicable to adhesive bonding. These are:

(1) Ultrasonic
(2) Acoustic impact (tapping)
(3) Liquid crystals
(4) Birefringent coatings
(5) Radiography
(6) Holography
(7) Infrared
(8) Proof test
(9) Leak test

Which methods can be used for a specific application will depend upon one or more of the following factors:

(1) Design and configuration of the structure
(2) Materials of construction
(3) Types of joints
(4) Material thicknesses
(5) Type of adhesive
(6) Accessibility to the joints

In some cases, it may be necessary to incorporate features in the component design or the adhesive to utilize an inspection process. For example, a filler may be required in the adhesive to increase its thermal or electrical conductivity or its density. To determine the applicability of a particular inspection method, the manufacturers of the particular type of equipment should be consulted.

SAFE PRACTICES

Adequate safety precautions must be observed with adhesives. Corrosive materials, flammable liquids, and toxic substances are commonly used in adhesive bonding. Therefore, manufacturing operations should be carefully supervised to ensure that proper safety procedures, protective devices, and protective clothing are being used. All federal, state, and local regulations should be complied with, including OSHA Regulation 29CFR 1900.1000, Air Contaminants.

GENERAL REQUIREMENTS

Flammable Materials

All flammable materials such as solvents should be stored in tightly sealed drums and issued in suitably labeled safety cans to prevent fires during storage and use. Solvents and flammable liquids should not be used in poorly ventilated, confined areas. When solvents are used in trays, safety lids should be provided. Flames, sparks, or spark-producing equipment must not be permitted in the area where flammable materials are being handled. Fire extinguishers should be readily available.

Toxic Materials

Severe allergic reactions can result from direct contact, inhalation, or ingestion of phenolics and epoxies as well as most catalysts and accelerators. The eyes or skin may become sensitized over a long period of time even though no signs of irritation are visible. Once a worker is sensitized to a particular type of adhesive, he may no longer be able to work near it because of allergic reactions. Careless handling of adhesives by production workers may expose others to toxic materials if proper safety rules are not observed: for example, coworkers may touch tools, door knobs, light switches, or other objects contaminated by careless workers.

For the normal individual, proper handling methods that eliminate skin contact with the adhesive should be sufficient. It is mandatory that protective equipment, protective creams, or both be used to avoid skin contact with certain types of formulations.

Factors to be considered in determining the extent of precautionary measures to be taken include:

(1) The frequency and duration of exposure

(2) The degree of hazard associated with the specific adhesive

(3) The solvent or curing agent used

(4) The temperature at which the operations are performed

(5) The potential evaporation surface area exposed at the work station

All these elements should be evaluated in terms of the individual operation.

PRECAUTIONARY PROCEDURES

A number of measures are recommended in the handling and use of adhesives and auxiliary materials.

Personal Hygiene

Personnel should be instructed in the proper procedures to prevent skin contact with solvents, curing agents, and uncured base adhesives. Showers, wash bowls, mild soaps, clean towels, refatting creams, and protective equipment should be provided.

Curing agents should be removed from the hands with soap and water. Resins should be removed with soap and water, alcohol, or a suitable solvent. Any solvent should be used sparingly and be followed by washing with soap and water. In case of allergic reaction or burning, prompt medical aid should be obtained.

Work Areas

Areas in which adhesives are handled should be separated from other operations. These areas should contain the following facilities in addition to the proper fire equipment:

(1) A sink with running water

(2) An eye shower or rinse fountain

(3) First aid kit

(4) Ventilating facilities

Ovens, presses, and other curing equipment should be individually vented to remove fumes. Vent hoods should be provided at mixing and application stations.

Protective Devices

Plastic or rubber gloves should be worn at all times when working with potentially toxic adhesives. When contaminated, the gloves must not contact objects that others may touch with their bare hands. Contaminated gloves should be discarded or cleaned using procedures that will remove the particular adhesive. Cleaning may require solvents, soap and water, or both. Hands, arms, face, and neck should be coated with a commercial barrier ointment or cream. This type of material may provide short term protection and facilitate removal of adhesive components by washing.

Full face shields should be used for eye protection whenever the possibility of splashing exists, otherwise glasses or goggles should be worn. In case of irritation, the eyes should be immediately flushed with water and then promptly treated by a physician.

Protective clothing should be worn at all times by those who work with the adhesives. Shop coats, aprons, or coveralls may be suitable and they should be cleaned before reuse.

Metric Conversion Factors

1 psi = 6.895 kPa

1 in. = 25.4 mm

$t_c = 0.556(t_F - 32)$

SUPPLEMENTARY READING LIST

Bethune, A. W., Durability of bonded aluminum structure. SAMPE *Journal,* 1975 July-Aug-Sept: 4-10.

Bodner, M. J., ed., *Symposium on Adhesives for Structural Applications,* New York: Interscience Publishers (Div., John Wiley and Sons), 1962 and 1968.

Bruno, E. J., ed., *Adhesives in Modern Manufacturing,* Dearborn, MI: Soc. of Manufacturing Engineers, 1970.

Cagle, C. V., *Handbook of Adhesive Bonding,* New York: McGraw-Hill, 1973.

De Lollis, N. J., *Adhesives For Metals—Theory And Technology,* New York: Industrial Press Inc., 1970.

Houwink, R. and Solomon, G., eds., *Adhesive and Adhesives,* New York: Elsevier Publishing Co., Vol. 1, 1965; Vol. 2, 1967.

Kantner, R. and Litvak, S., *Adhesives for Bonding Large High Temperature Sandwich Structures,* Warrendale, PA: Society of Automotive Engineers, 1967; Report No. 670858.

Katz, I., *Adhesive Materials, Their Property and Usage,* Long Beach, CA: Foster Publishing Co., 1971.

Minford, J. D., Evaluating adhesives for joining aluminum. *Metals Engineering Quarterly,* 1972 Nov.

Patrick, R. L., ed., *Structural Adhesives—With Emphasis On Aerospace Applications,* Vol. 4, New York: Marcel Dekkar, Inc., 1976.

Peters, R. A. and Logan, T. J., Microvoid epoxy adhesives for high peel and shear strength. *Adhesive Age,* 1975 Apr.

Rider, D. K., How to prepare surfaces for bonded metal assembly. *Product Engineering,* 35 (12), 1965: 75-78.

Rogers, N. L., Surface Preparation of metals for adhesive bonding, Applied Polymer Symposia No. 3., *Structural Adhesive Bonding,* New York: Interscience Publishers; 1966: 327-340.

Rolf, R. L., et al., Adhesive bonded structural joints in aluminium. *Adhesive Age,* 1971 July.

Schneberger, G. L., Polymer structures and adhesive behavior. *Adhesive Age,* 1974 April: 17-23.

Skeist, I., ed., *Handbook of Adhesives,* 2nd ed., New York: Van Nostrand Reinhold; 1976.

12

Thermal Spraying

Chapter Committee

W. B. MEYER, *Chairman**
St. Louis Metallizing Company

F. W. GARTNER, JR.
F. W. Gartner Company

E. S. HAMEL
Norton Company

J. D. HAYDEN
Hayden Corporation

F. J. HERMANEK
General Electric Company

G. M. HERTERICK
Bay State Abrasives Division

K. LEOVICH
Barrington Manufacturing, Inc.

M. A. LEVENSTEIN
Consultant

F. N. LONGO
Metco, Inc.

R. E. MAHOOD
St. Louis Metallizing Company

D. R. MARANTZ
Flame Spray, Inc.

J. RITCHIE
Bender Machine, Inc.

T. J. ROSEBERRY
Battelle Columbus Laboratories

N. L. RUNDLE
Union Carbide Corporation

J. WATSON
Hardfacing Welding and
Machine Company

Welding Handbook Committee Member

J. R. CONDRA
E. I. du Pont de Nemours and Company

*Deceased

12
Thermal Spraying

FUNDAMENTALS OF THE PROCESS

DEFINITION AND GENERAL DESCRIPTION

Thermal spraying (THSP) is a process in which a metallic or nonmetallic material is heated and then propelled in atomized form onto a substrate. The material may be initially in the form of wire, rod, or powder. It is heated to the plastic or molten state by an oxyfuel gas flame, by an electric or plasma arc, or by detonation of an explosive gas mixture. The hot material is propelled from the spray gun to the substrate by a gas jet. Most metals, cermets, oxides, and hard metallic compounds can be deposited by one or more of the process variations. The process can also be used to produce free-standing objects using a disposable substrate. It is sometimes called "metallizing" or "metal spraying."

When the molten particles strike a substrate, they flatten and form thin platelets that conform to the surface. These platelets rapidly solidify and cool. Successive layers are built up to the desired thickness. The bond between the spray deposit and substrate may be mechanical, metallurgical, chemical, or a combination of these. In some cases, a thermal treatment of the composite structure is used to increase the bond strength by diffusion or chemical reaction between the spray deposit and the substrate.

The density of the deposit will depend upon the material type, method of deposition, the spraying procedures, and subsequent processing. The properties of the deposit may depend upon its density, the cohesion between deposited particles, and its adhesion to the substrate.

Thermal spraying is widely used for surfacing applications to attain or restore desired dimensions; to improve resistance to abrasion, wear, corrosion, oxidation, or a combination of these; and to provide specific electrical or thermal properties. Frequently, thermal sprayed deposits are applied to new machine elements to provide surfaces with desired characteristics for the application.

PROCESS VARIATIONS

There are four variations of thermal spraying:
(1) Flame spraying (FLSP)
(2) Plasma spraying (PSP)
(3) Arc spraying (ASP)[1]
(4) Detonation flame spraying

These variations are based on the method of heating the spray material to the molten or plastic state and the technique for propelling the atomized material to the substrate.

In flame spraying, the surfacing material is continuously fed into and melted by an oxyfuel gas flame. The material may be initially in the form of wire, rod, or powder. Molten particles are projected onto a substrate by either an air jet or the combustion gases.

In plasma spraying, the heat for melting the surfacing material is provided by a non-transferred plasma arc.[2] The arc is maintained be-

1. This method is commonly called "electric arc spraying."

2. For information on plasma arc systems, refer to Chapter 9, Plasma Arc Welding, *Welding Handbook*, Vol. 2, 7th ed.

tween an electrode, usually tungsten, and a constricting nozzle which serves as the other electrode. An inert or reducing gas, under pressure, enters the annular space between the electrodes where it is heated to a very high temperature (above about 15 000° F). The hot plasma gas passes through and exits from the nozzle as a very high velocity jet. The surfacing material in powder form is injected into the hot gas jet where it is melted and projected onto the substrate.

The surfacing materials used with arc spraying are metals or alloys in wire form. Two continuously fed wires are melted by an arc operat-

ing between them. The molten metal is atomized and projected onto a substrate by a high velocity gas jet, usually air. This method is restricted to spraying of metals that can be produced in continuous wire form.

The detonation flame spraying method operates on principles significantly different from the other three methods. This method repeatedly heats charges of powder and projects the molten particles onto a substrate by rapid, successive detonations of an explosive mixture of oxygen and acetylene in a gun chamber. The particles leave the gun at much higher velocity than with the other methods.

FLAME SPRAYING

Flame spraying is used to deposit surfacing materials in the forms of metal wires, ceramic rods, and metallic and nonmetallic powders. A wide variety of materials in these forms can be sprayed with this method. Two exceptions are materials that cannot be melted with an oxyfuel gas flame and those that will burn when melted in the presence of an air jet.

Materials are normally deposited in multiple layers, each of which is less than 0.010-in. thick. The total thickness of a material that should be deposited will depend upon several factors including the following:

(1) Type of surfacing material and its properties

(2) Type and properties of the workpiece material

(3) Service requirements of the composite

(4) Postspray treatment of the composite

A typical flame spraying arrangement consists of the following:

(1) A flame spray gun

(2) A source of surfacing material and the associated feeding equipment

(3) Oxygen and fuel gas supplies, pressure regulators, and flowmeters

(4) A compressed air source and control unit, when required

(5) Workpiece holding device

The gun design depends upon the type of material to be sprayed and its physical form (wire, rod, or powder). When automated, the gun or the workpiece, or both, are driven by mechanisms designed to produce the desired deposit configuration.

Three common fuel gases are used for flame spraying: acetylene, propane, and methylacetylene-propadiene (MPS).[3] Acetylene in combination with oxygen produces the highest flame temperature. The distinct characteristics of an oxyacetylene flame make it rather easy to adjust the combustion ratio to produce oxidizing, neutral, or reducing conditions. The significant changes in flame appearance are not so evident with the other two gases. Hydrogen, which is used occasionally, and propane are suitable for flame spraying metals with low melting points such as aluminum, tin, zinc, and babbitt metal.

GAS CONTROLS

Oxygen and fuel gas flowmeters should be used to provide good control of the flame mixture and intensity. Their use usually permits higher spraying rates than with valve control of

---

3. Properties of these and other fuel gases are discussed in Chapter 13, Arc and Oxygen Cutting, *Welding Handbook,* Vol. 2, 7th ed.

gas flows. Since the molten particles are exposed to oxygen, an oxide film will form on them, even when a reducing gas mixture is used. The thickness of the oxide film does not appear to vary greatly with minor changes in the fuel gas-to-oxygen ratio.

COMPRESSED AIR SUPPLY

The cleanliness of the compressed air, when used to atomize and propel the molten surfacing material, is important in producing a quality deposit. Oil or water in the compressed air will cause a fluctuation in the flame, produce poor or irregular atomization of the spray material, and ultimately affect the quality of the deposit. In most cases, aftercoolers and chemical filters should be installed between the air source and the spray unit. Sometimes, filters on both sides of the pressure regulator are advantageous. Accurate regulation of the air pressure is important for uniform atomization.

WIRE FLAME SPRAYING

With wire flame spraying, the metal wire to be deposited is normally supplied to the gun continuously from a coil or spool. In some cases, cut lengths of metal rods are used.

A typical wire flame spray system is shown in Fig. 12.1, and a cross section of a gun is shown in Fig. 12.2. A wire type gun consists essentially of two subassemblies: a drive unit which feeds the wire, and a gas head which controls the flow of fuel gas, oxygen, and compressed air. The principles of operation of all wire type guns are very similar.

The wire drive unit consists of a motor and drive rolls. They may be air or electrically powered with adjustable speed controls. Speed controls may be mechanical, electromechanical, electronic, or pneumatic, depending upon the type of power.

The gas head consists of valves to control the fuel gas, oxygen, and compressed air, and a gas nozzle and air cap. The wire is fed through a central orifice in the nozzle. Various nozzles and air caps are used to accommodate different wire sizes or metals. The arrangement of the oxyfuel gas jets and compressed air orifices differs with the various manufacturers, as do the mechanisms for feeding the wire through the flame.

If the wire feed rate is excessive, the wire tip will extend beyond the hot zone of the flame and it will not melt and atomize properly. This will produce very coarse deposits. If the feed is too slow, the metal will oxidize badly, and the wire may fuse in the nozzle. The spray deposit will have a high oxide content.

Wire spraying units vary in size. Small hand-held units are manipulated in much the

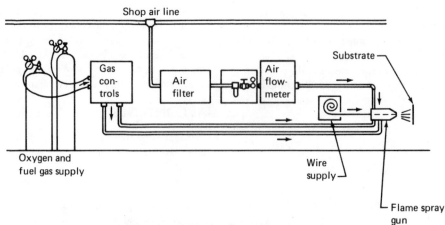

Fig. 12.1—Typical wire flame spray system

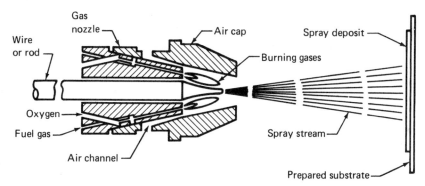

Fig. 12.2 — Cross section of a typical wire flame spray gun

same manner as paint spray guns. They are often used to apply protective coatings of aluminum or zinc to large objects such as tanks, ship hulls, and bridges. Larger units are usually designed to be mechanically manipulated for spraying moving parts.

CERAMIC ROD FLAME SPRAYING

Ceramic rod flame spraying is similar to wire flame spraying. Straight lengths of ceramic rod are successively fed into the flame by driven plastic rollers in the gun. The ceramic material is melted in the flame, then atomized and propelled from the nozzle by a compressed air jet. The molten particles flatten on impact with the substrate and interlock to form a rather dense, coherent deposit.

The bond between the ceramic deposit and the substrate is primarily mechanical in nature. The molten particles deform and take the shape of the prepared surface. Proper surface preparation is a prerequisite for a firmly bonded deposit.

Equipment and Materials

The equipment for ceramic rod spraying is similar to wire spraying equipment (Fig. 12.1) except that the wire feed unit is replaced by a mechanism to drive the rollers in the gun. The gun design is somewhat different with respect to the gas nozzle, air cap, and feed system. This equipment requires greater care in adjusting the spraying variables than does wire spraying equip-

ment because of the higher melting points and lower thermal conductivities of ceramics compared to metals. Some ceramic materials that are applied by thermal spraying are:

(1) Alumina-titania
(2) Alumina
(3) Zirconia
(4) Rare earth oxides
(5) Zirconium silicate
(6) Magnesium zirconate
(7) Barium titanate
(8) Chromium oxide
(9) Magnesia-alumina
(10) Mullite
(11) Calcium titanate

Each ceramic surfacing material has specific characteristics, economics, advantages, and limitations. The material is usually selected to provide specific properties for the expected service with proper consideration of the following factors:

(1) Thermal, electrical, and chemical characteristics
(2) Melting point
(3) Adherence or bond strength
(4) Density
(5) Cost

Some of the important characteristics of ceramic spray deposits are:

(1) Good adherence to a variety of substrate materials
(2) Economically applied in controlled thicknesses
(3) Good physical and chemical stabilities

(4) Low thermal and electrical conductivities

(5) High wear resistance

(6) Good finishing characteristics

POWDER FLAME SPRAYING

With this method, the material to be sprayed is supplied to the gun in powder form from a hopper. The hopper may be remote from the gun or mounted onto it. The powder may be aspirated or carried into the flame by an air feed system, by the oxygen stream, or by gravity. The powder is melted by the flame and propelled onto the substrate by either a compressed air jet or the combustion gases.

Powder flame spraying equipment is more simple and less costly than plasma spray equipment. However, the spray rate with flame spraying is lower. The equipment can be designed for easy portability.

A special case is a powder flame spray gun which is similar to an oxyacetylene welding torch. Powder to be sprayed is metered into the gas stream before it leaves the tip. Compressed air is not used. The torch can be used for preheating or fusing spray deposits when powder is not being injected into the gas stream.

Metals, ceramics, and ceramic-metal mixtures can be flame sprayed by the powder method. The metals are usually hard alloys designed for specific wear or corrosion resistant applications. Very hard metallic compounds, such as carbides and borides, can be blended with metal powders to form a composite, wear-resistant coating. The degree of melting of the particles of spray powder depends upon both the melting point of the material and the time that the particles are exposed to the heat of the flame (called the dwell time). Powders with low melting points will become completely molten, and those with high melting points, such as ceramics, may melt only on the particle surface.

PLASMA SPRAYING

GENERAL DESCRIPTION

Plasma spraying utilizes the heat of a plasma arc to melt the surfacing material. The term "plasma arc" is used to describe a family of metal working processes that use a constricted electric arc to provide high density thermal energy. Arc constriction is usually accomplished by forcing the arc plasma through a water-cooled copper orifice. The purpose of this is to control and increase the energy density of the arc stream. Plasma arc processes are employed for welding, cutting, and surfacing of metals. In plasma spraying, a non-transferred arc is constricted between an electrode and a constricting nozzle. A plasma spray unit will consist of a gun, power source, gas source, spray material supply, and associated fixturing and traversing devices.

EQUIPMENT

Torch Design

Several types of plasma spray torches are available. In each instance, the arc is generated between an electrode and a water-cooled chamber into which a plasma gas is injected.The gas picks up arc energy and exits from a nozzle in a configuration that resembles an open welding flame. A typical gun is shown in Fig. 12.3.

Two types of gas-stabilized torches are used for spraying. One type is vortex-stabilized and the other is gas-sheath stabilized. With vortex stabilization, gas is introduced into the chamber with a swirling motion and produces an intense vortex at the exit through the nozzle. This causes the arc plasma to travel from the electrode out through the nozzle and back to the face of the

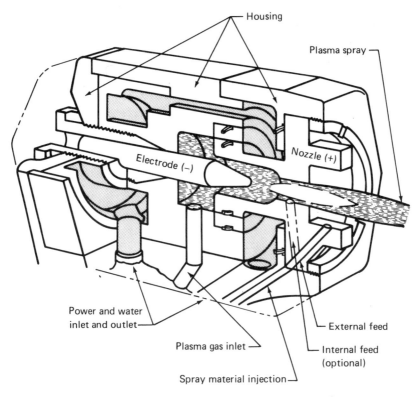

Housing

Plasma spray

Nozzle (+)

Electrode (−)

Power and water
inlet and outlet

Plasma gas inlet

Spray material injection

External feed

Internal feed
(optional)

Fig. 12.3—A typical plasma spray gun

chamber. In a gas-stabilized torch, the arc remains within the nozzle and is prevented from striking the chamber wall prematurely by a sheath of gas passing through the nozzle. The arc, when properly adjusted, penetrates through this gas sheath some distance down the nozzle. Arc positioning with respect to the nozzle is accomplished by control of the gas flow pattern and turbulence.

In reference to Fig. 12.3, the rear electrode may be fixed or adjustable, but it must be properly aligned with the nozzle or front electrode during operation. The flow of gas through the orifice is governed by the gas pressure. Several nozzle configurations can be used to accommodate different plasma gases and to spray different types of powders.

An important factor in producing quality deposits is the introduction of powder at the proper point in the arc plasma and at the correct feed rate. Since the particles are in the plasma for only about 2×10^{-5}s, slight variations in the location of the feed point may significantly change the amount of heat transmitted to the powder.

Current torch designs have power capacities of from 40 to 100 kW. Direct current of 100 to 1100 A is used at 40 to 100 V. High power is necessary when spraying with high particle velocities. Particle velocity is an important variable with respect to bond strength, deposit density, and deposit integrity.

Power Supply

Power supplies for plasma spraying should have the following characteristics:

(1) Constant current output at 100 percent duty cycle

(2) Variable open-circuit and load voltages

(3) Variable current control

(4) Good regulation

(5) Arc starting capability, either high-frequency or high voltage dc pulse

Rectifier type, solid-state units can generally meet the above requirements. Units are easily operated in parallel for high power operations. In general, they resemble arc welding power sources.[4]

Powder Feed Devices

Powder feed mechanisms are of two types: aspirator and mechanical. The latter is the more popular type. It utilizes the metering action of a screw or wheel to deliver powder at a constant rate to a mixing chamber. Here, the powder is introduced into the carrier gas stream.

Units are available to cover a wide range of spray rates. The range of a particular design is determined by the specific gravity of the surfacing material. Modifications are usually available to meet specific spray rate requirements.

System Control

A complete system, including the spray unit, can be operated from a control console. The console provides adjustment of the plasma gas flow rates, plasma current, starting and stopping functions and, in some cases, operation of the powder fed unit. These functions are common to all plasma spray systems.

Controlled Environment

Plasma spraying can be done in a sealed chamber containing an inert gas atmosphere at or below atmospheric pressure. This technique provides improved oxidation protection for the molten particles and the substrate. The substrate can then be preheated to a higher temperature without excessive oxidation. Deposition efficiency and deposit quality are better than when spraying in air.

GASES

Gases are utilized for three purposes in plasma spray systems: the primary plasma gas; a secondary gas mixed in small volumes with

the plasma gas; and a powder-carrying gas, usually the same as the primary plasma gas.

Monatomic and diatomic gases can be used for plasma spraying. Argon and helium are the two most frequently used monatomic gases. With monatomic gases, it is possible to attain plasma temperatures sufficiently high for most applications. Plasmas generated with polyatomic gases have greater heat contents. They not only release ionization energy but also the energy of molecular recombination. The choice of gas determines the quality of the plasma. The gases should be welding grade with low moisture and oxygen contents.

The four gases commonly used for plasma spraying and their important characteristics are as follows:

(1) Nitrogen is widely used because it is inexpensive, diatomic, and permits high spraying rates and deposit efficiencies. Nozzle life will be shorter than with monatomic gases, but this factor can be offset by the lower cost of this gas.

(2) Argon provides a high velocity plasma. It is used to spray materials that would be adversely affected if hydrogen or nitrogen were used.. Carbides and high temperature alloys are most commonly sprayed with argon.

(3) Hydrogen may be used as a secondary gas with nitrogen or argon in amounts of 5 to 25 percent. Hydrogen addition raises the arc voltage and, thus, the power. It may be detrimental with certain metals.

(4) Helium is usually used as a secondary gas mixed with argon. It will also tend to raise the arc voltage.

SURFACING MATERIALS

It is usually necessary to use plasma spraying equipment for powders with melting points above 5000° F. Because this method is well qualified for depositing refractory metals and ceramics, it can also deposit powdered materials that are normally applied by flame spraying but at a higher rate.

A partial list of surfacing materials applied by this method is given in Table 12.1. It does not include all available materials. Many commercial compositions are proprietary and de-

---

4. Arc welding power sources are discussed in Chapter 1, *Welding Handbook*, Vol. 2, 7th ed.

Table 12.1
Materials commonly applied by plasma spraying

Metals	Carbides[a]	Oxides	Cermets
Aluminum	Chromium carbide	Alumina	Alumina-nickel
Chromium	Titanium carbide	Chromium oxide	Alumina-nickel
Copper	Tungsten carbide	Magnesia	aluminide
Molybdenum		Titania	Magnesia-nickel
Nickel		Zirconia	Zirconia-nickel
Nickel-chromium alloys			Zirconia-nickel
Tantalum			aluminide
Tungsten			

a. Normally combined with a metal powder that serves as a binder.

signed for specific applications.

Plasma sprayed ceramic deposits exhibit higher densities and hardnesses than flame sprayed deposits. High density, plasma sprayed deposits can be thinner in some cases, but may be more susceptible to cracking. Deposition procedures can be designed to overcome differences in coefficients of thermal expansion of a ceramic and a metal substrate. This can be achieved by spraying mixtures of the ceramic and a suitable metal in various proportions to produce graded (layered) deposits.

ARC SPRAYING

GENERAL DESCRIPTION

Arc spraying (ASP) utilizes an arc between two wires to melt their tips. A jet of compressed gas, normally air, atomizes the molten metal and projects the particles onto a prepared substrate. A simple version of arc spraying is shown in Fig. 12.4.

The apparatus simultaneously and continuously feeds two wires through electrical contacts at uniform speeds. The contacts also serve as guides to direct the wires toward a point of intersection.

Direct current power is normally used for arc spraying; one wire is positive (anode) and the other is negative (cathode). The cathode wire tip is heated to a higher temperature than the anode wire tip and melts at a faster rate. Consequently, the particles atomized from the cathode wire are much smaller than those from the anode wire when the two wires are identical. When dissimilar metals are sprayed, the deposit is a heterogeneous blend of the two metals with some alloying.

The velocity of the gas as it leaves the gun nozzle can be regulated over a broad range to control the deposit characteristics. Molten metal particles are ejected from the arc at the rate of several thousand particles per second, some of which are in the superheated state. These molten particles will bond to minute protrusions of a properly cleaned and roughened substrate. The

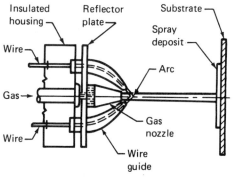

Fig. 12.4—Arc spraying

resulting deposit is highly adherent to the substrate and has strong interparticle cohesive strength.

In addition to the strong bond strength, other advantages of this process over wire flame spraying are:

(1) The quantity of metal oxides in the deposit can be controlled by spray conditions.

(2) Spray rates are generally higher.

(3) It may be more economical.

Arc spraying may be more economical than other spraying methods in some cases. Energy and labor costs may be lower because of the higher spray rates and deposition efficiencies available with this method. One adverse effect of the high energy state of the atomized particles is their tendency to change composition through selective oxidation or vaporization, or both. The nature of these effects is complex, but they can be minimized by judicious selection of wire composition.

The arc method is less versatile than the plasma method because only wire can be sprayed. Particle temperatures and velocities are generally lower than with plasma spraying but higher than with wire flame spraying.

EQUIPMENT

There are several equipment designs that may be used for arc spraying. They generally consist of four major components:

(1) A solid-state dc power source, usually of the constant voltage type

(2) A dual wire feeding system

(3) A control assembly with compressed gas attachments

(4) An arc spray gun

The dc power source and the control system supply the correct amount of electrical energy, atomizing air, and wire to the arc spray gun. Wire sizes of from 0.057 to 0.125 in. are normally employed. The wires are first drawn through straighteners and then passed through insulated, flexible conduits into the spray gun. The wires may be pushed through the conduits to the gun by wire feeders or pulled into the gun by a drive unit incorporated in it.

SYSTEMS OPERATION

Arc spray systems can be operated from a console control or the spray unit. Controls are provided for atomizing gas pressure and flow, wire feed rate, and arc power. On most equipment, switches are provided at the spray gun to energize the wire feed and the atomizing gas flow.

Arc voltage is adjusted at the power source. The arc is self-igniting as the two wires are advanced to their intersecting point. In some cases, special circuitry provides high voltage (approximately 75 volts) to initiate the arc. Correct and proper preparation of the base material is one of the most important parts of the spraying procedure. A partial metallurgical bond can sometimes be produced between the surface and the deposit. To achieve this, the substrate surface must be chemically clean.

High-strength bonds can be achieved by spraying the first pass using a high arc voltage, a low gas flow, and a short gun-to-work distance. These conditions will produce coarse, hot particles that will adhere well to the substrate. To avoid overheating of the substrate, the traversing speed should be fast, especially with manual spraying.

After the first pass has been uniformly applied over the entire surface, subsequent spraying is done using the standard gas pressure, the lowest possible arc voltage, and the normal spray distance. The shortest arc length consistent with arc stability is desired. This ensures:

(1) Fine spray particle size

(2) Minimum loss of alloy constitutents

(3) A concentrated spray pattern

(4) High melting capacity for the amount of electrical energy consumed.

DETONATION FLAME SPRAYING

Detonation flame spraying[5] is done with a specially designed gun, such as the one shown in Fig. 12.5. The gun contains a chamber into which are injected metered quantities of oxygen, acetylene, and a powdered surfacing material suspended in nitrogen. The oxyacetylene mixture is detonated by an electric spark several times per second. This creates a hot, high velocity gas stream that heats the powder to its plastic state and then accelerates the particles to a speed of about 2500 ft/s as they leave the gun barrel. The molten particles impinge on the surface of the workpiece. Successive detonations in the gun build up the deposit to the desired thickness.

The high bond integrity of detonation sprayed deposits is due to the high velocity at which the particles strike the substrate. It is roughly 2.5 to 5 times faster than the velocities of particles from a plasma or powder flame spraying gun. Since kinetic energy is a function of the square of the velocity, particles from a detona-

tion flame spray gun strike the surface with an energy at least 25 times greater than those from a flame spray gun. As a result, coatings with only 0.25 to 1 percent porosity are commonly achieved with this equipment. These include the following materials:

(1) Alumina

(2) Alumina-titania

(3) Chromium carbide with a nickel-chromium alloy binder

(4) Tungsten carbide with a cobalt binder

(5) Tungsten carbide-tungsten chromium carbide mixture with a nickel-chromium alloy binder

These are primarily wear resistant coatings for elevated temperature service. Sprays of liquid carbon dioxide are used to cool the workpiece during spraying. Consequently, metallurgical changes or warpage of the workpiece does not normally occur with this spraying method.

Since the spraying operation is a series of rapid detonations, the noise level is quite high. Spraying is normally done in a specially constructed application room to contain this noise.

---

5. This is a proprietary process.

Fig. 12.5—Detonation flame spraying equipment

PREPARATION FOR SPRAYING

The type of bonding between the spray deposit and the substrate may be mechanical, metallurgical, chemical, or combinations of these. It depends upon the substrate material, surfacing material and form, the thermal spraying method and procedures, and surface preparation. Of these, suitable surface preparation is most important to obtain a strong bond.

First, the surface and adjacent areas must be free of all oil, grease, oxides, dust, and other foreign matter. Cleanliness must be maintained before and during the spraying operation. Specific handling and storage procedure should be established to ensure this.

Some materials, such as cast iron, may absorb lubricants or other chemicals in service. Prior to surfacing, workpieces of these materials should be heated in air to a temperature between 400° and 600° F for 4 to 6 hours to evaporate these contaminants. A hot caustic soaking treatment followed by hot water rinsing may also be effective in removing these types of contaminants. Nonporous workpieces may be cleaned by vapor degreasing, steam, hot detergent washing, or manually applied solvents.

SURFACE ROUGHENING

Since mechanical bonding is the most prevalent mechanism with thermal spraying, some form of mechanical roughening is usually employed. For applications where the surface must be machined undersize to accommodate the spray deposit, a rough surface can be produced during this operation by such techniques as threading or grooving.

In any case, the surface should be blasted with a suitable hard, sharp abrasive or hardened and crushed steel grit. The type and size of blasting particles depend upon the composition and the size or thickness of the workpiece. Coarse particles can be used on heavy or thick sections, but only fine abrasives should be used on thin sections or small diameter parts. The blasting operation itself should not contaminate the substrate with dirty blasting materials, embedded particles, or oil or water from the compressed air supply. Any residue from the blasting operation must be removed prior to spraying because it will interfere with good bonding.

BOND LAYER

Thermal sprayed molybdenum, nickel-chromium alloy, or an exothermically reactive nickel-aluminum composite is widely used as a base for ceramic spray deposits, particularly on mechanical components. These materials are self-bonding to most clean, blasted metal surfaces, except copper and its alloys. Molybdenum, tungsten, exothermic materials, and certain other metals with high melting points can form a metallurgical bond with a metal substrate.

Nickel-chromium alloy is mostly used as a bond layer for ceramic deposits to obtain a strong mechanical bond and to provide an oxidation resistant layer with some applications. Molybdenum is not recommended for applications where operating temperatures will exceed 750° F in an oxidizing atmosphere. Nickel-aluminum composites may be used at temperatures up to 1500° F. These bond layers can be applied directly to most machined, sanded, or abrasive blasted surfaces. The desired spray deposit is applied directly to the bond layer.

A 9 percent aluminum-bronze alloy is a very reliable bond layer for copper, copper alloys, and aluminum. It also has good adhesion to steel. When the spray deposit will be less than about 0.05-in. thick, the bond layer for most metal deposits can be the same composition as the spray deposit. For deposit thicknesses greater than about 0.05 in., Ni-Cr or Ni-Cr-Fe alloy should be used for a bond layer for all ferrous deposits and aluminum-bronze for all nonferrous deposits.

PREHEATING

For most base metals, the part should be preheated to within the 200° to 300° F range and maintained at temperature during the thermal spraying operation. This prevents the formation of surface condensate, expands the substrate, and reduces stresses in the deposit as it cools. Since heat is being transferred to the workpiece during spraying, some form of cooling may be needed to prevent overheating and resulting undesirable deposit properties. Jets of air, inert gas, or liquid CO_2 may be used for cooling. In the case of brittle deposits, wide temperature excursions may cause them to crack or spall during the spraying operation.

FUSED SPRAY DEPOSITS

GENERAL DESCRIPTION

A fused spray deposit is a self-fluxing alloy deposited by thermal spraying and subsequently heated to coalescence within itself and with the substrate. The materials will wet the substrate without the addition of a fluxing agent, provided the substrate is properly cleaned and prepared to receive it. The materials are basically alloys based on nickel or cobalt in powder form, and are normally applied with powder flame spraying equipment. Plasma spraying is also a suitable method.

The application of a fused deposit involves four operations:

(1) Surface preparation

(2) Spraying the self-fluxing alloy

(3) Fusing the coating to the substrate

(4) Finishing the coating to meet surface and dimensional requirements

A finishing operation is not always required if the as-fused surface is suitable for the application. Pump packing sleeves, pump plungers, piston rods, and process rolls are examples of surfaced machine parts that require a subsequent finishing operation on fused deposits. Centrifuge screw flutes, buffing fixtures, and process piping are examples of components that may be used in the as-fused condition.

A properly sprayed and fused deposit will be nearly homogeneous, metallurgically bonded to the substrate, and have no open or visible porosity. It normally will have a higher hardness than an equivalent mechanically bonded deposit, and will withstand pressures and environments better than non-fused deposits.

SELF-FLUXING ALLOYS

Most self-fluxing alloys fall into two general groups:

(1) Nickel-chromium-boron-silicon alloys

(2) Cobalt-chromium-boron-silicon alloys

In some cases tungsten carbide or chromium carbide particles are blended with an alloy from one of the above groups.

The boron and silicon additions are crucial elements that act as fluxing agents and as melting point depressants. They permit fusing at temperatures compatible with steels, certain cast irons, and some nickel base alloys.

The hardness of fused deposits will range from about 20 to 60 HRC, depending upon alloy composition. It is virtually unaffected by the thermal spraying procedures since there is almost no dilution with the base metal.

Selection of an alloy composition for a particular application should be based on certain considerations including the following:

(1) Fusion temperature of the alloy and the thermal effects on the base metal

(2) Relative difference in the coefficients of thermal expansion of the base metal and the alloy deposit

(3) Service requirements of the part

(4) Finish requirements of the fused deposit and available finishing equipment

EQUIPMENT

In addition to cleaning, blasting, thermal spraying, and work-handling equipment, some device or method is needed to fuse the spray deposit. Fusing can be done with an oxyfuel gas torch or in a furnace.

Fusing Torches

A fusing torch can have a single or multiple jet tip, depending upon the mass of the workpiece. The fusing gas is usually acetylene, and a neutral or reducing flame can be used. Other fuel gases may be used except for cobalt-base alloys. An oxyacetylene torch with a reducing flame is recommended for these alloys. A combination spraying and fusing torch is available for applying these types of deposits. The coating is alternately deposited and fused with it. This type of equipment is particularly suited for repair work, but not for large workpieces nor for production work.

Fusing Furnaces

Spray deposits can be fused by placing the sprayed workpiece in an atmosphere furnace operating at fusing temperature. Argon, dry hydrogen, or vacuum atmospheres may be used. Furnace fusing is advantageous for high production applications, intricate part geometries, or parts with significant variations in section thickness.

BASE METALS

Fused thermal sprayed deposits can be applied to a wide variety of metals. However, varying degrees of skill, technique, and procedures are required. Some base metals are decidedly easier to surface than are others. Those which can be readily thermal sprayed with one or more self-fluxing alloys and then fused are as follows:

(1) Carbon and low alloy steel with less than 0.25 percent carbon
(2) AISI 300 series stainless steels, except Types 303 and 321
(3) Certain grades of cast iron
(4) Nickel and nickel alloys that are free of titanium and aluminum

Metals that require special procedures to avoid undesirable metallurgical changes are car-

bon and low alloy steels with more than 0.25 percent carbon and AISI 400 series stainless steels, except Types 414 and 431. Types 414, 431, and the precipitation hardening stainless steels are not recommended as base metals for self-fluxing alloys.

Cracking of some types of fused spray deposits on hardenable steels can be avoided by isothermal annealing of the parts from the fusing temperature to avoid martensitic transformation. Fused deposits with hardnesses above 25 HRC will likely crack when a martensitic steel is hardened because of the expansion that takes place during transformation. However, there are applications in which cracks in the fused deposit are not detrimental to service requirements.

FUSING

Fusing of a sprayed deposit is accomplished by heating the workpiece to the fusing temperature range for the particular self-fluxing alloy. The fusing temperatures of nickel-chromium-boron-silicon alloys range from 1875° to 2150° F. The cobalt-chromium-boron-silicon alloys fuse in the range of 2150° to 2250° F. The actual fusing temperature depends upon the composition of the alloy.

The most common method of fusing is with one or more oxyfuel gas heating torches using a reducing flame. A typical torch fusing operation is shown in Fig. 12.6. First, the torch is directed on the workpiece which is heated to a dull red color (about 1400° to 1600° F). Then, the torch is moved across the spray deposit to gradually increase the surface temperature until the deposit shows a glossy or greasy appearance. This indicates that the deposit has fused. Overheating should be avoided to prevent flow of the molten alloy. The temperature of the workpiece and deposit should be maintained as uniform as possible.

The fusing operation may also be done with other methods of heating including furnace and induction. With such processes, heating should be done in a neutral or reducing atmosphere to avoid oxidation of both the deposit and the base metal before the fusing temperature is attained.

Fig. 12.6—Fusing a deposit on a large roll with oxyacetylene torches

POST-TREATMENTS

SEALING

Sealing of sprayed deposits is used occasionally to either lengthen their lives or prevent corrosion of the substrate, or both. Thermal sprayed deposits of aluminum or zinc may be sealed with vinyl coatings, either clear or aluminum pigmented. The sealer may be applied to fill only subsurface pores in the deposit or both subsurface pores and surface irregularities. The latter technique will provide a smooth coating to resist industrial atmospheres. The vinyl coatings can be applied with a brush or spray gun.

Sealing is also sometimes used on surfaced machine parts. Where the spray deposit will be exposed to acids, it is often advisable to seal the surface with either a high-melting-point wax sealer or a phenolic plastic solution. Spray deposits on high-pressure hydraulic rams, pump shafts, and similar parts should be sealed with air-drying phenolics to prevent the seepage of the liquid through the deposit around the packing.

Pressure cylinders of all types are reclaimed by thermal spraying. Prior to finish grinding, the cylinder bore is sealed with a phenolic. This prevents grinding wheel particles from embedding in the pores of the sprayed metal and causing premature wear.

Epoxies, silicones, and other similar materials may be used as sealants for certain corrosive conditions. Vacuum impregnation with plastic solutions is also possible.

DIFFUSING

A thin layer of aluminum may be diffused into a steel or silicon bronze substrate at 1400° F. The diffused layer can provide corrosion protection against hot gases up to 1600° F. After depositing the aluminum, the part should be coated with an aluminum pigmented bitumastic sealer or other suitable material to prevent oxidation of the aluminum during the diffusion heat treatment.

SURFACE FINISHING

Techniques for surface finishing of thermal sprayed deposits differ somewhat from those commonly used for metals. Most sprayed deposits are primarily mechanically bonded to the substrates, except for fused coatings. Excessive pressure or heat generated in the deposit during the finishing operation can cause damage such as cracking, crazing, or separation from the substrate.

Since an as-sprayed deposit is composed of an aggregation of individual particles, improper finishing techniques can dislodge particles singly or in clusters. This may cause a severely pitted surface. The deposited particles should be cleanly sheared and not pulled from the surface. Even so, an ideally finished surface will probably not be shiny, but may have a matte finish due to porosity in the deposit.

The selection of a finishing method depends upon the type of deposit material, its hardness, and its thickness. Consideration should be given to the properties of the substrate material as well as dimensional and surface roughness requirements. Spray deposits of soft metals are often finished by machining, especially those applied to machine components. A good finish can be obtained using high cutting speeds and carbide tools for most such applications. More often, however, sprayed deposits are finished by grinding, particularly the hardfacing and ceramic types.

Various other finishing methods are occasionally used. These include buffing, tumbling, burnishing, belt polishing, lapping, and honing.

MACHINING

Tungsten carbide tools are commonly used for machining sprayed metal deposits and fused coatings. Proper tool angles play a critical role in the success of machining these coatings. The surface speeds and the depth of cut are of equal importance. Improper tool angle and tool pressure can result in excessive surface roughness and in the destruction of the bond between the deposit and the substrate.

A cutting tool with a slightly rounded tip and a rake angle of three degrees should be used. On outside diameters, the tip of the tool should be set three degrees below center; in bores, this should be three degrees above center. This will help to limit the stress on the deposit. Peripheral speed should not exceed 75 ft/min. The feed should be slow with light cuts for best surface finish.

Special cutting tools, such as oxide-coated carbide, cubic boron-nitride, ceramics, cermets, and diamonds, may be used to machine very hard metal, ceramic, and cermet deposits. In many cases, machining with these tools is replacing grinding of intricate shapes and large pieces. Machining of flat deposits requires extreme care at the corners and edges to avoid damage. Depth-of-cut and feed rate should be low.

GRINDING

Metal Deposits

Wet grinding is the preferred method. Large, wide wheels can be used, and the required amount of stock can be removed with one operation. Wet grinding permits closer tolerances than does dry grinding. Grinding wheel manufacturers can recommend wheel types and grinding procedures for various metal deposits using a particular type of grinding machine.

If it is necessary to grind metal deposits dry, as is done with portable grinders mounted on a lathe, the major amount of materials should be removed first by machining. Then, the deposit can be ground to the required finish and dimensions.

Wheels used for dry grinding operations are either aluminum oxide or silicon carbide, depending upon the metal to be ground. The factors to be considered in selecting a wheel for a spray deposit are similar to those for grinding the same metal in wrought or cast form. The grinding technique should be designed to minimize heat buildup in the deposit. The structure of the wheel should be as open as possible and the grain size as coarse as possible, consistent with the finish requirements. The wheel should be narrow, the infeed light, and the traverse as fast as possible without spiraling.

When grinding equipment is not available, metal deposits can be machined to within 0.002 to 0.006 in. of final size. Then they can be finished to size with a belt polishing unit. Close tolerances and fine finishes are possible with belt polishing by proper selection of abrasive type and grit size.

Fused Deposits

Because most fused deposits are designed for hardfacing purposes, grinding is usually the most economical method for finishing them.

Although most fused deposits can be machined with the proper type of cutting tool, close tolerance work is difficult because of rapid tool wear and the large amount of heat generated. Dry grinding may be suitable for some operations but, in this case also, heat and fast wheel wear make close tolerance work difficult. Wet grinding can produce close tolerance parts, fine finishes, and economical stock removal rates. Nickel-base alloys are best ground with silicon carbide grinding wheels and cobalt-base alloys with aluminum oxide grinding wheels.

Grinding wheel manufacturers may recommend the appropriate type for the job. Good practice usually suggests a coarse wheel, consistent with finish requirements; an open structure or soft bond; as large a wheel as possible; and good wheel dressing techniques. Surface finish of fused coatings can often be improved after grinding by polishing with fine grit belts.

Ceramic Deposits

The as-coated surface finish of flame sprayed ceramics is, in general, in excess of 150 $\times 10^{-6}$ in. Many applications require a better finish, and this can be accomplished by grinding.

Although the individual particles of a ceramic deposit have extreme hardness, the deposit can be finished by conventional grinding techniques on standard equipment. However, it is necessary to use the proper grinding wheel and to follow correct procedures. General recommendations for grinding ceramic deposits are available from grinding wheel manufacturers.

Flood cooling should be employed during grinding. Water containing a rust inhibitor is best. Water-soluble coolants are likely to stain light-colored ceramic deposits.

OTHER METHODS

Other methods of surface finishing are sometimes used for as-sprayed and fused deposits. These include:

(1) Manual buffing or polishing
(2) Abrasive tumbling
(3) Honing
(4) Lapping

As-sprayed or machined deposits may be buffed or polished manually with abrasive stones, cloth, or paper. Abrasive tumbling of small parts will polish the surface by removing the "high spots." An abrasive medium, cleaners, and usually a liquid are vibrated or rotated in a drum in which the parts are placed.

Honing is done with abrasive stones mounted in a loading device. The part normally moves in one direction or rotates while the stones are oscillated under pressure, transverse to the work motion. Lapping is done with a fine, loose abrasive mixed with a vehicle, such as water or oil. The mixture is spread on lapping shoes or plates that are then rubbed against the spray deposit. The lap rides against the deposit, and their relative movements are continually changed.

QUALITY CONTROL

A properly designed quality control program can ensure consistent quality in thermal sprayed deposits. Proper quality control consists of more than just the examination of the workpiece after the spraying is complete. Each step in the operation should be monitored by an inspector. This includes not only the spraying and fusing steps but also the preparation of the substrate and the various stages of handling and storage of the workpiece between operations. In addition, the quality of the spray materials must be controlled. Since bond strength and spray deposit soundness are difficult to determine by nondestructive techniques, the procedures for accomplishing each step of the thermal spraying operation should be documented. The procedures should be qualified by appropriate destructive tests of sample parts.

In general, sprayed deposits are inspected visually for quality and soundness. With fused deposits, lack of bonding may be detected by localized torch heating of the suspected area. Lack of bonding will be indicated by a hot spot or spalling of the deposit material. Ultrasonic techniques may also be used to detect lack of bonding. Penetrant or magnetic particle inspection can detect exposed cracks or surface connected pores.

PROPERTIES

The quality and the properties of thermal sprayed deposits are largely determined by (1) the size, temperature, and velocity of the spray droplets as they impinge on the substrate and (2) the degree of oxidation of both the droplets and the substrate during spraying. These factors will vary with the method of spraying and the procedures employed.

Metal alloys deposited by the thermal spray process do not retain their original chemical composition unless special techniques are used. Their properties may change significantly because of this, depending upon the spray method used. With the plasma and arc methods, appreciable amounts of low melting point constituents may be lost by vaporization. Oxidation of the droplets may also be significant when air is used as the propellent.

The physical and mechanical properties of a spray deposit normally differ greatly from those of the original material. The deposit structure is lamellar and non-homogeneous. Its cohesion is generally the result of mechanical interlocking, some point-to-point fusion, and sometimes oxide-to-oxide bonding. The tensile strengths of these structures are low compared to those of the same materials in wrought or cast form. Sometimes, the compressive strength is quite high but the ductility is low. Deposits from wire or rod are less dense than the original material. In any case, spray deposits should be considered as a separate and distinct form of fabricated material.

Oxide spray deposits tend to retain their physical properties with only some decomposition. In many cases, the deposit will have a crystalline structure. Alumina, which may be alpha originally, may deposit with a metastable gamma structure. The chemical compositions of reactive type ceramics, such as carbides, silicides, and borides, normally change when the materials are sprayed in air with the flame or plasma methods.

MICROSTRUCTURE

The microstructure of a transverse section through a flame sprayed metal deposit will show a heterogeneous mixture of layered metal particles (white), metal oxide inclusions (gray), and pores (black). Figure 12.7 is a photomicrograph of a transverse section through a flame sprayed deposit of AISI 1080 steel. The light layered particles are bonded to one another by chemical and mechanical interactions. Figure 12.8 is a photomicrograph of a transverse section through a copper deposit and its substrate at the bond line. The roughness of the prepared substrate surface is apparent.

The microstructure of the polished and etched surface of the AISI 1080 steel deposit is shown in Fig. 12.9. It has an emulsified appearance because the flattened steel particles (light) are separated by the oxide (gray).

Fig. 12.7—Transverse section through a flame sprayed AISI 1080 steel deposit (×500 reduced on reproduction)

Fig. 12.8—Transverse section through a thermal sprayed copper deposit (top) and the substrate (bottom) (×500 reduced on reproduction)

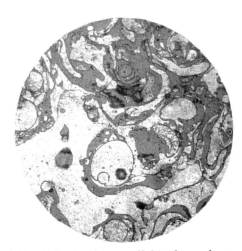

Fig. 12.9—Section parallel to the surface of a flame sprayed deposit of AISI 1080 steel (×500 reduced on reproduction)

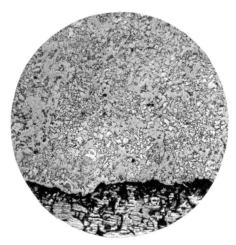

Fig. 12.10—Microstructure of a fused coating of a self-fluxing nickel-chromium alloy (top) on a substrate (bottom) (×250 reduced on reproduction)

As-sprayed, self-fluxing alloy deposits are similar in appearance to any typical metal deposit, except that there is significantly less oxide. These materials are oxidation resistant in nature. After fusing, the deposit will have an equiaxed cast structure with some porosity and inclusions. The microstructure of a fused nickel-chromium self-fluxing alloy deposit is shown in Fig. 12.10. The roughness of the prepared substrate is evident here also.

HARDNESS

The heterogenous structures of spray deposits generally have a lower macrohardness than the original rod or wire supplied to the gun. However, the hardness of individual deposit particles (microhardness) may be much higher than that of the overall deposit. The type of hardness test should be selected to give either the overall deposit hardness or the particle hardness. The thickness of the deposit must also be considered in selecting the type of test. If the deposit is too thin, the indenter may penetrate through it and into the substrate. This would obviously give a false reading.

The Brinell and Rockwell hardness tests can be used to determine the hardness of fairly thick metallic deposits. Superficial Rockwell and Vickers hardness tests may be suitable for thin metallic deposits. Requirements for various hardness tests are covered in the appropriate ASTM Standards. Table 12.2 relates the minimum spray deposit thicknesses to the various Rockwell hardness tests.

Hardness tests with diamond indenters are

Table 12.2
Minimum deposit thicknesses for
Rockwell hardness tests

Rockwell scale	Minimum thickness, in.
15N	0.015
30N	0.025
45N	0.035
A	0.040
B	0.060
C	0.070
D	0.050

not entirely satisfactory for determining the true hardness of heterogeneous spray deposits, but they can be used for spot checks and shop guides. Microhardness tests can be used to determine the hardness of individual particles. Since the deposited particles are relatively thin, hardness impressions should be taken on a transverse section. The Knoop indentation hardness test is best suited for this.

BOND STRENGTH

The strength of the bond between a spray deposit and the substrate depends upon many factors including the following:

(1) Substrate material
(2) Preparation of the substrate surface
(3) Preheat
(4) Bond layer material and its application method and procedures
(5) Deposit material and its application method and procedures
(6) Thickness of the deposit
(7) Postspraying thermal treatment
(8) Design of the bond strength test
(9) Technique of attachment to the deposit surface
(10) Effect of the attachment on the deposit properties

A standard test for determining the bond or cohesive strength of thermal spray deposits is described in ASTM C633, *Standard Test Method for Adhesion or Cohesive Strength of Flame Sprayed Coatings*. The arrangement for this test is shown in Fig. 12.11.

In this test, each specimen is an assembly of a substrate block to which the deposit is applied and a loading block. The flat end of the substrate block is prepared, and the deposit is applied to it. Then, the deposit is machined or ground flat and uniform in thickness using procedures appropriate for the deposit material. The loading block is then adhesive bonded to the flat deposit surface to produce a tension specimen. The specimen is loaded in tension at a constant rate using the self-aligning device. The maximum load is recorded. From this, the bond strength or the cohesive strength of the deposit can be calculated, depending upon the fracture location. The bond strengths of several self-bonding spray materials are presented in Table 12.3

This test method is limited to deposit thicknesses greater than 0.015 in. because adhesive bonding agents tend to infiltrate porous deposits. If the bonding agent penetrates to the substrate, it will affect the test results.

DENSITY

Thermal sprayed deposits have densities less than 100 percent because they are porous and contain some oxide. The densities of the flame sprayed deposits and the original wire for several metals are given in Table 12.4. In these cases, the deposits have densities greater than 80 percent.

Porosity in spray deposits consists of isolated and interconnected pores. It is difficult to determine the amount accurately. However, it can be estimated by several methods. The simplest one is to superimpose a grid over the microstructure of a prepared surface and then count the number of grid squares occupied by pores. Other methods include water or toluene immersion and paraffin absorption. Because of lack of total interconnection of the pores, however, no method is perfect.

The porous nature of spray deposits can be used to advantage, especially for bearing surfaces. The porosity permits oil retention and provides an escape for foreign material from actively loaded areas. Where corrosion is a factor, porosity is a disadvantage. It limits the use of deposits to those that are anodic to the base material unless special overcoatings of paints or sealers are used.

SHRINKAGE

Spray deposits contract upon cooling. The amount of shrinkage varies widely with different materials and spraying methods, but it will not be the same as that of the original material in cast or wrought form. Contraction sets up tensile stresses in the deposit as well as shear stresses across the bond between the de-

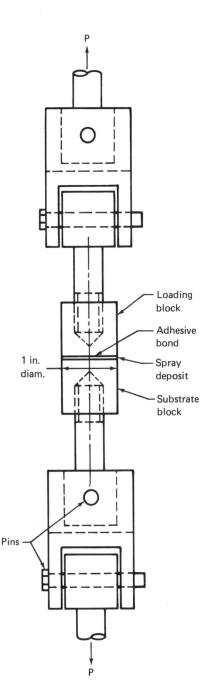

Fig. 12.11—Standard test arrangement for determining the bond or cohesive strength of thermal spray deposits

Table 12.3
Typical bond strengths of self-bonding materials[a]

Material	Bond strength, psi		
	Plasma sprayed	Flame sprayed	
		Powder	Wire
Columbium	2400		
Molybdenum	3200	3600	3300
Nickel-aluminide	3000	2750	3150
Tantalum	2750		

a. Applied to a smooth, unprepared metal surface.

posit and substrate. These stresses tend to crack the deposit or spall it from the substrate. Surface preparation and the selection of material for spraying are important factors in preventing these problems. Metals having low coefficients of thermal expansion should be used wherever possible, especially for thick deposits and buildup of internal surfaces.

APPLICATIONS

CORROSION AND OXIDATION PROTECTION

Thermal spray deposits can provide protection against many types of corrosive attack on iron and steel. Zinc, aluminum, stainless steel, bronze, hard alloys, and ceramics are used as surfacing materials. Service conditions determine both the material type and its application procedures. Undercoatings for organic materials, such as paints and plastic finishes, can be applied by this process. A thick layer of zinc or aluminum can protect steel against oxidation and provide a strong bond for an organic coating.

The porosity of anodic deposits is not important because they protect steel by electrochemical action. Sealing these deposits with paint or plastic solutions can impart a longer life.

Nickel, a nickel-copper alloy, stainless steels, and bronzes are some metals that are cathodic to steel. They should be used as a deposit on steel only if they are made impermeable to corrosive agents by sealing. The sealer is likely to entrap air bubbles in the pores of the spray deposit. The component should not be heated because expansion of the air bubbles may rupture the sealer.

Hard alloy deposits are often used on machine components such as pump plungers, pump rods, hydraulic rams, packing sections of steam turbine shafts, and valves. When sealed, these materials provide both corrosion and wear resistance.

Several different materials may be used to give oxidation protection. The choice depends upon the operating temperature. For applications up to 1600° F, the part can be aluminized by depositing a thin layer of aluminum. The aluminum is then diffused into the surface by a suitable heat treatment. For temperatures above 1600° F, a nickel-chromium alloy deposit may be used. This is followed by a spray deposit of aluminum. Often, this combination deposit is then covered with an aluminum pigmented bitumastic sealer. The part is then diffusion heat-treated in a furnace or it is placed directly in service if the operating temperature is above 1600° F. Such deposits are sometimes used for cyanide pots, furnace kiln parts, annealing boxes, and furnace conveyors.

Zirconia and alumina ceramics are sometimes used for thermal barrier layers. When the workpiece will be exposed to thermal cycling, a bond layer of nickel aluminide or nickel-chromium alloy may help to minimize thermal stresses in the ceramic deposit.

MACHINE ELEMENTS

In the mechanical field, thermal spray hardfacing materials can be used to combat many types of wear. The ability of metal spray deposits to absorb and maintain a film of lubricant is a distinct advantage in many applications. Spray deposits often give longer life than the original surfaces, except where severe conditions of shock loading or abrasion are encountered. A low cost base metal can be protected on only the areas of wear with a high quality, wear resistant deposit.

Some metal deposits, such as nickel-copper alloy, nickel, and stainless steel, are virtually impervious to penetration by corrosives when they are applied in sufficient thickness and are exposed only to moderate pressure. These surfaces can be vacuum impregnated with various phenolic or vinyl solutions or with fluorocarbon resins for high-pressure operation. For applications where extreme wear or corrosion resistance, or both, are encountered, fused spray coatings may be used.

ELECTRICAL

The electrical resistance of a metal spray deposit may be 50 to 100 percent higher than that of the same metal in cast or wrought form. This should be taken into consideration in the design of spray deposits for electrical conductors. Such applications include spraying of copper on electrical contacts, carbon brushes, and glass in automotive fuses as well as silver on copper contacts.

In the field of electrical insulation, various ceramic deposits can be used for insulators. Magnetic shielding of electrical components may be done with deposits of zinc applied to electronic cases and chassis. Condenser plates can be made by spraying aluminum on both sides of a cloth tape.

FOUNDRY

Changes in contour of expensive patterns and match plates can be readily accomplished by the application of thermal spray deposits followed by appropriate finishing. Patterns and molds can be repaired with wear resistant deposits. Blow holes in castings that appear during machining can be filled to salvage the parts.

BRAZING AND SOLDERING

Thermal spraying is frequently used for the preplacement of soldering or brazing filler metals. The usual practice is to apply the filler metal in place using standard thermal spraying techniques.

AIRCRAFT AND MISSILES

The process may be used for air seals and wear resistant surfaces to prevent fretting and galling at elevated temperatures. Deposits of alumina and zirconia may be used for thermal insulation.

Table 12.4
Comparison of the densities of flame sprayed metal deposits and the wire

| Metal | Density, lb/in.^{3} | |
	Flame sprayed deposit (wire)	Wire
Type 1100 Aluminum	0.087	0.098
Copper	0.271	0.324
Molybdenum	0.326	0.369
AISI 1025 steel	0.244	0.284
Type 304 stainless steel	0.249	0.29
Zinc	0.229	0.258

SAFETY

The potential hazards to the health and safety of personnel involved in thermal spraying operations and to persons in the immediate vicinity can be grouped as follows:

(1) Electrical shock
(2) Fire
(3) Gases
(4) Dust and fumes
(5) Arc radiation
(6) Noise

These hazards are not unique to thermal spraying methods. For example, flame spraying has hazards similar to those associated with the oxyfuel gas welding and cutting processes. Likewise, arc spraying and plasma spraying are similar in many respects to gas metal arc and plasma arc welding respectively. However, thermal spraying does generate dust and fumes to a greater degree.

EQUIPMENT

Gas Systems

Local, state, and federal regulations relative to the storage of gas cylinders should be investigated and their rules complied with. Storage, handling, and use of gas cylinders should be done in accordance with ANSI Z49.1, *Safety in Welding and Cutting,* and CGA P-1, *Safe Handling of Compressed Gases.* Improper storage, handling, and use of these cylinders constitute safety hazards in thermal spraying operations. Oil or grease must not be used on oxygen equipment; only special oxidation resistant lubricants may be used.

Acetylene pressures in excess of 15 psi (103 kPa) are dangerous and should not be used. When acetylene pressure of 15 psi is too low for the application, another fuel gas should be used. Alloys containing more than 67 percent copper or silver must not be used in acetylene systems because dangerous explosive compounds may be formed .

Flame Spray Guns

Flame spray guns should be maintained in accordance with the manufacturer's recommendations. Each operator should be familiar with the operation of the flame spray gun and should read the instruction manual thoroughly before using it.

A friction lighter, a pilot light, or arc ignition should be used to ignite the gun. Matches are not safe. A flame spray gun and its hoses should not be hung on gas regulators or cylinder valves because of the danger of fire or explosion.

Plasma and Arc Spraying Equipment

The plasma and arc methods of thermal spraying use equipment where high voltages and amperages present a hazard. Operators should be thoroughly instructed and trained in the operation of the unit. They should be familiar with the operating and safety recommendations and, at the same time, observe proper safety precautions for electrical equipment.

The plasma or arc spraying equipment itself should be safe to operate. Exposed electrodes of plasma guns should be grounded or adequately insulated. Periodic inspections should be made of cables, insulation, hoses, and gas lines. Faulty equipment must be repaired or replaced immediately. The entire system, including the power supply, must be shut down before repairing any part of the power supply, console, or gun.

Arc spray guns should be cleaned frequently according to the manufacturer's operation manual to prevent the accumulation of metal dust. If the plasma or arc spray gun is suspended, the suspension hook must be insulated or grounded. Contact between any ungrounded portion of the plasma or arc spray gun and the spray booth or chamber must be avoided.

FIRE PREVENTION

Airborne finely divided solids, especially metal dusts, must be treated as explosives. To minimize danger from dust explosions, adequate ventilation should be provided to spray booths.

A wet collector of the water-wash type is recommended to collect the spray dust. Bag or filter type collectors are not recommended. Good housekeeping in the work area should be maintained to avoid accumulation of metal dusts, particularly on rafters, tops of booths, and in floor cracks.

Paper, wood, oily rags, and other combustibles in the spraying area can cause a fire and should be removed before the equipment is operated.

PROTECTION OF PERSONNEL

The general requirements for the protection of thermal spray operators are the same as for welders as set forth in ANSI Z49.1, *Safety in Welding and Cutting;* ANSI Z87.1, *Practices for Occupational and Educational Eye and Face Protection;* ANSI Z88.2, *Practices for Respiratory Protection;* and ANSI Z89.1, *Safety Requirements for Industrial Head Protection.*

Eye Protection

Helmets, hand shields, face shields, or goggles should be used to protect the eyes, face, and neck during all thermal spraying operations. These are described in ANSI Z87.1 and Z89.1. Safety goggles should be worn at all times. Helmets, hand shields, or goggles must be equipped with suitable filter plates to protect the eyes from excessive ultraviolet, infrared, and intense visible radiation. Table 12.5 is a guide for the selection of the proper filter shade number.

Respiratory Protection

Most thermal spraying operations require that respiratory protective devices be used by the operator. The nature, type, and magnitude of the fume and gas exposure determine which respiratory protective device should be used. The selection of these devices should be in accordance with ANSI Z88.2. This Standard contains descriptions, limitations, operational procedures, and maintenance requirements for standard respiratory protective devices. All devices selected should be of a type approved by the U.S. Bureau of Mines, National Institute for

Table 12.5
Recommended eye filter plates for thermal spraying operations

Operation	Filter shade numbers
Wire flame spraying (except molybdenum)	2 to 4
Wire flame spraying of molybdenum	3 to 6
Flame spraying of metal powder	3 to 6
Flame spraying of exothermics or ceramics	4 to 8
Plasma and arc spraying	9 to 12
Fusing operations	4 to 6

Occupational Safety and Health, or other approving authority for the purpose intended.

Ear Protection

Ear protectors or properly fitted soft rubber ear plugs should be worn to protect the operator from the high-intensity noise from the gun. Such protection should reduce the noise level to below 80 decibels. Cotton wads should not be used for ear protection as they are ineffective against high-intensity noise. Federal, state, and local codes should be checked for noise protection requirements.

Protective Clothing

Appropriate protective clothing required for a thermal spraying operation will vary with the size, nature, and location of the work to be performed. When working in confined spaces, flame resistant clothing as well as leather, rubber, or asbestos gauntlets should be worn. Clothing should be fastened tightly around the wrists and ankles to keep dusts from contacting the skin.

For work in the open, ordinary clothing such as overalls and jumpers may be used. However, open shirt collars and loose pocket flaps are potential hazards. High-top shoes should be worn, and cuffless trousers should cover the shoe tops.

The intense ultraviolet radiation of plasma arc spraying can cause skin burns through normal clothing. When using this process, the clothing should provide protection against such radiation. Thick, tightly-woven wool clothing may be sufficient. For exposure to more intense radiation, leather capes may be necessary. Protection against radiation with arc spraying is practically the same as that used for normal electric arc welding.

Metric Conversion Factors

$t_C = 0.56\,(t_F - 32)$

1 in. = 25.4 mm

$1\ lb/in.^3 = 2.77 \times 10^4\ kg/m^3$

1 psi = 6.89 kPa

1 ft/s = 0.305 m/s

SUPPLEMENTARY READING LIST

AWS C2.1-73, *Recommended Safe Practices for Thermal Spraying,* Miami: American Welding Socirty, 1973.

AWS C2.2-67, *Recommended Practices for Metallizing With Aluminum and Zinc for Protection of Iron and Steel,* Miami: American Welding Society, 1967.

AWS C2.13-70, *Flame Spraying of Ceramics,* Miami: American Welding Society, 1970.

AWS C2.15-75, *Recommended Practices for Fused Thermal Sprayed Deposits,* Miami: American Welding Society, 1975.

AWS C2.16-78, *Guide for Thermal Spray Operator and Equipment Qualification,* Miami: American Welding Society, 1978.

AWS C2.17-78, *Recommended Practices for Electric Arc Spraying,* Miami: American Welding Society, 1978.

Hermancek, F. J., Determining the adhesive/cohesive strength of thin thermally sprayed deposits. *Welding Journal,* 57 (11): 31-35; 1978 Nov.

Hermancek, F. J., Thermal conductivity and thermal shock qualities of zirconia coatings on thin gage Ni-Mo-C metal. *Metal Progress,* 97 (3): 104; 1970 March.

Ingham, H. S., Jr. and Fabel, A. J., Comparison of plasma flame spray gases. *Welding Journal,* 54 (2): 101-5; 1975 Feb.

Irons, G. C., Laser fusing of flame sprayed coatings. *Welding Journal,* 57 (12): 29-32; 1978 Dec.

Longo, F. N., Use of flame sprayed bond coatings. *Plating,* 61(10): 306-11; 1974 Oct.

Longo, F. N. and Durmann, G. J., Corrosion prevention with thermal sprayed zinc and aluminum coatings. *Welding Journal,* 53 (6): 363-70; 1974 June.

Papers of the Eighth International Thermal Spraying Conference, Miami Beach, FL: Sept. 27 to Oct. 1, 1976; American Welding Society, 1976.

Phelps, H. C., Fuel gas additive helps solve problem of rejects during thermal spraying operation. *Welding Journal,* 56 (7): 32-35; 1977 July.

13
Other Welding Processes

Chapter Committee

L. J. PRIVOZNIK, *Chairman*
    Westinghouse Electric Corporation
P. B. DICKERSON
    Aluminium Company of America

H. D. FRICKE
    U.S. Thermit, Inc.
A. N. KUGLER
    Consulting Engineer

T. J. MOORE
    *National Aeronautics and
    Space Administration*

Welding Handbook Committee Member

L. J. PRIVOZNIK
    Westinghouse Electric Corporation

13

Other Welding Processes

THERMIT WELDING

FUNDAMENTALS OF THE PROCESS

Definition

Thermit[1] welding is a process that produces coalescence of metals by heating them with superheated molten metal from an aluminothermic reaction between a metal oxide and aluminum. Filler metal is obtained from the liquid metal. The process had its beginning at the end of the 19th century when Hans Goldschmidt discovered that the exothermic reaction between aluminum powder and a metal oxide can be initiated by an external heat source. Once the reaction is started, it is self-sustaining because of its exothermic nature.

Principles of Operation[2]

The thermochemical reaction takes place according to the following general equation:

Metal oxide + aluminum (powder)→
aluminum oxide + metal + heat

The reaction can only be started and completed if the oxygen affinity of the reducing agent (aluminum) is higher than the oxygen affinity of the metal to be reduced. The heat generated by this exothermic reaction results in a liquid product consisting of metal and aluminum oxide. If the density of the slag is lower

1. Thermit is the term commonly used to identify this welding process even though it is a registered trademark.
2. See also Ch. 2, "Physics of Welding," Vol. 1, 7th ed., 43-44.

than that of the metal, as in the case of steel and aluminum oxide, they separate immediately. The slag floats to the surface and the molten steel can be used for welding purposes.

Typical thermochemical reactions and the thermal energies produced are as follows:

$$3\,Fe_3O_4 + 8\,Al \rightarrow 9\,Fe + 4Al_2O_3 : \Delta H = 3350\,kJ$$
$$3\,FeO + 2\,Al \rightarrow 3\,Fe + Al_2O_3 : \Delta H = 880\,kJ$$
$$Fe_2O_3 + 2\,Al \rightarrow 2\,Fe + Al_2O_3 : \Delta H = 850\,kJ$$
$$3\,CuO + 2\,Al \rightarrow 3\,Cu + Al_2O_3 : \Delta H = 1210\,kJ$$
$$3\,Cu_2O + 2\,Al \rightarrow 6\,Cu + Al_2O_3 : \Delta H = 1060\,kJ$$

In the above reactions, aluminum is the reducing agent. Theoretically, the elements magnesium, silicon, and calcium can be used as well; but, for general applications, magnesium and calcium have found limited use because of technical disadvantages. Silicon is often used in Thermit mixtures where a liquid product is not wanted, such as for heat treatment. In other cases, an alloy of equal parts of aluminum and silicon is used as a reducing agent.

The first of the reactions above is the one most commonly used as a basis of mixtures for Thermit welding. The proportions of such mixtures are usually about three parts by weight of iron oxide to one part of aluminum. The theoretical temperature created by this reaction is about 5600° F. Additions of nonreacting constituents, as well as heat loss to the reaction vessel and radiation, reduce this temperature to about 4500° F. This is about the maximum temperature that can be tolerated, since aluminum vaporizes at 4530° F. On the other hand, the maximum temperature should not be much lower because the aluminum slag (Al_2O_3) solidifies at 3700° F.

The heat loss depends very much upon the quantity of Thermit being reacted. With large quantities, the heat loss per pound of Thermit is considerably lower and the reaction more complete when compared with small quantities of Thermit.

Alloying elements can be added to the Thermit compound in the form of ferroalloys to match the chemistry of the parts to be welded. Other additions are used to increase the fluidity and lower the solidification temperature of the slag.

The Thermit reaction is nonexplosive and requires less than one minute for completion, regardless of quantity. To start the reaction, a special ignition powder or ignition rod is required; both can be ignited by a regular match. The ignition rod or powder will produce enough heat to raise the Thermit powder in contact with it to its ignition temperature, which is about 2200° F.

The parts to be welded must be lined up properly; the faces to be joined must be free of rust, loose dirt, and grease. A proper gap must be provided between the faces, the size depending upon the cross section of the joint. A mold, which is either built up on the parts or premanufactured to conform to the parts, is placed aound the joint to be welded.

To accomplish a butt weld, the ends of the parts must be preheated sufficiently to provide conditions for complete fusion between the molten steel and the base metal. Even though it is called a welding process, Thermit welding resembles metal casting where proper gates and risers are needed to:

(1) Compensate for volume shrinkage during solidification

(2) Eliminate typical defects that appear in castings

(3) Provide proper flow of the molten steel

(4) Avoid turbulence as the metal flows into the joint

APPLICATIONS

Rail Welding

The most common application of the process is the welding of rail sections into continuous lengths. It is an effective means of minimizing the number of bolted joints in the track structure. In coal mines, the main haulage track is often welded to minimize maintenance and to reduce excessive coal spillage caused by uneven track. Crane rails are regularly welded to minimize joint maintenance and vibration of the building as heavily loaded wheels pass over the joint.

Thermit mixtures are available for all types of rail steels. The majority of rails are C-Mn steels, but Cr, Cr-Mo, Cr-V, and Si alloy rail steels are manufactured abroad. Addition of rare earth metals or alloys may decrease the amount of sulfur and phosphorous in steel made by the Thermit process, resulting in an improvement in mechanical properties.

Welding with Preheat. Premanufactured molds of split design are generally used for welding standard rail sizes. The mold should be aligned so that its center coincides with the center of the gap between the rail ends. The rail ends are preheated in the range of 1100° F to 1800° F with a gas torch flame directed into the mold as shown in Fig. 13.1. The refractory-lined crucible containing the Thermit charge is positioned above the mold halves after preheating is completed. The charge is then ignited and the molten steel pours into the joint. In some procedures, the metal is fed into the middle of the joint gap (top-poured); in other procedures the metal enters the bottom of the mold at the outer leg of the rail base and rises vertically in the middle (bottom-poured).

A self-tapping seal (disk) is used in the bottom of the crucible. A few seconds after the Thermit reaction is complete, the molten metal melts the seal and pours out of the bottom of the crucible into the gap between the two rail sections. The liquid slag separates from the steel within the crucible because of its lower density. It does not reach the mold cavity until all of the molten steel has entered and filled

***Fig. 13.1—Preheating rail ends with a gas torch
prior to Thermit welding***

both the cavity between the rail sections and the mold itself. The slag remains on top of the weld and solidifies there. When the metal has solidified, the mold halves are removed and discarded. The excess metal is removed by hand grinding or by hydraulic or manual shearing devices.

Thermit welding with short preheating time usually requires a larger Thermit charge than is needed with long preheating time. The heat dissipated into the work during welding has to be provided by a larger mass of molten steel.

Welding without Preheat. The self-preheating method is designed to eliminate the variables associated with torch preheating and the equipment needed to perform that operation.

The rail ends are preheated by a portion of the molten metal produced by the Thermit reaction. The crucible and mold are a one-piece design as shown in Fig. 13.2. The molds, commonly known as shell molds, are premanufactured of sand bonded with phenolic resins. They are very light, non-hygroscopic, and moisture-free with unlimited shelf life. After the Thermit reaction is completed, the molten steel automatically flows from the crucible into the joint rather than passing through the atmosphere, as is the case with a separate crucible.

Figure 13.3, a section through a mold, shows the shape of the cavity in which the molten filler metal flows. There is a hollow chamber in the mold underneath the weld area

Preparing the Joint. The pieces to be joined should be properly positioned in contact and aligned for welding. Firm marks should then be made on the pieces outside the area to be covered by the mold box. They will be used to reposition the pieces after the groove faces are prepared for welding. The metal may then be cut with a cutting torch along the line of fracture to provide a parallel-sided gap. The gap width depends upon the size of the section to be welded as shown in Table 13.1. All loose oxide and slag from torch cutting, as well as dirt and grease, must be removed from the pieces where the mold will be located.

To allow for contraction of the weld during cooling, the pieces are initially spaced 1/16 to 1/4 in. further apart than their original positions, using the markers on the pieces for reference. The exact increase depends upon the size of the weld and the gap length. The amount of contraction allowance required can be judged quite

Fig. 13.2—Combination mold-crucible in position for Thermit welding without preheat

that receives the first molten metal to allow it to preheat the rail ends. This metal is called the preheat metal. By the time the chamber is filled, sufficient molten metal should have passed over the rail ends to preheat them to the required temperature to assure complete fusion with the base metal. Thermit portions for this process are about twice the size of those used for the external preheat method.

The heat-affected zones in the adjacent rail sections are considerably smaller than when external preheating is used. Figure 13.4 shows a typical section through a Thermit rail weld made with the self-preheating process.

Repair Welding

Repair welds are normally nonrepetitive and, therefore, premanufactured molds are not used. A mold must be made for each weld so that it will conform to the shape of the part.

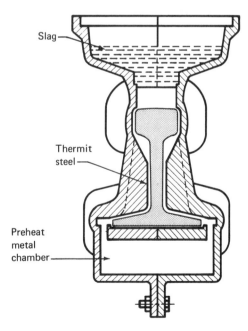

Slag

Thermit steel

Preheat metal chamber

Fig. 13.3—Section through a mold-crucible having a preheat metal chamber

Table 13.1
Examples of Thermit weld dimensions and mold requirements

Section size or diam., in.	Gap, in.	Collar, in.	Risers		Pouring gates		Heating gates		Connecting gates		Thermit req'd,[a] lb.
			No.	Diam., in.	No.	Diam., in.	No.	Diam., in.	No.	Diam., in.	
Rectangular sections											
2x2	7/16	1-1/2x7/16	1	3/4	1	3/4	1	1-1/4			6
2x4	9/16	1-5/16x9/16	1	3/4	1	1	1	1-1/4			12
4x4	11/16	2-5/8x11/16	1	1	1	1	1	1-1/4			25
4x8	7/8	3-7/16x7/8	1	1	1	1	2[b]	1-1/4			50
8x8	1- 1/8	4-5/8x1-1/8	1	1-3/4	1	1-1/4	2[b]	1-1/4	—	—	125
8x12	1- 1/4	5-1/2x1-1/4	1	1-3/4	1	1-1/4	1	1-1/4	1	1-1/4	175
12x12	1- 7/16	6-1/2x1-7/16	1	2-1/2	1	1-1/2	2[b]	1-1/2	1	1-1/2	300
12x18	1-11/16	7-3/4x1-11/16	1	2-1/2	1	1-1/2	2[b]	1-1/2	1	1-1/2	500
16x16	1- 3/4	8-15/16x1-3/4	1	2-3/4	2	2	2	1-1/2	2	1-1/2	700
16x24	2	9-15/16x2	1	2-3/4	2	2	2	1-1/2	2	1-1/2	1150
24x24	2- 5/16	11-13/16x2-5/16	2	2-1/2	2	2	2	1-3/4	2	1-3/4	1875
24x36	2- 5/8	14-1/8 x2-5/8	2	2-1/2	2	2	2	2	4	2	3125
Round sections											
2	7/16	1-3/8x7/16	1	3/4	1	3/4	1	1-1/4			5
4	5/8	2-3/8x5/8	1	1	1	1	1	1-1/4			25
8	1	4-3/16x1	1	1-1/2	1	1-1/4	1	1-1/4			75
12	1- 5/16	5-7/8x1-5/16	1	1-3/4	1	1-1/2	1	1-1/112	1	1-1/2	200
16	1- 5/8	7-1/2x1-5/8	1	2	1	1-1/2	1	1-1/2	1	1-1/2	425

a. Thermit required includes provision for a 10% excess of steel in slag basin for a single pour and a 20% excess for a double pour.
b. Includes one separate back heating gate.

accurately with experience.[3]

Applying the Mold. When a single large weld is to be made, a wax pattern is used to shape the mold cavity at the joint, similar to the lost-wax casting process. The wax is placed in the gap and on the surfaces of the parts to produce the exact shape desired for the finished weld, including the collar of weld reinforcement. A sand mold is then built up around the pattern, using a suitable mold box to contain the mold sand.

Wood patterns of pouring and heating gates and risers are positioned within the mold as it is being rammed. Where two pieces of the same size are to be welded together, the heating gate is centered directly on the wax pattern. If unequal sections are being welded together, the heating gate is directed toward the larger section to provide somewhat uniform heating of the two parts. Where there are one or more high points on a joint of complex cross section, riser gates will be required at all of them. The top of the mold is hollowed out to provide a basin for the slag produced by the thermit reaction. The mold must be adequately vented to facilitate the escape of moisture during the preheating process. Finally, the wood patterns are removed. Figure 13.5 shows a section through a completed Thermit weld mold with the crucible in position.

The quality of the molding sand requires special attention. It must have high refractoriness, high permeability, and adequate shear strength. It should be free of clay components with low melting points.

Preheating. Preheating is accomplished by directing a gas flame into the chamber through

3. Transverse shrinkage in butt welds is discussed in Ch. 6, "Residual Stresses and Distortion," Vol. 1, 7th ed: 250-255.

Upon completion of the preheating, the heating gate must be blocked. A short length of steel rod of appropriate diameter is pushed into the gate against a shoulder and then backed with molding sand.

Charging the Crucible. The Thermit reaction, as in the case of preheated type rail welding, takes place in a refractory-lined, cone-shaped crucible as shown in Fig. 13.6. A hard refractory (magnesite) stone at the bottom of the crucible holds a replaceable refractory orifice or thimble. The thimble is plugged by inserting a tapping pin through it and then placing a metal disk on top of the pin. The disk is covered with a layer of refractory sand. The Thermit mixture should be placed in the crucible in a manner that will not dislodge the sand layer.

Low carbon steel punchings are sometimes added to the Thermit mixture to augment the metal produced. The quantity of Thermit mixture required for a joint can be calculated by the following equation:

$$X = \frac{E}{0.5 + 0.01S}$$

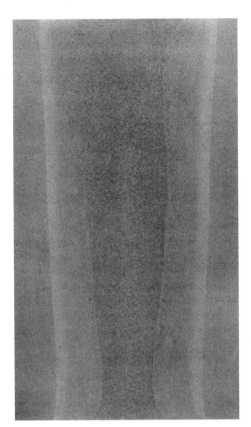

Fig. 13.4—Photomacrograph of a vertical section through a typical thermit rail weld

the heating gate. A torch designed specifically for the purpose may burn propane, natural gas, kerosene, or gasoline.

The initial purpose of preheating is to remove the wax. The heat is applied gradually, and the torch is frequently removed from the heating gate to allow the melted wax to flow out. After the wax has been removed, the heat is gradually increased to preheat the faces of the base metal and thoroughly dry the mold. The mold must be completely dried to avoid weld porosity generated by residual moisture in the molding sand. Preheating is continued until the ends of the parts to be welded are cherry red in color, an indication that their temperature is between 1500° F and 1800° F.

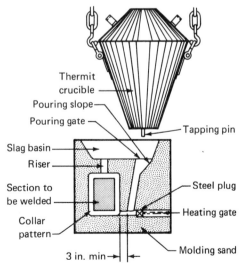

Fig. 13.5—Cross section of a typical thermit mold for repair welding with external preheat

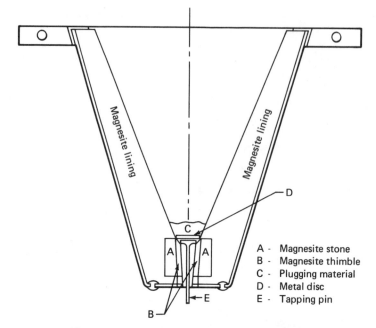

A - Magnesite stone
B - Magnesite thimble
C - Plugging material
D - Metal disc
E - Tapping pin

Fig. 13.6 — Cross section of a thermit crucible

where:

X = quantity of Thermit required, lb

E = quantity of molten steel required to fill the gap, including 10 percent for losses, lb

S = percent of steel punchings to be included in the charge

As a rough rule, the amount of Thermit is 25 pounds for each pound of wax in the pattern.

Making the Weld. The reaction can be initiated by two methods: (1) starting powder that can be ignited by a match or regular gas striker, or (2) an ignition rod.

After the reaction is completed and the action of the molten steel subsides, the crucible is tapped by striking the tapping pin with a sharp upward blow. The molten steel flows into the mold and fills the joint.

The mold is stripped away after the weld metal has solidified. Whenever possible, the entire weldment should be annealed to stress relieve it.

If required, the collar around the weld may be removed by machining or grinding. Risers and the gate are removed with an oxyfuel gas cutting torch.

Applications of Repair Welding. Thermit welding is employed in the marine field for the repair of heavy sections of ferrous metal such as broken stern frames, rudder parts, shafts, and struts.

Broken necks, pinions, and pinion teeth of sheet and plate rolls are replaced with entirely new pieces, cast or forged slightly oversize to permit machining. They are Thermit welded to the main section.

Badly worn wobblers on the ends of steel mill rolls may similarly be replaced with a sufficiently tough Thermit metal deposit that is machinable. The method is particularly applicable for repairs involving large volumes of metal, where the heat of fusion cannot be raised satisfactorily or efficiently by other means or where fractures or voids in large sections require a large quantity of weld metal.

Thermit welding can be used for the repair of ingot molds at significant savings over replacement: the bottom of the mold can be cut off and completely rebuilt with Thermit metal, or an eroded cavity in the bottom can be filled with Thermit metal. The first method of repair is more sophisticated and requires larger quantities of Thermit, but the lifetime of the ingot mold will be more than doubled. The latter type of repair has to be repeated after every second or third pour.

With large dredge cutters, the blades may be Thermit welded to a center ring. Quantities up to several thousand pounds are poured at one time. In this case, Thermit welding is a production tool rather than a repair method.

Reinforcing Bar Welding

Thermit welding without preheat is one way of splicing concrete reinforcing steel bars. Continuous reinforcing bars permit the design of concrete columns or beams smaller in section than when the bars are not welded together.

The two mold halves, which are premanufactured by the CO_2 or shell-mold process, are positioned at the joint in the aligned bars and sealed to them with asbestos and sand to avoid loss of molten metal. The arrangements for horizontal and vertical welding are shown in Fig. 13.7. A closure disk is located in a well at the base of the Thermit crucible section of the

mold. The Thermit powder is placed in the crucible and the reaction initiated. After completion of the reaction, the molten steel melts through the closure disk and fills the gap between the bars. The initial molten steel entering the mold chamber preheats the bar ends as it flows over them into a preheat metal chamber. The molten steel fills the joint gap and consummates a weld. Reinforcing bars can be welded by this process in any position with properly designed molds. Thermit welded reinforcing bar specimens tested in tension and bending are shown in Fig. 13.8.

As an alternative to welding, reinforcing bars can be joined end-to-end by casting steel between a steel sleeve and the enclosed bars. The joint is primarily mechanical. Arrangements for horizontal and vertical connections are shown in Fig. 13.9. The sleeve is placed around the abutted bars. Its inside diameter is somewhat larger than the diameter of the bars to provide space for the cast steel. Both the inner surface of the sleeve and the surfaces of the bars are serrated. A graphite or CO_2 mold is mounted over an orifice in the sleeve through which the molten steel flows into the annular space between the bars and sleeve. The Thermit

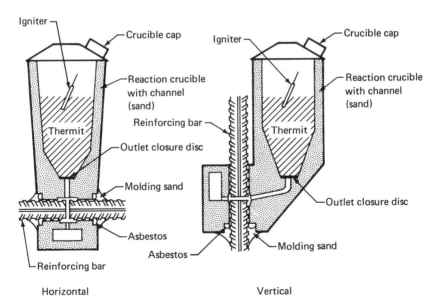

Fig. 13.7—Thermit welding concrete reinforcing steel bars

mix is placed above a metal closure disk within the mold.

Upon ignition of the Thermit powder, molten steel is produced. It melts the metal disk, flows into the annular space between the bars

Fig. 13.8—Thermit welded reinforcing steel bars after tensile and bend testing

the sleeve, and then solidifies. About five minutes are required to make a joint with this technique.

Electrical Connections

A Thermit mixture of copper oxide and aluminum is used for the welding of joints in copper conductors. The reaction between the two materials produces superheated molten copper and slag in 1 to 5 seconds. Other metals in the form of slugs or powder can be added to produce alloys for particular applications of the process.

The process is primarily used for welding copper bars, cables, and wires together, as well as copper conductors to steel rails for grounding. For the latter application, a graphite mold is clamped to the rail section at the joint. As soon as the Thermit reaction is complete, the molten copper melts the disk and flows into the joint cavity. It solidifies in a few seconds and creates a weld between the base metal and the copper cable. Then the mold is removed. It can be used again after removing the slag from the reaction chamber.

Heat Treatment of Welds

The development of high alloy steels and

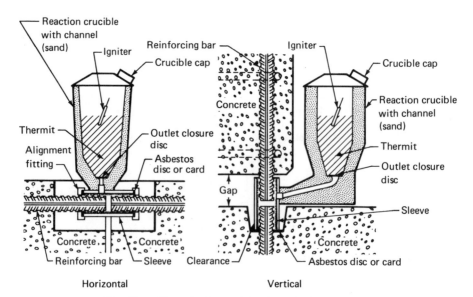

Fig. 13.9—Thermit sleeve joint for reinforcing bars

the application of the Thermit process for welding them created a need for special types of Thermit mixtures that produce heat only. Molten metal is not produced. This particular type of Thermit is designed to create sufficient heat for heat-treating purposes.

Using special binders, the Thermit mixture itself is formed to the configuration of the parts to be heat treated. It keeps its exact shape during and after the Thermit reaction. The maximum temperature produced in the part can be adjusted by the design of the Thermit compound. Figure 13.10 shows Thermit blocks in place for heat treating a rail section.

SAFETY

The presence of moisture in the Thermit mix, in the crucible, or on the workpieces can lead to rapid formation of steam when the Thermit reaction takes place. This may cause ejection of molten metal from the crucible. Therefore, the Thermit mix should be stored in a dry place, the crucible should be dry, and moisture should not be allowed to enter the system before or during welding.

The work area should be free of combustible materials that may be ignited by sparks or small particles of molten metal. The area should be well ventilated to avoid the buildup of fumes and gases from the reaction. Starting powders and rods should be protected against accidental ignition.

Personnel should wear appropriate protec-

Fig. 13.10—Bonded Thermit blocks in place for heat treating a rail section

tion against hot particles or sparks. This includes full face shields with filter lenses for eye protection and headgear. Safety boots are recommended to protect the feet from hot sparks. Clothing should not have pockets or cuffs that might catch hot particles.

Preheating should be done taking safety precautions applicable to all oxyfuel gas equipment and operations.

COLD WELDING

FUNDAMENTALS OF THE PROCESS

Definition and General Description

Cold welding (CW) is a solid-state process in which pressure is used at room temperature to produce coalescence of metals with substantial deformation at the weld. A characteristic of the process is the absence of heat, either applied externally or generated by the welding process itself. A fundamental requisite for satisfactory cold welding is that at least one of the metals to be joined is highly ductile and does not exhibit extreme work-hardening. Both butt and lap joints can be cold welded. Typical joints are shown in Fig. 13.11.

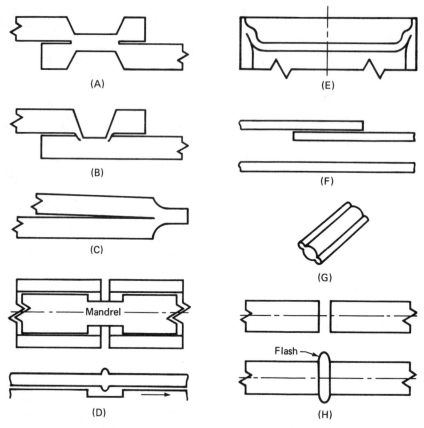

(A) lap weld, both sides indented; (B) lap weld, one side indented; (C) edge weld, both sides indented; (D) butt joint in tubing, before and after welding; (E) draw weld; (F) lapped wire, before and after welding; (G) mash cap joint; (H) butt joint in solid stock, before and after welding

Fig. 13.11—Contour of typical cold welded joints

Materials Welded

Metals with face-centered-cubic (FCC) lattice structure are best suited for cold welding, provided they do not work-harden rapidly. Soft tempers of metals such as aluminum and copper are most easily cold welded. It is more difficult to weld cold-worked or heat-treated alloys of these metals. Work-hardening at the interface becomes pronounced or fracture occurs elsewhere in the piece when ductility becomes exhausted. Other FCC metals that may be cold welded readily are gold, silver, palladium, and platinum.

The joining of copper to aluminum by cold welding is a good application of the process, especially where aluminum tubing or electrical conductor grade aluminum is joined to short sections of copper to provide transition joints between the two metals. Such cold welds are characterized by substantially greater deformation of the aluminum than the copper because of the difference in the yield strengths and work-hardening behaviors of the two metals.

Dissimilar Metal Welds

Numerous dissimilar metals may be joined by cold welding whether they are quite soluble or relatively insoluble in one another. In some cases, the two metals may combine to form intermetallic compounds. Since cold welding is carried out at room temperature, there is no significant diffusion between dissimilar metals during welding. The alloying characteristics of the metals being joined do not affect the manner in which the cold welding operation is carried out. However, interdiffusion at elevated temperatures can affect the choice of postweld thermal treatments and the performance of the weld in service.

Welds made between metals that are essentially insoluble in each other are usually stable. Diffusion can form an intermetallic compound at elevated service temperatures. In some cases, this intermetallic layer can be brittle and cause a marked reduction in the ductility of the weld. Such welds are particularly sensitive to bending or impact loading after an intermetallic layer has formed.

The rate at which intermetallic compounds form depends upon the specific diffusion constants for the particular metals in the weld as well as the time and temperature of exposure. Thus, bimetal cold weldments require careful consideration of the diffusion couple and the service environment. For example, a layered structure forms at the interface in an aluminum-copper weldment at elevated temperatures as shown in Fig. 13.12. The layered structure contains a brittle Al-Cu intermetallic compound that weakens the weldment. Figure 13.13 shows how rapidly the thickness of the diffused zone increases at high service temperatures. Mechanical tests have shown that the strength and ductility of the joint decrease when the thickness of the interfacial layer exceeds about 0.002 inch. Consequently, aluminum-copper cold welds should be used only in applications where service temperatures are low and peak temperatures seldom, if ever, exceed 150° F.

Metallurgical Structure

In butt joints, the lateral flow of metal between the dies during upset produces a cross-grained structure adjacent to the interface of the

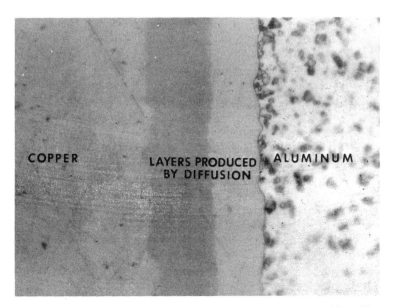

Fig. 13.12—Layered structure in an aluminum-copper cold weld after exposure at 500° F for 60 days

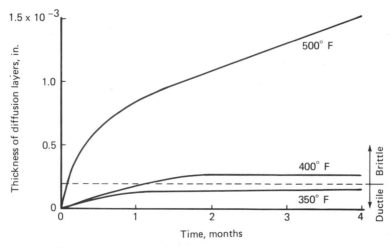

Fig. 13.13 — Change in thickness of the diffusion layer in aluminum-copper cold welds with time at three elevated temperatures

weld, as shown in Fig. 13.14. This cross-grained material is essentially a narrow transverse section in the weldment. The presence of this section is not important in metals that are essentially isotropic, such as aluminum and some aluminum alloys. In nonisotropic metals, fatigue or corrosion resistance may be substantially lower at the welded joint.

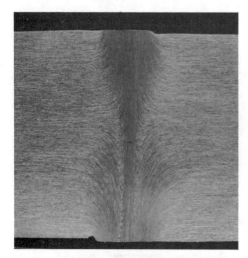

Fig. 13.14 — Transverse flow lines in a cold welded butt joint

Surface Preparation

Cold welding requires that clean metal faces come into intimate contact for a strong joint. Proper surface preparation is very important to assure joints of maximum strength. Dirt, adsorbed gas, oils, or oxide films on the surface interfere with metal-to-metal contact and must be removed to obtain strong welds.

The best method of surface preparation for lap welds is wire brushing at a surface speed of about 3000 ft per minute. A motor-driven rotary brush of 0.004-in. diameter stainless steel wire is commonly used. Softer wire brushes may burnish the surface; coarser types may remove too much metal and roughen the surface. The surfaces should be degreased prior to brushing to avoid contamination of the wire brushes. It is important that the clean surface not be touched with the hands because grease or oil on the faying surfaces impairs the formation of a strong joint. Welding should take place as soon as practical after cleaning to avoid interference with bonding from oxidation. In the case of aluminum, for example, welding should be done within about 30 minutes.

Chemical and abrasive cleaning methods have not proved satisfactory for cleaning surfaces to be joined by cold welding. The

The bar type indentation, Figs. 13.17(A) and (B), causes metal deformation along both of its sides. Indentations in the form of a ring, Fig. 13.17(C), may cause undesirable curvature of the sheet surfaces. A ring weld may have a smooth convex dome of metal within the ring. This dome is formed when the metal is forced from between the dies as pressure is applied. Continuous seam welding, Fig. 13.17(E), can be employed in the manufacture of thin-wall tubing or lap welds in sheet.

Symmetrical dies that indent both sides of the joint, Figs. 13.11(A) and (C), are generally used. If one surface must be free of indentation, a flat plate or anvil may be used on one side to produce the weld shown in Fig. 13.11(B). Thinner gages of sheet metal or wire can be cold

Fig. 13.15—A manually operated cold welding machine

residue from chemical cleaning or abrasive particles embedded in or left on the surface may prevent the formation of a sound weld.

EQUIPMENT

Pressure for welding may be applied to overlapped or butted surfaces with hydraulic or mechanical presses, rolls, or special manual or pneumatically operated tools. A hand tool of the toggle cutter type is suitable for very light work. A common manually operated press, as shown in Fig. 13.15, may be used for medium size work. Very heavy work requires power operated machines such as the one in Fig. 13.16. The rate of pressure application does not usually affect the strength or quality of the weld.

Regardless of how pressure is applied, the proper indentation for lap welds is important. The indentation may take the form of a narrow strip, a ring, or a continuous seam. Typical lap weld indentor configurations used for cold welding are presented in Fig. 13.17. The selection of indentation configuration is largely determined by the desired appearance and performance characteristics.

Fig. 13.16—A hydraulically operated cold welding machine

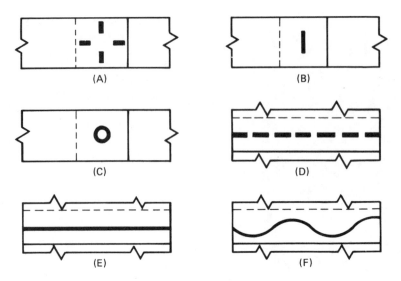

(A) and (B), bar type; (C) ring type; (D), (E), and (F), intermittent and continuous seam types

Fig. 13.17—Typical lap weld indentor configurations used in cold welding

welded using simple dies mounted in hand-operated tools.

Draw welding is a form of lap welding used to seal containers. Both the lid and the can are flared before welding. The components are placed in a close fitting die. A punch forces the components into the die which cold welds the flared metal as it is drawn down over the punch. Figure 13.11(E) illustrates such a joint.

Dies are usually subjected to high pressures and should be made of tool steel hardened to about 60 Rockwell C. Pressures of from 150 to 500 ksi are required to weld aluminum depending upon alloy composition and temper. Copper requires pressures that may be two to four times greater than those required for aluminum. Aluminum can be cold welded to copper using specially designed dies to compensate for the difference in yield strengths.

BUTT JOINTS

Cold welding is commonly used to produce butt joints in wire, rod, tubing, and simple extruded shapes of like and unlike metals. Figures 13.11(D) and (H) illustrate butt joints in tubing and solid forms. The weld is frequently as strong as the base metal when correct procedures are used.

Preparation for Welding

A short section is usually sheared from the ends of the parts to be joined to expose fresh, clean surfaces. The shear should be designed to produce square ends so that the parts do not deflect from axial alignment as welding force is applied. During shearing, a thin film of a particular metal being cut can accumulate on the shear blades. If the shear is then used to cut a different metal, accumulated metal on the blade may transfer to the cut surface and inhibit welding. Therefore, the shear blades should be cleaned before shearing parts of another metal for cold welding.

It is not usually necessary to degrease parts to be cold welded before shearing if the residual film of lubricant is very thin. However, degreasing may be necessary if there is a heavy oil film on the metal to avoid contamination of the cut surfaces and to prevent the parts from slipping in the clamping dies.

Welding Procedure

The parts must be positioned in the clamping dies with sufficient initial extension of each part between the dies to ensure adequate upset to produce a weld. However, extension of the parts should not be excessive to avoid bending of the parts. The upsetting force will cause the parts to bend or assume an S-shaped curve, as shown in Fig. 13.18, if the initial die opening is too large. The ends can deflect and slide past one another when force is applied if the projecting length of each part exceeds about twice the thickness or diameter of the parts. In other words, the initial opening between the dies should be no greater than four times the diameter or thickness of the parts. This distance is the maximum total upset that can be used to effect welding. The minimum upset distance varies with the alloy being joined.

The welding dies must firmly grip the parts to prevent slippage when upset force is applied. Any slippage will reduce the amount of upset. For a firm grip, the dimensions of the parts must be within close tolerance so that the dies can nearly close to hold each part securely. The allowable tolerance depends on (1) the design of the die and die holder, and (2) the gripping surface finish. Deep knurling on the gripping surfaces will indent into the part. In most cases, the allowable tolerance for a round part is about 3 percent of the diameter.

Somewhat wider tolerances are permissible for rectangular-shaped parts because the dies usually bear on only two sides. The gap between the closed grips must, however, be

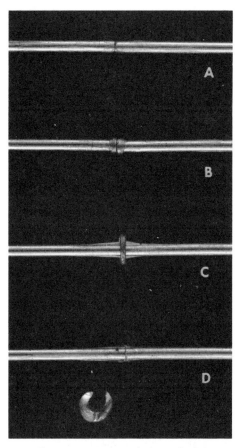

Fig. 13.19—States of upset during butt cold welding

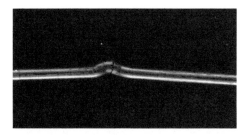

Fig. 13.18—Bending produced during upset from excessive projecting lengths

small to obtain uniform upsetting of metal. It should be no more than about 10 percent of the part thickness.

The application of upset force causes the metal between the dies to upset laterally as illustrated in Fig. 13.19. This lateral flow of metal

(1) Breaks up the oxide film present on the abutting surfaces and carries most of it out of the joint.

(2) Enables oxide-free metal on one side of the interface to achieve intimate contact with oxide-free metal on the other side

(3) Provides the energy that enables the

contacting surfaces to achieve a submicroscopic bond registry with one another

Thus, all requirements needed to form a metal bond are fulfilled and a metallurgical union forms. The flash formed by the lateral flow of metal can be pinched off by the dies as they close together.

Cold welds are usually insensitive to the rate of upsetting of the metal, within limits. Regardless of upset speed, welding will take place if there is sufficient upset.

Multiple Upset

The amount of upset required to produce a full-strength weld in some alloys sometimes exceeds that which can be provided in one step because of a limitation on part extension. If an initial upset will produce a bond of sufficient strength to hold the parts together, additional upset can be applied to produce a full-strength weld by repositioning the weldment in the dies. Surface preparation prior to welding is relatively unimportant when a multiple upsetting technique is used. This technique will completely displace contaminants from the interface.

Fig. 13.21—Multiple upset cold weld in Type 1100 aluminum wire using an offset flash technique

Single upset welding is not normally used for welding butt joints in wires smaller than 3/16-inch in diameter. A cross section of a single upset butt joint in type 1100 aluminum wire is shown in Fig. 13.20. Compare this weld with the multiple upset weld in Type 1100 aluminum alloy illustrated in Fig. 13.21. The multiple upset, offset-flash technique is commonly used in wire drawing to splice 0.025 to 0.128-in. diameter wires as well as those aluminum alloy wires that cannot be welded effectively with single upset. Figure 13.22 illustrates the various stages involved in making a multiple upset butt weld between strips.

Offset Welds

Figure 13.23 illustrates a cold weld being made in a die designed to produce an offset flash. This technique will produce a discontinuous flash that is easy to remove as well as a weld joint that is at an angle to the wire axis. The weld joint being at an angle to the axis will be less influenced by discontinuities in the weld.

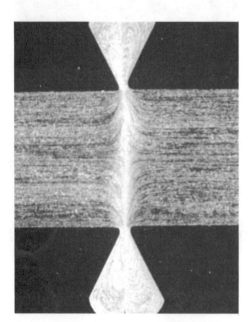

Fig. 13.20—Single upset cold weld in Type 1100 aluminum wire

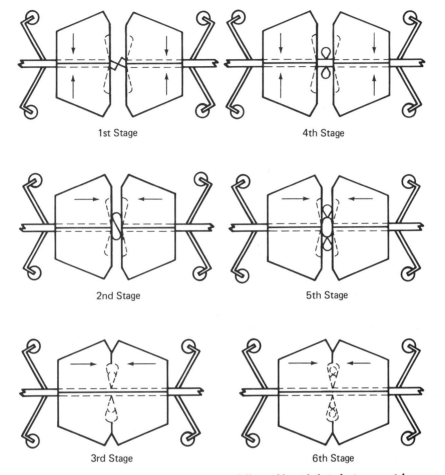

1st Stage   4th Stage

2nd Stage   5th Stage

3rd Stage   6th Stage

Fig. 13.22—Stages in multiple upset welding of butt joints between strips

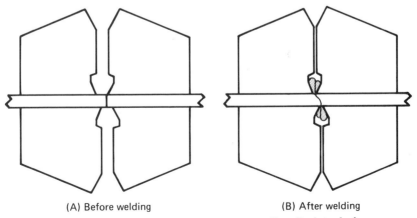

(A) Before welding   (B) After welding

Fig. 13.23—Cold welding of wire with an offset flash technique

Applications

Cold welded butt joints are used in the manufacture of aluminum, copper, gold, silver, and platinum wire. The most common use is to join successive reels of wire for continuous drawing to smaller diameters. Butt joints are also used to repair breaks in the wire that occur during the drawing operation. Diameters ranging from 0.0025 to 0.50-in. have been welded successfully. The aluminum alloys that have been welded with good results include Types EC, 1100, 2319, 3003, 4043, all of the 5000 series, 6061, and 6201. With most of these alloys, the as-welded wire can be drawn successfully to smaller diameters after removal of the flash. Cold welded joints in wire of very high-strength alloys, such as Types 2014 and 7178, usually must be annealed to prevent breaks during subsequent drawing operations. Aluminum alloys that contain lead and bismuth (Types 2011 and 6262) are difficult to cold weld.

Where welding is permitted by ASTM Specifications, cold welding is used to join successive lengths of wire to permit stranding of long lengths of multiple-strand electrical conductors. The weld flash is removed and the weld is dressed with a file or suitable abrasive to obtain a smooth, uniform appearance.

Welds in annealed wire of any of the weldable aluminum alloys exhibit tensile strengths exceeding 95 percent of the base metal. In cold worked Type EC or 5005 aluminum wire and heat-treated Type 6201 wire, weld efficiencies of 92 to 100 percent are attained. In bend testing, the welded joint can be bent or twisted about half as many times before failure as unwelded wire of the same alloy.

Seven-strand No. 4 AWG aluminum alloy conductors with a cold weld in one strand have shown the same breaking strength as similar conductors without welded strands. Types EC, 5005, and 6201 aluminum alloys were used in making these tests.

For copper wire, work hardening at the cold weld increases the metal strength to that of the drawn wire.

LAP WELDING

Procedures

Lap welds can be used for joining aluminum sheet or foil to itself and also to copper sheet or foil. Pressure is applied to the lapped parts by dies that indent the metal and cause it to flow at the interface. This pressure ranges between 150 and 500 ksi for aluminum, depending upon the compressive yield strength of the alloy being welded. Excellent lap welds can be produced in non-heat-treatable aluminum alloys, such as Types EC, 1100, and 3003. However, aluminum alloys containing more than 3 percent magnesium, the 2000 and 7000 series of wrought alloys, and castings are not readily lap welded.

Table 13.2 gives recommended deformations and typical joint efficiencies for lap welds in several common aluminum alloys. For most alloys, joint strength is maximum when deformations between 60 and 70 percent are used. It is apparent that the intrinsic strength of the weld may increase at deformations exceeding 70 percent, but the overall strength of the assembly is decreased. Lap welds exhibit good shear and tensile strengths but have poor resistance to bending or peel loading.

Equal deformations can be achieved when lap welding aluminum to copper or welding dissimilar aluminum alloys together by using dies with the bearing areas approximately in inverse proportion to the compressive yield strength of each metal.

Applications

Commercial applications of lap welding are shown in Figs. 13.24 and 13.25. They include packaging as well as electrical applications. The latter is probably the field in which lap welding finds the greatest application. It is especially useful in the fabrication of electrical devices in which a transition from aluminum windings to copper terminations is required. The range of electrical applications covers large distribution transformers to small electronic devices.

two across the midpoint of the weld. As with sheet, the interior of the tubing must be clean to accomplish a leak-free weld across the flattened tube.

The die face radius and width must be designed to accomplish cold welding and ultimately cut the tubing in two. The opposing dies must be carefully aligned for welding. The face radius is the key to successful cold welding and must be determined experimentally for the metal and tube wall thickness.

Fig. 13.24—Application of cold welding in the manufacture of an electrical component

A variation of cold lap welding is the sealing of commercially pure aluminum, copper, or nickel tubing by pinching it in two between two dies. The tubing is placed transversely between two radius-faced linear dies. As the dies are forced together, the tubing is pinched flat against itself. As the force on the dies is increased, the metal between the dies is upset and extruded from between them as in lap welding. The force is increased until the tube walls are cold welded together and then finally parted in

Fig. 13.25—Application of cold welding in the manufacture of industrial packaging

Table 13.2
Lap type cold welds in selected aluminum alloys

Alloy	Temper	Recommended deformation, %	Joint efficiency, %
3003	O	50	85
3003	H14	70	70
3003	H16	70	60
3003	H18	60	55
3004	O	60	60
3004	H34	55	40
5052	O	60	65
5052	H34	60	45
6061	T6	60	50
7075 Alclad	T6	40	10

HOT PRESSURE WELDING

PRESSURE GAS WELDING

Definition and General Description

Pressure gas welding is an oxyfuel gas welding process which produces coalescence simultaneously over the entire area of abutting surfaces by heating them with oxyfuel gas flames and then applying adequate pressure.[4] No filler metal is used. The two variations of the process are the closed joint and open joint methods. In the closed joint method, the clean faces of the parts to be joined are abutted together under moderate pressure and heated by gas flames until a predetermined upsetting of the joint occurs. In the open joint method, the faces to be joined are individually heated by the gas flames to the melting temperature and then brought into contact for upsetting. Both methods are easily adapted to mechanized operation. Pressure gas welding can be used for welding low and high carbon steels, low and high alloy steels, and several nonferrous metals and alloys.

In the closed joint method, since the metal along the interface does not reach the melting point, the mode of welding is different from that of fusion welding. Generally speaking, welding takes place by the action of grain growth, diffusion, and grain coalescence across the interface under the impetus of high temperature (about 2200° F for low carbon steel) and upsetting pressure. The welds are characterized by a smooth-surfaced bulge or upset, as shown in Fig. 13.26, and by the general absence of cast metal at the weld line.

In the open joint method, the joint faces are melted, but the molten metal is squeezed from the interface to form a flash when the joint is upset. These welds resemble flash welds in general appearance.

Principles of Operation

Closed Joint Method. The faces to be welded are butted together under initial pressure to as-

sure intimate contact. The metal at the joint is then heated to welding temperature with a gas flame. Finally, the metal is upset sufficiently to consummate a weld.

Heating is generally done with water-cooled, multiflame oxyacetylene torches. These torches are designed to generate sufficient heat and distribute it uniformly throughout the entire section to be welded. For sections over 1-in. thick, it is advisable to heat the joint uniformly from all sides as shown in Fig. 13.27.

Solid or hollow round sections, such as shafts or piping, are usually welded with circular ring torches. The torch head may be a split type for easy loading and removal of work from the welding machine. A typical head of this type for welding 2.5-in. diameter tubing with a 0.25-in. wall thickness is shown in Fig. 13.28. More elaborate heating heads are required for more complicated shapes. They should conform to the shape of the part to provide uniform heating.

The apparatus for pressure gas welding must be designed to apply the desired pressure and maintain alignment during welding. Provision for maintaining uniform pressure is essential.

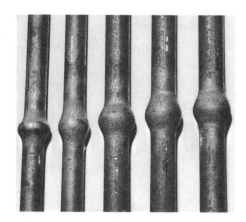

Fig. 13.26—Typical pressure gas welds in 1 and 1.25-in. diameter steel bars

---

4. This process is covered more fully in Ch. 22, "Pressure Gas Welding," Section 2, 4th ed., 1958.

Fig. 13.27—Torch arrangement for pressure gas welding a Type 321 stainless steel ring with 1.88 by 2.63-in. cross section

The quality and type of end preparation of the parts to be welded depend upon the type of steel. In general, the abutting ends should be machined or ground to a smooth, clean surface. Freedom from oil, rust, grinding dust, and other foreign material is of great importance.

The geometry of the abutting faces depends upon the application and the alloy. Some control of the shape of the upset metal can be obtained by beveling one or both of the parts. Figure 13.29 illustrates typical joint preparations for pressure gas welding and the effect of beveling on the shapes of the completed welds.

For illustration, assume that two 5-in. diameter by 1/4-in. wall steel pipes are to be pressure gas welded end-to-end using a butt joint. The general procedures are as follows: A split torch head that will provide small oxyacetylene flames for the full circumference of the joint is selected. The head is mounted in the same plane as the interface with provision for axial oscillation. The abutting ends of the pipe are beveled to an included angle of 6 to 10 degrees with a smooth, clean finish. The pipes are placed in the machine and aligned. Then a force of 5850 lbs is applied to produce a low compressive pressure of 1500 psi. While this force is maintained, the torch flames are oscil-

Fig. 13.28—Split annular torch for pressure welding piping, tubing, or solid rounds

lated axially a short distance across the weld joint. As the joint heats up, the metal will upset. The joint faces will close together preventing oxidation at higher temperatures. As the metal temperature increases, the compressive strength of the steel decreases and the joint begins to upset uniformly. At this time, the metal is at welding temperature throughout its full thickness and an upsetting pressure of 4000 psi (15 600 lb force) is applied until the weld zone is upset for a distance of 3/16 inch. The torches may then be extinguished and the assembly removed from the machine.

There are a few variations of this basic procedure, the principal one being the sequence of pressure application. These variations are introduced to meet special requirements of certain metals such as high carbon steels, high chromium steels, and nonferrous metals. For example, the constant pressure method is recommended for welding high carbon steel parts.

Another variation of the basic method is applicable to high chromium steels and some nonferrous metals. A high initial pressure in the range of 6000 to 10 000 psi is applied before heating is started and is maintained until the metal in the weld zone starts to upset. This high pressure forces the joint faces together to prevent oxidation. Pressure is then decreased until welding temperature is reached when the high pressure is again applied to upset the joint.

Examples of typical pressure cycles used for pressure gas welding several metals are given in Table 13.3. Table 13.4 gives the average dimensions of closed joint pressure welds in parts of various thicknesses.

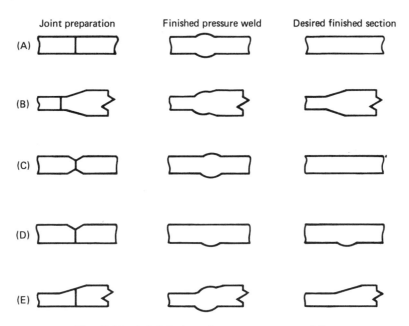

Joint preparation   Finished pressure weld   Desired finished section

(A)

(B)

(C)

(D)

(E)

Fig. 13.29—Joint designs for pressure gas welding

The quality of a weld depends to an important degree upon proper upsetting during the welding operation. The upset distance or shortening of the weld zone increases with metal thickness. Recommended amounts of upset are given in Table 13.4. These values are usually measured from fixed points on the parts or the welding machine.

Open Joint Method. Machines for open joint

pressure gas welding must provide more accurate alignment and be of rugged construction to withstand rapidly applied upset forces. Machines similar to those used for flash welding are suitable.

The most satisfactory heating head is a flat, multiflame type burner, such as the one shown in Fig. 13.30, that produces a uniform flame pattern conforming to the cross section of the members to be welded. Good alignment of the heat-

Table 13.3
Typical upset pressure cycles for pressure gas welds

Type of metal	Method	End pressure, psi		
		Initial	Intermediate	Final
Low carbon steel	Closed joint	500-1,500	. . .	4,000
High carbon steel	Closed joint	2,700	. . .	2,700
Stainless steel	Closed joint	10,000	5,000	10,000
Monel alloy	Closed joint	6,500	. . .	6,500
Steel (carbon and alloy)	Open joint	. . .	. . .	4,000-5,000

Table 13.4
Joint dimensions of pressure gas welds, squared end preparation, closed joint method

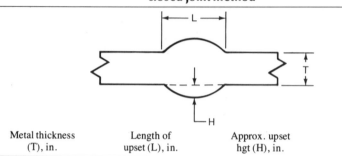

Metal thickness (T), in.	Length of upset (L), in.	Approx. upset hgt (H), in.	Total upset, in.
1/8	3/16-1/4	1/16	1/8
1/4	5/16-1/2	3/32	1/4
3/8	9/16-5/8	1/8	5/16
1/2	3/4-7/8	3/16	3/8
3/4	1-1/16-1-3/16	1/4	1/2
1	1-1/4-1-1/2	3/8	5/8

ing head with the faces of the joint is important to minimize oxidation and to obtain uniform heating and subsequent upsetting. A removable spacer block can be used during alignment.

Saw cut surfaces are satisfactory for welding since the ends are thoroughly melted before the weld is consummated. A thin layer of oxide on the joint faces has little effect on weld quality, but major amounts of foreign substances, such as rust or oil, should be removed before welding.

The general procedure for open joint pressure gas welding is to align the parts with a suit-able torch tip properly spaced between the joint faces (Fig. 13.30). When thin sections are welded, the torch tip is placed just outside the joint with the flames directed at the joint faces. The tip is designed to heat the full cross section of the faces. The flames are maintained in this position until a molten film entirely covers both faces. The torch is then withdrawn, and the parts are rapidly brought together with a force that will produce a constant pressure of 4000 to 5000 psi at the interface. This step is shown in Fig. 13.31. Pressure is maintained until upsetting of

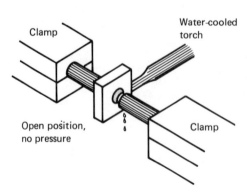

Fig. 13.30—Torch and general setup for open joint pressure gas welding

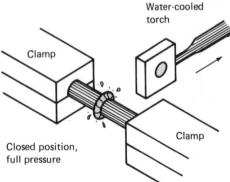

Fig. 13.31—An open joint pressure gas weld as upsetting starts

the metal ceases. The total upset is controlled by both the applied pressure and the temperature of the hot metal. It is not preset on the equipment.

Equipment

Machines. The apparatus for pressure gas welding comprises:

(1) Equipment for applying upsetting force

(2) Suitable heating torches and tips designed to provide uniform and controlled heating of the weld zone

(3) Necessary indicating and measuring devices for regulating the process during welding.

The complexity of the machine depends upon the configuration and size of the parts to be welded and the degree to which the process is mechanized. In most cases, it is advisable to use special heating torches and tips as well as special apparatus for gripping and applying force to the parts.

Figure 13.32 illustrates a simple, manually operated gas pressure welding machine capable of welding bars and tubes up to 3 inches in diameter.

Fig. 13.32—A small pressure gas welding machine for rods and tubes up to 3-in. diameter

Auxiliary Equipment. Some auxiliary equipment is necessary. The gas supply must be adequate for the maximum flow requirement, and the gas regulators must be capable of maintaining a uniform flame adjustment. Quick-acting gas shut-off valves are very desirable. In many instances, needle valves are advantageous for fine adjustment of the flame. The best control of gas flow and heat input is obtained when the pressure gages are located close to the torch. This permits the operator to check gas pressures readily. Flowmeters may be used to assure uniform gas flow.

An ample supply of water is needed for cooling the torches and, in some cases, the clamps and parts of the press. Adequate jigs for aligning and supporting the parts are usually needed. Automatic control units for regulating the upset force and heating cycles and then terminating the operation can be incorporated in a machine.

Applications

Metals Welded. Pressure gas welding has been successfully applied to plain carbon, low alloy, and high alloy steels, and to several nonferrous metals, including nickel-copper, nickel-chromium, and copper-silicon alloys. It has been very useful for joining dissimilar metals.

Rails. The first commercial application of this process was the welding of railroad rails, and thousands of joints were made. However, this process has largely been replaced by the flash welding process. When pressure gas welding is used, the closed butt method is usually employed with equipment specifically designed for this application.

The rail ends are carefully prepared by power sawing and then are cleaned. The rails are gripped by special clamps and a force applied to produce of about 2800 psi on the joint. The joint is then heated with specially shaped heating tips or heads that are oscillated automatically across the joint until the metal reaches welding temperature. Adequate pressure is then applied to produce the required upset. Typical rail welding equipment is shown in Fig. 13.33. Most of the upset metal on the ball and edges of the base is removed by oxygen cutting. In the next position, the weld zone is normalized

Fig. 13.33—Pressure gas welding railroad rails

to refine the grain size and restore normal hardness. Finally the weld is ground to rail contour, examined by magnetic particle inspection, and oiled for protection against rusting. Rails that have been welded, normalized, and inspected in this manner have given satisfactory service under both heavy and fast traffic for extended time periods.

Other Applications. Pressure gas welding has been largely superseded by other welding processes. The basic elements of the process assisted in the development of similar processes, such as flash and friction welding, that use other sources of energy. Automatic welding of pipe, a former application, is now accomplished using automatic gas metal arc welding.

Properties and Heat Treatment

In general, pressure gas welding has a minimum effect on the mechanical and physical properties of the base metals. Because of the relatively large mass of hot metal in the weld zone, its cooling rate is usually quite low. Tests have shown that with steel over 1-in. thick, the cooling rate approximates that at the far end of a Jominy end quench specimen.

In the closed joint method, the maximum temperature of the metal is below the temperature at which overheating and rapid grain growth occur. In the open joint method, the melted metal film is squeezed out of the joint during upset. These characteristics are advantageous for welding high carbon steels and some nonferrous alloys that are hot-short or affected by overheating.

Another important factor is the absence of deposited metal. The entire weld zone is base metal, and, hence, it responds to heat treatment in the same manner. This, of course, includes the effect of the heat of welding on the corrosion resistance of welded stainless steels. If unimpaired corrosion resistance is desired, stabilized stainless steel must be used or the welded assembly must be given a stabilization heat treatment after welding.

Pressure gas welds in low carbon steels seldom require heat treatment or stress relief since the heat-affected zone in such steels is usually normalized and relatively stress-free. Pressure gas welding has been used with low alloy and high carbon steels for fabricating assemblies subject to high service stresses, and postweld heat treatment was necessary. Heat treatment may frequently be done with the same heating heads used for welding. In rails, for example, the annealed zone on each side of the weld may be too soft. To overcome this problem, the weld zone can be heated to normalizing temperature using heating heads and then air cooled to restore the desired hardness. Similarly, heat treatment with the welding flame may be suitable for developing desired mechanical properties in welded joints in some low alloy steels such as those used for oil well drilling tools. Such heat treatment, which is essentially a normalizing operation, will refine the grain size in the weld zone and improve ductility and toughness. For highly hardenable steels, annealing or slow cooling after the welding operation may be necessary to prevent hardening or surface cracking in the weld zone. To develop optimum properties in welds in heat-treatable steels, furnace heat treatment is commonly used.

Weld Quality

Mechanical Properties.[5] Since there is no deposited metal in pressure welds, the mechanical properties of welds will depend upon the composition of base metals, the cooling rate, and the quality of the weld. When dissimilar steels are joined, the properties of the welded joint will be more nearly those of the weaker member.

Metallurgical Structure. The location of the original interface in pressure gas welds in plain carbon steels and many alloy steels is very difficult to detect in a metallographic cross section using normal etchants. It is possible to locate the weld line with special polishing and etching techniques. A typical photomicrograph of a pressure gas weld in steel is shown in

5. Mechanical properties of typical pressure gas welds are given in Ch. 22, "Pressure Gas Welding," Section 2, 4th ed., 1958.

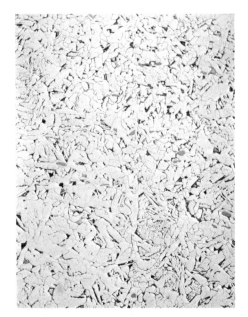

Fig. 13.34—Photomicrograph of a pressure gas weld in 1020 steel, as-welded

Fig. 13.34. Although not apparent, the interface extends vertically through the center of the photomicrograph.

Quality Control

Process Control. Successful pressure gas welding by the closed joint method requires positive and continuous control of the variables that influence the quality of a weld. These variables include the following:

(1) Degree of roughness and cleanliness of end preparation

(2) Pressure cycle

(3) Alignment of the parts

(4) Welding cycle time

(5) Performance of the heating torches

(6) Desired upset or shortening

(7) Cooling time in the machine after upset

Examination of the pressure-upsetting cycle can indicate if the welding conditions are conforming to the prescribed procedure. With constant heat input into the weld zone, constant width of heated zone, and uniform pressure sequence, the entire cycle of heating and up-setting should be completed with a variation in welding time of not more than 10 percent. On this basis, if a weld requires an unduly long or short time, the conditions that prevailed during welding should be evaluated. A large time variation from nominal would indicate that (1) some factor other than time was controlled improperly, and (2) the weld might be of questionable quality. Malfunctioning of the pressure system or the heating heads and slipping of the parts in the clamps are examples of the conditions that might cause poor quality welds.

Autographic or other records of the following variables have proved of value in maintaining good control:

(1) Pressure cycle

(2) Total time or times for certain stages of the procedure

(3) Gas flow rates

(4) Total upset distance

Some conditions of importance in the closed joint method are not so important with the open joint method. With the open joint method, the cleanliness of the joint faces is not critical except for excessive amounts of foreign matter. Melting of the faces offsets a need for thorough surface preparation. The amount of upset or shortening is not necessarily constant, and, therefore, it is not an index of weld quality. However, due attention must be given to the pressure cycle and the performance of the heating torches.

Inspection. The first inspection is usually a visual one to evaluate the following general characteristics:

(1) Presence or absence of excessive melting

(2) Contour and uniformity of upset

(3) Position of the weld line with respect to the midpoint of the upset zone

If there is no appreciable variation from an accepted standard and the controls were adequate, it may usually be concluded that the pressure weld is of normal quality.

In many highly stressed assemblies, added assurance of weld consistency and quality may be needed. Sample welds selected either at random or at fixed intervals should be destructively tested. This procedure will serve as a positive and continuous check on the welding cycle and process controls as well as the properties of the welded assemblies.

Magnetic particle inspection can be used for nondestructive inspection of pressure welded rails. The nick-break test[6] can be used as a convenient quality check for soundness. Fracture of a sample weld along the weld line will show the extent of metallic bonding, grain size, and evidence of overheating of the faces. Changes in the welding cycle can be checked quickly by this test. Experience has proved that when the nick-break tests show satisfactory crystalline fracture throughout the weld cross section, all other tests will usually prove satisfactory.

Proof testing may be used as an alternative to destructive testing. The test is designed to disclose defective welds and pass acceptable welds. A welded joint is subjected to either a tensile or a bending load, or both, to produce a maximum tensile stress just below the yield strength of the metal. A poor quality weld will fail in this test.

FORGE WELDING

Fundamentals of the Process

Forge welding is a form of hot pressure welding which produces coalescence of metals by heating them in air in a forge or other furnace and then applying sufficient pressure or hammer blows to cause permanent deformation at the interface. Forge welding was the earliest welding process and the only one in common use until well into the nineteenth century. Blacksmiths used this process. Pressure vessels and steel pipe were among the industrial items once fabricated by forge welding.

The process still finds some application with modern methods of applying the heat and pressure necessary to achieve a weld. The chief present-day applications are in the production of tubing and clad metals.

Principles of Operation

The sections to be joined by forge welding may be heated in a forge, furnace, or by other appropriate means until they are very malleable. A weld is accomplished by removing the parts from the heat source, superimposing them, and then applying pressure or hammer blows to the joint.

Heating time is the major variable that affects joint quality. Insufficient heat will fail to bring the surfaces to the proper degree of plasticity, and welding will not take place. If the metal is overheated, a brittle joint of very low strength may result. The overheated joint is likely to have a rough, spongy appearance where the metal is severely oxidized. The temperature must be uniform throughout the joint interfaces to yield a satisfactory weld.

Process Modes

Hammer Welding. In hammer welding, coalescence is produced by heating the parts to be welded in a forge or other furnace and then applying pressure by means of hammer blows. Manual hammer welding is the oldest technique. Pressure is applied to the heated members by repeated high velocity blows with a comparatively light sledge hammer. Modern automatic and semiautomatic hammer welding is accomplished by blows of a heavy power-driven hammer operating at low velocity. The hammer may be powered by steam, hydraulic, or pneumatic equipment.

The size and quantity of parts to be fabricated will determine the choice of either manual or power-driven hammer welding. This process may still be used in some maintenance shops, but it largely has been replaced by other welding processes.

Die Welding. This is a forge welding process where coalescence is produced by heating the parts in a furnace and then applying pressure by means of dies. The dies also form the work to shape while it is hot.

Metals Welded

Low carbon steels are the metals most

---

6. For a description of this test, refer to *Welding Inspection,* Miami: American Welding Society, 1968: 113-114.

commonly joined by forge welding. Sheets, bars, tubing, pipe, and plates of these materials are readily available.

The major influences on the grain structure of the weld and heat-affected zone are the amount of forging required and the temperature at which the forge welding takes place. A high temperature is generally necessary for the production of a sound forge weld. Annealing can refine the grain size in a forge welded steel joint and improve joint ductility.

Thin extruded sections of aluminum alloy are joined edge-to-edge by a forge welding process with automatic equipment to form integrally stiffened panels. The panels are used for lightweight truck and trailer bodies. Success of the operation depends upon the use of correct temperature and pressure, effective positioning and clamping devices, edge preparation, and other factors. Although the welding of aluminum for this application is called forge welding, it could be classified as hot pressure welding because the edges to be joined are heated to welding temperature and then upset by the application of pressure.

Joint Design

The five joint designs applicable to manual forge welding are the lap, butt, cleft, jump, and scarf types shown in Fig. 13.35. The joint surfaces for these welds are slightly rounded or crowned. This shape ensures that the center of the pieces will weld first so that any slag, dirt, or oxide on the surfaces will be forced out of the joint as pressure is applied. Lap, pin, and butt joints used for automatic forge welding are shown in Fig. 13.36.

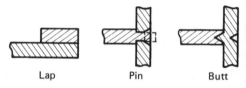

Fig. 13.36—Typical joint designs used for automatic forge welding

Scarfing is the term applied to the preparation of the workpieces of forge welding. Similarly, the prepared surface is referred to as a scarfed surface. Each workpiece to be welded must be upset sufficiently for an adequate distance from the scarfed surface to provide metal for mechanical working during welding.

Fluxes

In the forge welding of certain metals, a flux must be used to prevent the formation of oxide scale. The flux and the oxides present combine to form a protective coating on the heated surfaces of the metal. This coating prevents the formation of additional oxide and lowers the melting point of the existing oxide.

Two commonly used fluxes for steels are silica sand and borax (sodium tetraborate). Flux is not required for very low carbon steels (ingot iron) and wrought iron because their oxides have low melting points. The flux most commonly used in the forge welding of high carbon steels is borax. Because it has a relatively low fusion point, borax may be sprinkled on the metal while it is in the process of heating. Silica sand is suitable as a flux in the forge welding of low carbon steel.

ROLL WELDING

Roll welding is a solid-state welding process where coalescence of metals is produced by heating them to a proper temperature and then applying sufficient pressure with rolls to cause deformation at the faying surfaces. This process is used mostly for the manufacture of clad steel plates. In the fabrication of clad plates, two sheets of steel are thoroughly cleaned and alternately sandwiched with two

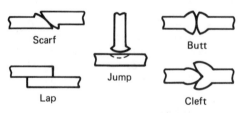

Fig. 13.35—Typical joint designs employed for manual forge welding

clean sheets of cladding metal. A nonfusing parting compound separates the two sets of metal sheets, each of which is welded around the edges to exclude air and prevent slippage during the roll welding operation.

The sandwiched material is heated uniformly to a temperature between 2100 and 2350° F in a soaking pit. The assembly is then rolled until the cladding and the base plate are welded together. Besides welding the cladding to the base plate, the rolling reduces the thickness of the composite. After rolling, the sandwiched assembly is separated into two clad sheets.

CARBON ARC WELDING

DEFINITION AND GENERAL DESCRIPTION

Carbon arc welding is a process in which an arc is established between a nonconsumable carbon (graphite) electrode and the work, or between two carbon electrodes. The latter is a variation known as twin carbon arc welding. Two other variations known as shielded and gas carbon arc welding no longer have commercial significance.

Although carbon arc welding has been superseded to a great extent by other welding processes, there are many applications for which it can be used to good advantage. In operation, the carbon arc is used only as a source of heat. In this respect, it resembles the gas tungsten arc welding process.

Figure 13.37 shows the carbon arc welding process. The arc stream usually develops a temperature of from 7000 to 9000° F, depending upon the amount of current used. Because the electrode burns off very slowly, it does not have an appreciable effect on the composition of the deposited metal, provided filler metal is added.

PRINCIPLES OF OPERATION

The application of heat and the filler metal feed are controlled separately in carbon arc welding. With some welds, the carbon arc is used to fuse the edges together without the addition of filler metal. A carbon arc can produce the high heat needed· for welding metals that have high heat conductivity, such as copper.

In carbon arc welding, direct current electrode negative (straight polarity) should be used. The arc is formed between the tip of the carbon electrode and the base metal. Welding current is adjusted to provide sufficient heat to melt the base metal and welding rod uniformly as welding progresses. Recommended current ranges for carbon and graphite electrodes are given in Table 13.5. Amperages are recommended on the basis of maximum electrode life. Higher amperages can be used, but the electrode will be consumed faster.

The properties of welds made with the carbon arc in mild steel may be adequate for noncritical applications. The process does not provide as much shielding from the atmosphere

A - Base metal
B - Penetration
C - Deposited metal
D - Carbon electrode
E - Arc flame
F - Arc stream
G - Welding rod

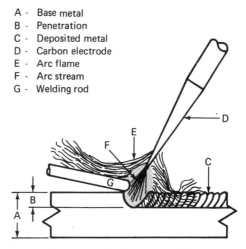

Fig. 13.37—Carbon arc welding

Table 13.5
Recommended current ranges for carbon and graphite electrodes[a]

Electrode diameter, in.	Current, A[b]	
	Carbon electrodes	Graphite electrodes
1/8	15-30	15-35
3/16	25-55	25-60
1/4	50-85	50-90
5/16	75-115	80-125
3/8	100-150	110-165
7/16	125-185	140-210
1/2	150-225	170-260
5/8	200-310	230-370
3/4	250-400	290-490
7/8	300-500	400-750

a. Recommended with regard to maximum electrode life. Where electrode cost is not a factor, higher amperages may be used.
b. Direct current electrode negative (straight polarity)

as the shielded metal arc or gas metal arc welding processes.

EQUIPMENT

Carbon electrodes range in size from 1/8 to 7/8-in. diameter. Baked carbon electrodes last longer than graphite electrodes. Figure 13.38 shows typical air-cooled carbon electrode holders. Water-cooled holders are available for use with the larger sizes of electrodes, or adapters can be fitted to regular holders to permit accommodation of the larger electrodes. Direct current welding machines of either the rotating or rectifier type are excellent power sources for the carbon arc welding process.

WELDING TECHNIQUE

The workpieces must be free from grease, oil, scale, paint, and other foreign matter. The two pieces should be clamped tightly together with no root opening. They may be tack welded together.

Carbon electrodes, 1/8 to 5/16-in. diameter, may be used depending upon the current required for welding. The end of the electrode should be prepared with a long taper to a point. The diameter of the point should be about half that of the electrode. For steel, the electrode should

protrude about 4 to 5 in. from the electrode holder.

A carbon arc may be struck by bringing the tip of the electrode into contact with the work and immediately withdrawing it to the correct length for welding. In general, an arc length between 1/4 and 3/8 in. will be best. If the arc length is too short, there is likely to be excessive carburization of the molten metal resulting in a brittle weld.

When the arc is broken for any reason, it should not be restarted directly upon the hot weld metal as this is likely to cause a hard spot in the weld at the point of contact. The arc should be started on cold metal to one side of the joint and then quickly returned to the point where welding is to be resumed.

When the joint requires filler metal, the welding rod is fed into the molten weld pool with one hand while the arc is manipulated with the other. The arc is directed on the surface of the work and gradually moved along the joint, constantly maintaining a molten pool into which the welding rod is added in the same manner as in gas tungsten arc welding. Progress along the weld joint and the addition of welding rod must be timed to provide the size and shape of weld bead desired. Welding vertically or overhead with the carbon arc is difficult because carbon arc welding is essentially a puddling process.

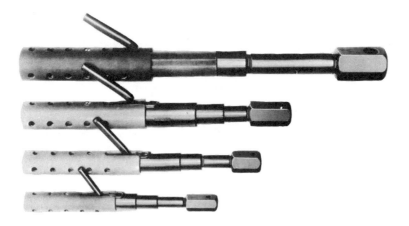

Fig. 13.38—Typical air-cooled carbon electrode holders

The weld joint should be backed up, especially in the case of thin sheets, to support the molten weld pool and prevent excessive melt-thru.

For outside corner welds in 14 to 18 gage steel sheet, the carbon arc can be used to weld the two sheets together without a filler metal. Such welds are usually smoother and more economical to make than shielded metal arc welds made under similar conditions.

METALS WELDED

The carbon arc can be used for welding steels and nonferrous metals. It can also be used for surfacing.

Steels

The principal use of carbon arc welding of steel is making edge welds without the addition of filler metal. This is done chiefly in thin gage sheet metal work, such as tanks, where the edges of the work are fitted closely and fused together using appropriate flux.

Galvanized sheet steel can be braze welded with the carbon arc.[7] A bronze welding rod is used. The arc is directed on the rod so that the galvanizing is not burned off the steel sheet. The arc should be started on the welding rod or a

---

7. Refer to Ch. 11, *Welding Handbook,* Vol. 2, 7th ed., p. 434-7 for a discussion of braze welding.

starting block. Low current, a short arc length, and rapid travel speed should be used. The welding rod should melt and wet the galvanized steel.

Cast Iron

Iron castings may be welded with the carbon arc and a cast iron welding rod. The casting should be preheated to about 1200° F and slowly cooled if a machinable weld is desired.

Copper

Straight polarity should always be used for carbon arc welding of copper. Reverse polarity will produce carbon deposits on the work that inhibit fusion.

The work should be preheated in the range of 300 to 1200° F depending upon the thickness of the parts. If this is impractical, the arc should be used to locally preheat the weld area. The high thermal conductivity of copper causes heat to be conducted away from the point of welding so rapidly that it is difficult to maintain welding heat without preheating.

A root opening of 1/8 in. is recommended. Best results are obtained at high travel speeds with the arc directed on the welding rod. A long arc length should be used to permit carbon from the electrode to combine with oxygen to form CO. This will provide some shielding of the weld metal.

Fig. 13.39—A twin carbon arc welding torch

TWIN CARBON ARC WELDING

With a twin carbon arc torch, the arc heat can be used for welding, brazing, surfacing, or soldering operations as well as for preheating or postheating the work. The heat is produced by an arc between two carbon electrodes. The work is not part of the electrical circuit. Twin carbon arc welding is used principally for maintenance operations. A twin carbon arc torch, shown in Fig. 13.39, has two adjustable arms in which the carbon electrodes are clamped. To maintain a constant distance between the electrodes (arc length) as they are consumed, adjustment of electrode position can usually be made while operating the torch.

Small ac arc welding machines are normally used with the twin carbon arc. Copper coated carbon electrodes are generally used in 1/4-in. to 3/8-in. diameter. The current should never be set so high that the copper coating is burned away over 1/2 in. ahead of the arc. Only enough current should be used to cause the filler material to flow freely on the work. This will avoid consuming carbons too rapidly.

SAFETY

Safety procedures and equipment normally used with other arc welding processes should also be used with this one. This includes welding helmets with appropriate filter lenses, protective clothing, and gloves. Adequate ventilation should be provided. The requirements of ANSI Z49.1, *Safety in Welding and Cutting*, latest edition, and appropriate federal, state, and local regulations should be followed when carbon arc welding.

BARE METAL ARC WELDING

Bare metal arc welding is an arc welding process that obtains coalescence by heating a joint with an electric arc between a bare or lightly coated metal electrode and the work. Neither shielding nor pressure is used, and filler metal is obtained from the electrode.

The chief disadvantage of welding with a bare electrode is that the molten filler and weld metal are exposed to the atmosphere. The molten metal transferring across the arc and the molten weld metal are both subjected to oxidation and nitrification. As a result, the molten metal oxidizes rapidly and the weld metal is likely to have unsatisfactory fusion with the base metal.

Formation of porosity in the weld will have a detrimental effect upon the strength and ductility of the welded joint. Nitrogen in the form of nitrides tends to cause high hardness and poor ductility. Water vapor dissociates in the arc to produce hydrogen, which may cause hydrogen embrittlement of some metals and underbead cracking in some steels.

Covered electrodes have largely replaced bare and lightly covered electrodes. Most of the bare wire electrodes are used with one of the gas shielded welding processes. They are not used without some form of shielding for the arc and molten weld pool.

ATOMIC HYDROGEN WELDING

Atomic hydrogen welding[8] is a process that produces coalescence of a joint by heating it with an electric arc maintained between two tungsten electrodes in an atmosphere of hydrogen. Shielding is obtained from the hydrogen. Filler metal may or may not be added.

In this process, the arc is maintained entirely independent of the work or parts being welded. The work is a part of the electrical circuit only to the extent that a portion of the arc comes in contact with the work, at which time a voltage exists between the work and each electrode.

Historically, atomic hydrogen welding was the forerunner of the gas shielded arc welding processes. At that time, it was the best process for welding of metals other than carbon and low alloy steels. With the advent of low-cost inert gases, the gas shielded arc welding processes have largely replaced atomic hydrogen welding.

Hydrogen in its normal state is diatomic.

8. More detailed information is presented in Ch. 26, *Welding Handbook,* Section 2, 5th ed., 1963.

Each molecule consists of two atoms. When an arc is established in hydrogen between two electrodes, the temperature in the arc stream reaches approximately 11 000°F and the molecular hydrogen dissociates into its atomic form. In the process of dissociation, a large amount of heat is absorbed by the hydrogen from the arc. The heat is subsequently liberated on recombination of the hydrogen atoms at the surface of the work. A sudden decrease in the temperature of the hydrogen as it strikes a relatively cold surface (weld area) is accompanied by a rapid release of heat as the hydrogen atoms recombine to the molecular form. By varying the distance between the arc stream and the surface, the available energy can be varied over a wide range. The hydrogen also cools the electrodes and protects both the electrodes and the metal from oxidation.

Atomic hydrogen welding had one unique advantage: the ability to control heat input over a very wide range by manipulating the arc. It was widely used for tool and die repair and similar operations where very precise metal buildup with accurate alloy control was necessary.

Metric Conversion Factors

1 Btu = 1 J
1 in = 25.4 mm
1 ft/min = 5.08 mm/s
1 lbf = 4.45 N
1 psi = 6.89 kPa
$t_c = 0.556 (t_F - 32)$

SUPPLEMENTARY READING LIST

Thermit Welding

Ailes, A. S., Modern applications of Thermit welding. *Weld. Met. Fab.* 32 (9): 335-43, 414-19; 1964.

Cikara, M., Repair of rails by Thermit welding and some observations on the testing of welded joints, *Welding and Allied Processes in Maintenance and Repair Work,* New York: Elsevier Pub. Co., 1961: 318-34

Fricke, H. D., Guntermann, H., and Jacoby, N., Thermit welding process for rails of special quality. *ETR,* 25(4); 1976 (in German)

Guntermann, H., Thermit butt joints for concrete-steel construction. *Maschinernmarket 75,* No. 75; 1969 (in German)

Guntermann, H., The applications of the Thermit process in areas besides rail welding, *ZEV-Glaser Annalen;* 1975 (in German)

Jacoby, N., Special processes of the Thermit welding technique. *Der Eisenbahningenieur* No. 3: 1977 (in German)

Kubaschewski, E., Evans, L. L., and Alcock, C. B., *Metallurgical Thermochemistry,* 4th ed., London-New York: Pergamon Press, 1967

Rossi, B., E., *Welding Engineering,* New York: McGraw-Hill, 1954

Cold Welding

Houldcraft, P. T., *Welding Process Technology,* London: Cambridge University Press, 1977, 217-21

Milner, D. R. and Rowe, G. W., Fundamentals of solid phase welding. *Metallurgical Review,* 28 (7): 433-80; 1962

Mohamed, H. A. and Washburn, J., Mechanism of solid-state pressure welding. *Welding Journal,* 54(9): 302s-10s; 1975 Sept.

Tylecote, R. F., *The Solid-State Welding of Metals,* New York: St. Martin's Press, 1968

Hot Pressure Welding

Bryant, W. A., A method for specifying hot isostatic pressure welding parameters. *Welding Journal,* 54(12): 433s-35s; 1975 Dec.

Guy, A. G. and Eiss, A. L., Diffusion phenomena in pressure welding. *Welding Journal,* 36 (11): 473s-80s; 1957 Nov.

Hastings, D. C., An application of pressure welding to fabricate continuous welded rails. *Welding Journal.* 34 (11): 1065-69; 1955 Nov.

Lage, A. P., Application of pressure welding to the aircraft industry. *Welding Journal,* 35 (11): 1103-09; 1956 Nov.

Lessmann, G. G. and Bryant, W. A., Complex rotor fabrication by hot isostatic pressure welding. *Welding Journal,* 51 (12): 606s-14s; 1972 Dec.

McKittrick, E. S. and Donalds, W. E., Oxyacetylene pressure welding of high-speed rocket test track. *Welding Journal,* 38 (5): 469-74; 1959 May

Welding Handbook
Index of Major Subjects

Z

INDEX